METHODS IN MOLECULAR BIOLOGY

Series Editor
John M. Walker
School of Life and Medical Sciences
University of Hertfordshire
Hatfield, Hertfordshire, UK

For further volumes:
http://www.springer.com/series/7651

For over 35 years, biological scientists have come to rely on the research protocols and methodologies in the critically acclaimed *Methods in Molecular Biology* series. The series was the first to introduce the step-by-step protocols approach that has become the standard in all biomedical protocol publishing. Each protocol is provided in readily-reproducible step-by-step fashion, opening with an introductory overview, a list of the materials and reagents needed to complete the experiment, and followed by a detailed procedure that is supported with a helpful notes section offering tips and tricks of the trade as well as troubleshooting advice. These hallmark features were introduced by series editor Dr. John Walker and constitute the key ingredient in each and every volume of the *Methods in Molecular Biology* series. Tested and trusted, comprehensive and reliable, all protocols from the series are indexed in PubMed.

Neuronal Morphogenesis

Methods and Protocols

Edited by

Kazuhito Toyooka

Neurobiology and Anatomy, Drexel University, Philadelphia, PA, USA

Humana Press

Editor
Kazuhito Toyooka
Neurobiology and Anatomy
Drexel University
Philadelphia, PA, USA

ISSN 1064-3745 ISSN 1940-6029 (electronic)
Methods in Molecular Biology
ISBN 978-1-0716-3968-9 ISBN 978-1-0716-3969-6 (eBook)
https://doi.org/10.1007/978-1-0716-3969-6

This Humana imprint is published by the registered company Springer Science+Business Media, LLC, part of Springer Nature.
The registered company address is: 1 New York Plaza, New York, NY 10004, U.S.A.

Preface

This book provides basic understanding of the molecular mechanisms that are involved in neuronal morphogenesis, as well as detailed methods for analyzing these mechanisms. This book offers an overview of numerous methods for analyzing neuronal morphology in vitro and in vivo using a variety of model systems, including rodents and human-induced pluripotent stem cells (hiPSCs). The book is suitable for a wide range of readers, including students and independent researchers interested in the mechanisms of neuronal morphology and related neurological disorders such as psychiatric and neurodevelopmental disorders.

Philadelphia, PA, USA *Kazuhito Toyooka*

Contents

Contributors

PRASHANT ADHIKARI • *Center for Biomedical Engineering and Science, Department of Mechanical Engineering and Engineering Science, University of North Carolina at Charlotte, Charlotte, NC, USA*

JAYNE AIKEN • *Department of Physiology, Perelman School of Medicine, University of Pennsylvania, Philadelphia, PA, USA*

RAFAEL CARVALHO ALMADA • *Behavioural Neuroscience Institute (INeC), São Paulo, Brazil; Department of Biological Sciences, Laboratory of Neurobiology and Neurobiotechnology, São Paulo State University (UNESP), School of Science, Humanities and Languages, Assis, São Paulo, Brazil*

JESSIE R. BADLEY • *Department of Molecular Biosciences, School of Veterinary Medicine, University of California, Davis, CA, USA*

EMILY ANNE BATES • *Department of Pediatrics, Section of Developmental Biology, University of Colorado Anschutz Medical Campus, Aurora, CO, USA*

CLAUDIA LATTE BOVIO • *Tissue Electronics, Istituto Italiano di Tecnologia, Naples, Italy; Dipartimento di Chimica, Materiali e Produzione Industriale, Università di Napoli Federico II, Naples, Italy*

HOLLY B. BURBIDGE • *Division of Molecular and Cellular Function, Faculty of Biology, Medicine and Health, University of Manchester, Manchester, UK*

INLE BUSH • *Weill Institute for Cell and Molecular Biology and Department of Molecular Biology and Genetics, Cornell University, Ithaca, NY, USA*

WEIXUN CHEN • *College of Electrical and Information Engineering, Hunan University, Hunan, China; National Engineering Research Center for Robot Visual Perception and Control Technology, Changsha, China*

GISELLE CHEUNG • *Institute of Science and Technology Austria (ISTA), Klosterneuburg, Austria*

LAURA PULIDO CIFUENTES • *Department of Biological Sciences, Purdue University, West Lafayette, IN, USA*

NORBERTO CYSNE COIMBRA • *Department of Pharmacology, Laboratory of Neuroanatomy and Neuropsychobiology, Ribeirão Preto Medical School of the University of São Paulo (FMRP-USP), Ribeirão Preto, São Paulo, Brazil; Behavioural Neuroscience Institute (INeC), São Paulo, Brazil; Ophidiarium LNN-FMRP-USP/INeC, Ribeirão Preto Medical School of the University of São Paulo, Ribeirão Preto, São Paulo, Brazil; NAP-USP-Neurobiology of Emotions Research Centre (NuPNE), Ribeirão Preto Medical School of the University of São Paulo, Ribeirão Preto, São Paulo, Brazil*

RAMAN M. DAS • *Division of Molecular and Cellular Function, Faculty of Biology, Medicine and Health, University of Manchester, Manchester, UK*

TRACEY DAVEY • *Electron Microscopy Research Services, Newcastle University, Newcastle-upon-Tyne, UK*

JENNY R. DIAZ • *Departments of Neurology, New York University Grossman School of Medicine, New York, NY, USA*

MOHAMED H. FARAH • *Department of Neurology, Neuromuscular Division, Johns Hopkins University School of Medicine, Baltimore, MD, USA*

Maria de Fátima dos Santos Sampaio • *Department of Pharmacology, Laboratory of Neuroanatomy and Neuropsychobiology, Ribeirão Preto Medical School of the University of São Paulo (FMRP-USP), Ribeirão Preto, São Paulo, Brazil; Behavioural Neuroscience Institute (INeC), São Paulo, Brazil; Laboratory of Tissue and Cellular Biology, Darcy Ribeiro Northern Fluminense State University (UENF) Centre of Biosciences and Biotechnology, Rio de Janeiro, Brazil*

Renato Leonardo de Freitas • *Department of Pharmacology, Laboratory of Neuroanatomy and Neuropsychobiology, Ribeirão Preto Medical School of the University of São Paulo (FMRP-USP), Ribeirão Preto, São Paulo, Brazil; Behavioural Neuroscience Institute (INeC), São Paulo, Brazil*

Ketakee Ghate • *Indian Institute of Science Education and Research (IISER) Pune, Pune, India; INSERM UMR-S 1270/ Institut du Fer à Moulin/ Sorbonne University, Paris, France*

Aurnab Ghose • *Indian Institute of Science Education and Research (IISER) Pune, Pune, India*

Nanako Hamada • *Department of Molecular Neurobiology, Institute for Developmental Research, Aichi Developmental Disability Center, Kasugai, Aichi, Japan*

Chun Han • *Weill Institute for Cell and Molecular Biology and Department of Molecular Biology and Genetics, Cornell University, Ithaca, NY, USA*

Yasunori Hayashi • *Department of Pharmacology, Kyoto University Graduate School of Medicine, Kyoto, Japan*

Paloma Molina Hernandes • *Department of Pharmacology, Laboratory of Neuroanatomy and Neuropsychobiology, Ribeirão Preto Medical School of the University of São Paulo (FMRP-USP), Ribeirão Preto, São Paulo, Brazil; Department of Biological Sciences, Laboratory of Neurobiology and Neurobiotechnology, São Paulo State University (UNESP), School of Science, Humanities and Languages, Assis, São Paulo, Brazil*

Victoria E. Higgs • *Division of Molecular and Cellular Function, Faculty of Biology, Medicine and Health, University of Manchester, Manchester, UK*

Simon Hippenmeyer • *Institute of Science and Technology Austria (ISTA), Klosterneuburg, Austria*

Erika L. F. Holzbaur • *Department of Physiology, Perelman School of Medicine, University of Pennsylvania, Philadelphia, PA, USA*

Chia-Wei Huang • *Institute of Brain Science, College of Medicine, National Yang Ming Chiao Tung University, Taipei, Taiwan*

Mini Jose • *Centre for Neuroscience, Indian Institute of Science, Bangalore, India; Cell and Tissue Imaging Center, St. Jude Children's Research Hospital, Memphis, TN, USA*

Prashant Kaushal • *Mass Spectrometry & Proteomics Core Facility, University of Nebraska Medical Center, Omaha, NE, USA*

Vikas Kumar • *Mass Spectrometry & Proteomics Core Facility, University of Nebraska Medical Center, Omaha, NE, USA; Department of Genetics, Cell Biology and Anatomy, University of Nebraska Medical Center, Omaha, NE, USA*

Ross Laws • *Electron Microscopy Research Services, Newcastle University, Newcastle-upon-Tyne, UK*

Pamela J. Lein • *Department of Molecular Biosciences, School of Veterinary Medicine, University of California, Davis, CA, USA*

Zhuangdian Lin • *College of Electrical and Information Engineering, Hunan University, Hunan, China; National Engineering Research Center for Robot Visual Perception and Control Technology, Changsha, China*

MIN LIU • *College of Electrical and Information Engineering, Hunan University, Hunan, China; National Engineering Research Center for Robot Visual Perception and Control Technology, Changsha, China*
XIAONAN LIU • *Department of Pharmacology and Physiology, Drexel University College of Medicine, Philadelphia, PA, USA*
ANNA MARIANO • *Tissue Electronics, Istituto Italiano di Tecnologia, Naples, Italy*
KATHERINE L. MARSHALL • *Department of Neurology, Neuromuscular Division, Johns Hopkins University School of Medicine, Baltimore, MD, USA*
ERIK MEIJERING • *School of Computer Science and Engineering, University of New South Wales, Sydney, NSW, Australia*
GEORGE MERCES • *Biosciences Institute, Innovation, Methodology and Application (IMA) research Theme, Faculty of Medical Sciences, Newcastle University, Newcastle-upon-Tyne, UK; Image Analysis Unit, Faculty of Medical Sciences, Newcastle University, Newcastle-upon-Tyne, UK*
OSVALDO A. MIRANDA • *Institute of Science and Technology Austria (ISTA), Klosterneuburg, Austria*
VALENTINA MOLLO • *Tissue Electronics, Istituto Italiano di Tecnologia, Naples, Italy*
ALICE E. MORTIMER • *Division of Molecular and Cellular Function, Faculty of Biology, Medicine and Health, University of Manchester, Manchester, UK; Division of Cell Matrix Biology and Regenerative Medicine, Faculty of Biology, Medicine and Health, University of Manchester, Manchester, UK; Department of Plastic Surgery & Burns, Wythenshawe Hospital, Manchester University NHS Foundation Trust, Manchester, UK*
SAMPADA P. MUTALIK • *Indian Institute of Science Education and Research (IISER) Pune, Pune, India; Department of Cell Biology and Physiology, University of North Carolina, Chapel Hill, NC, USA*
KOH-ICHI NAGATA • *Department of Molecular Neurobiology, Institute for Developmental Research, Aichi Developmental Disability Center, Kasugai, Aichi, Japan*
NAOKO NIIMI • *Department of Diseases and Infection, Tokyo Metropolitan Institute of Medical Science, Setagaya-ku, Tokyo, Japan*
KYLE R. NORTHINGTON • *Department of Pediatrics, Section of Developmental Biology, University of Colorado Anschutz Medical Campus, Aurora, CO, USA*
ISABEL PÉREZ-FERRER • *Instituto centrifugede Neurociencias, Consejo Superior de Investigaciones Científicas-Universidad Miguel Hernández (CSIC-UMH), Alicante, Spain*
KHAYZARAN QUBBAJ • *Center for Biomedical Engineering and Science, Department of Mechanical Engineering and Engineering Science, University of North Carolina at Charlotte, Charlotte, NC, USA*
AMY K. REEVE • *Wellcome Centre for Mitochondrial Research, Translational and Clinical Research Institute, Faculty of Medical Sciences, Newcastle University, Newcastle-upon-Tyne, UK*
ADAM J. REID • *Division of Cell Matrix Biology and Regenerative Medicine, Faculty of Biology, Medicine and Health, University of Manchester, Manchester, UK; Department of Plastic Surgery & Burns, Wythenshawe Hospital, Manchester University NHS Foundation Trust, Manchester, UK*
MARTIN J. SADOWSKI • *Departments of Neurology, New York University Grossman School of Medicine, New York, NY, USA; Departments of Psychiatry, New York University Grossman School of Medicine, New York, NY, USA; Departments of Biochemistry and Molecular Pharmacology, New York University Grossman School of Medicine, New York, NY, USA*

CARLOS ERNESTO GARRRIDO SALMON • *Department of Physics, InBrain Lab, Ribeirão Preto School of Philosophy, Sciences and Literature of the University of São Paulo, Ribeirão Preto, São Paulo, Brazil*

CARLOS SÁNCHEZ-HUERTAS • *Instituto centrifugede Neurociencias, Consejo Superior de Investigaciones Científicas-Universidad Miguel Hernández (CSIC-UMH), Alicante, Spain*

TAKEO SANEYOSHI • *Department of Pharmacology, Kyoto University Graduate School of Medicine, Kyoto, Japan*

KAZUNORI SANGO • *Department of Diseases and Infection, Tokyo Metropolitan Institute of Medical Science, Setagaya-ku, Tokyo, Japan*

FRANCESCA SANTORO • *Tissue Electronics, Istituto Italiano di Tecnologia, Naples, Italy; Faculty of Electrical Engineering and IT, RWTH Aachen, Aachen, Germany; Institute of Biological Information Processing – Bioelectronics, IBI-3, Forschungszentrum Juelich, Jülich, Germany*

DAVI HENRIQUE SALVADOR PEREIRA DOS SANTOS • *Department of Pharmacology, Laboratory of Neuroanatomy and Neuropsychobiology, Ribeirão Preto Medical School of the University of São Paulo (FMRP-USP), Ribeirão Preto, São Paulo, Brazil; Darcy Ribeiro Northern Fluminense State University (UENF) Science and Technology Centre, Rio de Janeiro, Brazil*

KANIKA SHARMA • *Mass Spectrometry & Proteomics Core Facility, University of Nebraska Medical Center, Omaha, NE, USA*

YEN-LIN SU • *Institute of Brain Science, College of Medicine, National Yang Ming Chiao Tung University, Taipei, Taiwan*

DANIEL M. SUTER • *Department of Biological Sciences, Purdue University, West Lafayette, IN, USA; Weldon School of Biomedical Engineering, Purdue University, West Lafayette, IN, USA; Purdue Institute for Integrative Neuroscience, Purdue University, West Lafayette, IN, USA; Purdue Institute for Inflammation, Immunology, and Infectious Disease, Purdue University, West Lafayette, IN, USA; Bindley Bioscience Center, Purdue University, West Lafayette, IN, USA; Birck Nanotechnology Center, Purdue University, West Lafayette, IN, USA*

HIDENORI TABATA • *Department of Molecular Neurobiology, Institute for Developmental Research, Aichi Developmental Disability Center, Kasugai, Aichi, Japan*

SHIZUKA TAKAKU • *Department of Diseases and Infection, Tokyo Metropolitan Institute of Medical Science, Setagaya-ku, Tokyo, Japan*

GABRIELA TORO-TAPIA • *Division of Molecular and Cellular Function, Faculty of Biology, Medicine and Health, University of Manchester, Manchester, UK*

KAZUHITO TOYO-OKA • *Department of Neurobiology and Anatomy, Drexel University College of Medicine, Philadelphia, PA, USA*

JIN-WU TSAI • *Institute of Brain Science, College of Medicine, National Yang Ming Chiao Tung University, Taipei, Taiwan; Brain Research Center, National Yang Ming Chiao Tung University, Taipei, Taiwan; Department of Biological Science & Technology, College of Biological Science and Technology, National Yang Ming Chiao Tung University, Hsinchu, Taiwan*

JANE TWEEDY • *Wellcome Centre for Mitochondrial Research, Translational and Clinical Research Institute, Faculty of Medical Sciences, Newcastle University, Newcastle-upon-Tyne, UK*

FARHAD ULLAH • *Department of Pharmacology, Laboratory of Neuroanatomy and Neuropsychobiology, Ribeirão Preto Medical School of the University of São Paulo (FMRP-*

USP), Ribeirão Preto, São Paulo, Brazil; Department of Zoology, Islamia College Peshawar, Peshawar, Khyber Pakhtunkhwa, Pakistan

AMY E. VINCENT • *Wellcome Centre for Mitochondrial Research, Translational and Clinical Research Institute, Faculty of Medical Sciences, Newcastle University, Newcastle-upon-Tyne, UK; John Walton Muscular Dystrophy Research Centre, Translational and Clinical Research Institute, Faculty of Medical Sciences, Newcastle University, Newcastle-upon-Tyne, UK; Newcastle University Medical School, Framlington Place, Newcastle-upon-Tyne, UK*

YAONAN WANG • *College of Electrical and Information Engineering, Hunan University, Hunan, China; National Engineering Research Center for Robot Visual Perception and Control Technology, Changsha, China*

REBECCA J. WILSON • *Department of Molecular Biosciences, School of Veterinary Medicine, University of California, Davis, CA, USA*

YINENG XU • *Weill Institute for Cell and Molecular Biology and Department of Molecular Biology and Genetics, Cornell University, Ithaca, NY, USA*

HIDEJI YAKO • *Department of Diseases and Infection, Tokyo Metropolitan Institute of Medical Science, Setagaya-ku, Tokyo, Japan*

IN HONG YANG • *Center for Biomedical Engineering and Science, Department of Mechanical Engineering and Engineering Science, University of North Carolina at Charlotte, Charlotte, NC, USA*

Chapter 1

Culturing Primary Cortical Neurons for Live-Imaging

Kyle R. Northington and Emily Anne Bates

Abstract

Primary neuronal cultures allow for in vitro analysis of early developmental processes such as axon pathfinding and growth dynamics. When coupled with methods to visualize and measure microtubule dynamics, this methodology enables an inside look at how the cytoskeleton changes in response to extracellular signaling cues. Here, we describe the culturing conditions and tools required to extract primary cortical neurons from postnatal mouse brains and visualize cytoskeletal components.

Key words Primary cortical neuron, Neuron culture, Cytoskeleton, Microtubules, Live-imaging

1 Introduction

Changes to mammalian genes can result in differences in brain function and gross morphology, but it is difficult to determine how molecular changes affect neuronal cell biology to cause large scale changes. Primary cultured mammalian cortical neurons are ideal for studies defining the unique cell biology required for neurodevelopment. The recapitulation of neuronal processes in an in vitro setting allows for live-imaging and high-resolution imaging with controlled variables. These studies can define important aspects of neurodevelopment including cytoskeletal regulation of neuronal morphology, neuron polarization, protein localization, synapse formation, axonogenesis, dendrite branching, and axon growth and guidance.

Primary cortical neuron cultures provide an opportunity to study the basic biology behind microtubules and actin and how these changes can affect neuron and tissue morphology. The neuronal cytoskeleton is regulated through tight control of protein levels. While there are multiple mammalian isotypes of alpha and beta tubulin, they are highly conserved across species [1]. The majority of variance occurs in the carboxy terminal tails of the proteins, an area that is exposed on the microtubule lattice and

Kazuhito Toyooka (ed.), *Neuronal Morphogenesis: Methods and Protocols*, Methods in Molecular Biology, vol. 2831,
https://doi.org/10.1007/978-1-0716-3969-6_1,

undergoes a variety of posttranslational modifications [2, 3]. Cultured primary neurons can be used to determine how the microtubule cytoskeleton is regulated for axon pathfinding by reducing function of tubulin isotypes, microtubule-associated proteins (MAPs), motor proteins, and enzymes that remove or add posttranslational modifications. Additionally, cultured neurons can be used to assess features of the actin-rich growth cone [4–6]. Cultured neurons can be used to assess the consequences of mutations on the cell biology of the neuron. For example, mutations in tubulin are responsible for brain malformations [7–11]. These diseases termed tubulinopathies were believed to be due to haploinsufficiencies; however, recent work has shown that mutant tubulin can act in a dominant negative fashion [12]. Thus, wildtype neurons can be transfected with genes encoding tubulin isotypes with tubulinopathy-associated mutations to determine how these affect microtubule function [13, 14].

Cultured neurons can be used to mechanisms underlying axon guidance. During development, axons navigate through a complex changing environment. A variety of receptors and secreted cues allow for precise pathfinding [15–21]. Axon behavior (growth, retraction, and directed turning) can be assessed in a culture environment.

Cultured neurons can be used to study subcellular specialized features of the neuron like the axon initial segment (AIS), a hallmark feature of axons [22–25]. The AIS is a dense region of proteins located at the proximal end of the axon. This region contains Ankyrin proteins that anchor ion channels that are needed to initiate neuronal action potentials [22, 23].

Limitations While primary cultured neurons are a good model to investigate the internal mechanisms of neuronal development and function, they do not completely recapitulate brain development. Neurons on a dish do not migrate as they would be in the developing brain and thus the culture system cannot be used to study neuronal migration.

2 Materials

1. Dissection Media (DM) 250 mL.	
HBSS (Ca^{2+} and Mg^{2+} free)	247.5 mL.
1 M HEPES	2.5 mL.
Kynurenate solution	200 μL per 20 mL (to be added before using solution on dissection plates).
Notes: Solution should be sterile filtered before use.	

2. Plating Media (PM) 250 mL.	
DMEM w/glucose and sodium pyruvate	247.5 mL.
Glutamax (100X)	2.5 mL.
Pen/Strep (100X)	2.5 mL.
Notes: Solution should be sterile filtered before use.	

3. Maintenance Media (MM) 50 mL.	
Neurobasal A	247.5 mL.
B27 (50X)	1 mL.
Glutamax (100X)	500 µL.
B-FGF (0.1 mg/mL)	2.5 µL.
Notes: Solution should be sterile filtered before use. If culturing for more than 4 days, replace media with media containing 10 ng/mL B-FGF on DIV 3–4.	

4. Borate Buffer (0.1 M).	
Boric Acid	1.24 g.
Sodium tetraborate	1.90 g.
MilliQ Water (pH 8.5)	400 mL.
Notes: Solution should be sterile filtered using a 0.2 µm filter before use. Several hours of stirring may be required for these chemicals to dissolve. The solution can be stored at 4 °C for 3–4 weeks.	

5. Poly-D Lysine working solution.	
Borate Buffer	8 mL.
Poly-D Lysine Stock	2 mL.
Notes: Mix the stock Poly-D Lysine with the previously made Borate Buffer (1:5 dilution).	

6. Kynurenate Solution (100 mM).	
Kynurenic Acid	Batch specific MW—check bottle before use.
10 N NaOH	100 µL.
MilliQ Water	9.9 mL.
Notes: Add water and NaOH to beaker to make 100 mM NaOH, then slowly add kynurenic powder. Stir on hot plate at 45–60 °C. Make 1 mL aliquots and store at −20 °C.	

7. Papain Solution.	
Dissection Media	10 mL.
Kynurenate (100 mM)	100 μL.
Papain	Depends on batch you need 50 U/brain.
Cysteine (1 M)	2 mg.
DNAse I	500 μL.
Notes: Prepare this solution for the morning of dissection. You will need 5 mL per brain, so this recipe is designed for two mouse pups. Vortex the solution and aliquot into 15 mL conical tube with 5 mL per conical tube. Place tubes in the incubator for at least 30 min and leave in incubator until ready to add brain tissue. Make sure that the papain solution is within date; expired papain reduces efficacy of cell dissociation resulting in clumps on the plate.	

8. Miscellaneous supplies.
Small bore glass pipette.
Large bore glass pipette.
Sylgard-coated Petri dishes.
Dissection tools.
Trypan blue.

3 Methods

On the day of the dissection, begin by making the papain solution in a biosafety cabinet (BSC). Simultaneously, coat your cell culture vessel with Poly-D Lysine solution. This needs to sit at room temperature for at least 1 h before plating cells. For optimal cell viability it is important to take the mouse pups between P0 and P4 for dissection.

Next, prepare the materials for the dissection, which includes Kim wipes, ethanol, scissors, forceps, a Sylgard-coated Petri dish, Dissection Media (DM), sterile conical tubes, and a dissection scope. When ready to begin the dissection, spray a Kim wipe with ethanol and wipe the pup's head to both sterilize and anesthetize the pup.

Next, use sterile scissors to decapitate the mouse pup as close to the base of the skull as possible. Place the decapitated head into a petri dish filled with PBS to remove blood from the wound. Move the head into a gloved hand and cut down the midline to peel back the skin on the head. Carefully cut through the skull along the midline while avoiding cutting into the brain tissue. Next, peel the skull toward the side to expose the brain tissue. Then use a slotted scoop to remove the brain and place into a Sylgard-coated Petri dish filled with DM. Next, use fine-tip forceps to remove the olfactory bulb and cerebellum/hindbrain, and then press down on the brain tissue at midline to separate the cortices (Figs. 1 and 2). Then use fine-tip forceps to remove the meninges from the dissected cortex. Next, chop the cortex into small pieces and

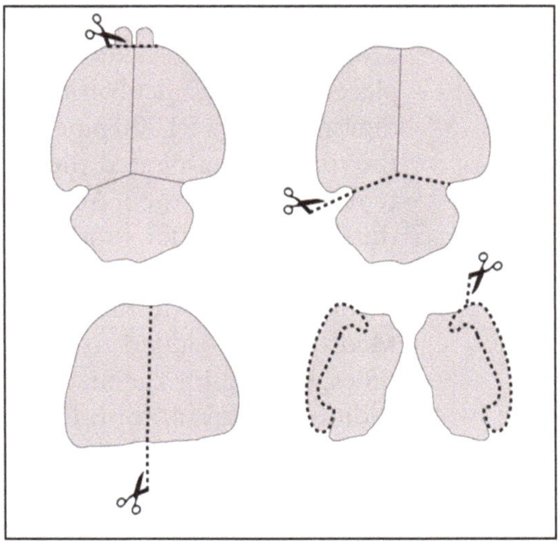

Fig. 1 Diagram of cortical brain isolation and removal. Olfactory bulbs should be removed first, followed by the hindbrain. Next, split the brain along the midline and lay it open like a book. The cortex can be visually identified by a subtle line that runs along the brain. Once cortical pieces are isolated, it is important to cut them into small pieces to increase the surface area in which papain can interact

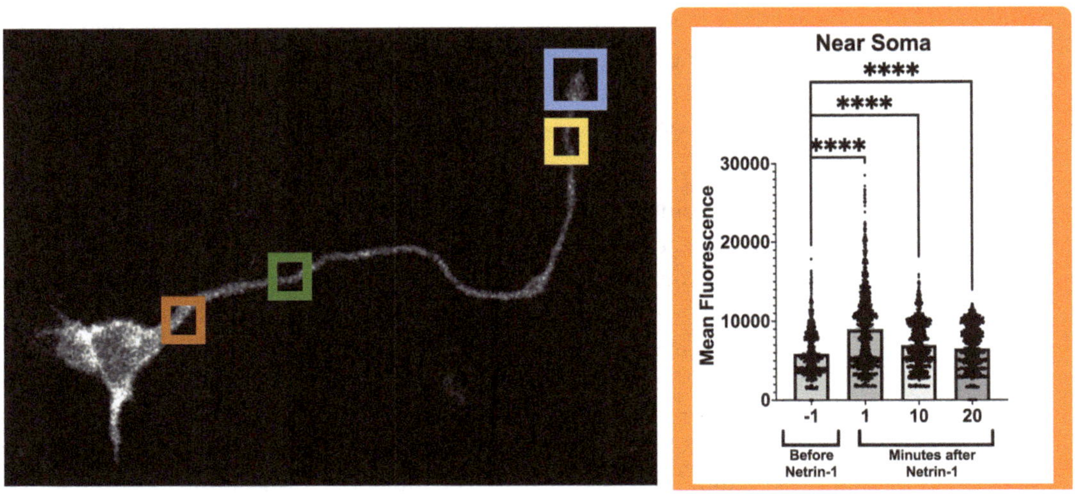

Fig. 2 DIV1 neurons nucleofected with GFP-CSAP and myrTdTomato. Neurons were imaged every 10 min before and after a bath application of Netrin-1. The axon and growth cone were measured for changes in fluorescence intensity using a Zeiss 900 microscope with Airyscan 2

transfer the tissue to a 15 mL conical tube with 4 mL fresh DM using forceps with textured grips, keeping the tissue at room temperature.

This process is repeated for each mouse, sterilizing the dissection instruments in between to maintain sterile conditions. To improve cell viability, only dissect two mice at a time, then move on to papain dissociation to ensure that the cells remained viable. Once the first two mice are placed in papain, the dissection can be continued with the next two mice.

Brain pieces were transferred to 5 mL of papain solution in 15 mL conical tubes using a P1000 pipette. The tubes were then placed in a 37 °C incubator for 40–45 min to digest the tissue. After digestion, use a P1000 pipette to remove any goopy meninges from the tissue and aspirated the papain, leaving approximately 100 μL remaining in the tube. At this stage the tissue is not a hard pellet at the bottom of the tube, so the liquid must be aspirated very carefully to avoid loss of tissue.

To remove the remaining papain, 5 mL of warmed Plating Media (PM) is added to each tube and gently inverted them 3–4 times, waiting for the tissue to settle before aspirating the PM. An additional 5 mL of fresh PM is added to each tube. Next, triturate the tissue with a wide-bore glass pipette approximately 10 times. Larger pieces of brain tissue will sink to the bottom of the tube. At that point, quickly transfer the supernatant using a narrow-bore glass pipette to a new tube. Add 1 mL of fresh PM and triturate the solution with the narrow-bore glass pipette approximately 10 times before moving on to downstream protocols.

Cells were counted and then diluted to the desired density in PM before plating them at a density of 300,000 cells per well. The next day or approximately 2–4 h later, aspirate the media from the wells and replace it with warm PM. Cells were taken for imaging at DIV1 (24 h after plating). If the cells will be grown longer than DIV1, media should be replaced with warmed Maintenance Media (MM) with araC at 1–5 μM at 24 h to inhibit the growth of nonneuronal cells.

Imaging

Using primary cortical neurons can be a useful medium to investigate aspects of axon extension and pathfinding. Live-imaging provides a means to visualize how cell biology processes can change during axon extension. To visualize these changes, neurons were nucleofected using an AMAXA Nucleofector 2b and a Lonza Mouse Neuron Nucleofector Kit. This step is performed after the dissociation of cells, but before plating. Primary cortical neurons were nucleofected using 4 μg of DNA per plasmid. First, the plasmids are mixed with the nucleofection solution with added supplement solution to create a total volume of 50 μL. Next, after centrifuging the cells at 400× *g* for 5 min, the media is carefully aspirated to avoid loss of the cell pellet. The cells will be resuspended in 50 μL of the nucleofection solution. Once resuspended, the 50 μL containing the plasmids will be added and the two solutions need to be mixed with a pipette. The 100 μL of total solution is transferred to the Amaxa cuvettes. Quickly take the cuvettes and place them in the Amaxa nucleofection device and electroporate the cells using setting O-03. For nucleofection a higher concentration of cells is desirable as there will be a loss in viability. As a result, it is important to increase the cell number to 500,000 cells per plate. Once the cells have been nucleofected, they need to be placed in a 15 mL conical tube with 2 mL of PM supplemented with 1 μL of L-glutamine. The tube will then be placed in a 37 °C incubator for 20 min to allow for recovery. After the 20-min incubation, the cells can be plated in your desired Poly-D-Lysine treated dish.

To image these cells, we use an inverted Zeiss 900 confocal microscope with Airyscan 2. This microscope allows for rapid image capture times that allow us to visualize microtubule and other cellular dynamics; however, any confocal or spinning disk microscope should be sufficient. Ideally, an incubation setup will be used with the microscope to ensure that the primary neurons are in optimal conditions. If this is unavailable, there are media alternatives that will be substituted for the Neurobasal-A to have better CO_2 buffering properties such as Hibernate-A media. However, studying microtubule dynamics in the absence of an incubation chamber setup is not recommended as microtubule dynamics can change based on ambient temperatures.

4 Notes

1. Prepare all solutions for the morning of dissection. PM and DM can be kept at 4 °C and used for subsequent dissections for approximately 1 month. MM should be made fresh each time it is used to preserve the efficacy of the B-FGF.
2. It is important to use a Sylgard-coated Petri dish for the brain dissection as it protects the fine-tip forceps from being damages. Damaged forceps make removing the meningeal tissue very difficult.
3. Remove as much of the meninges as possible as it inhibits downstream processing and can make the cells very hard to pipette during nucleofection.
4. Make sure to use the appropriate amount and concentration of DNase I. A reduction in the amount causes the tissue to be viscous after papain treatment.

5. It is possible to solely use narrow bore pipettes and skip the transferring of supernatant to a new tube. Doing so may result in a loss of cell viability as it will be necessary to triturate the cells more vigorously to ensure there are no clumps of tissue. If you use this method, it is important to stop triturating as soon as possible.
6. While using P0–P4 mice results in the greatest likelihood of a successful primary culture, it may be possible to go slightly beyond that age recommendation.
7. Once the cells have been electroporated, they should be quickly moved to the recovery media.
8. Ideally the time between dissecting the brain and placing it into papain should be around 20 min. This allows for an estimated 10 min per brain. Going significantly over that time can result in loss of viability.
9. Neurons can be visualized to determine the success of the experiment the next morning using a brightfield microscope if they have not been transfected with a fluorescent reporter; however, it is much easier to visualize them if they contain a fluorescent tag.

References

1. Little M, Krauhs E, Ponstingl H (1981) Tubulin sequence conservation. Biosystems 14:239–246
2. Janke C (2014) The tubulin code: molecular components, readout mechanisms, and functions. J Cell Biol 206:461–472
3. Janke C, Magiera MM (2020) The tubulin code and its role in controlling microtubule properties and functions. Nat Rev Mol Cell Biol 21:307–326
4. Minegishi T, Fujikawa R, Kastian RF, Sakumura Y, Inagaki N (2021) Analyses of actin dynamics, clutch coupling and traction force for growth cone advance. J Vis Exp. https://doi.org/10.3791/63227
5. Flynn KC, Pak CW, Shaw AE, Bradke F, Bamburg JR (2009) Growth cone-like waves transport actin and promote axonogenesis and neurite branching. Dev Neurobiol 69:761–779
6. Slater PG et al (2019) XMAP215 promotes microtubule-F-actin interactions to regulate growth cone microtubules during axon guidance in Xenopus laevis. J Cell Sci 132: jcs224311
7. Bahi-Buisson N et al (2014) The wide spectrum of tubulinopathies: what are the key features for the diagnosis? Brain J Neurol 137: 1676–1700
8. Fertuzinhos S, Legué E, Li D, Liem KF (2022) A dominant tubulin mutation causes cerebellar neurodegeneration in a genetic model of tubulinopathy. Sci Adv 8:eabf7262
9. Hebebrand M et al (2019) The mutational and phenotypic spectrum of TUBA1A-associated tubulinopathy. Orphanet J Rare Dis 14:38
10. Gartz Hanson M et al (2016) Novel α-tubulin mutation disrupts neural development and tubulin proteostasis. Dev Biol 409:406–419
11. Buscaglia G, Northington KR, Aiken J, Hoff KJ, Bates EA (2021) Bridging the gap: the importance of TUBA1A α-tubulin in forming midline commissures. Front Cell Dev Biol 9: 789438
12. Aiken J, Buscaglia G, Aiken AS, Moore JK, Bates EA (2019) Tubulin mutations in brain development disorders: Why haploinsufficiency does not explain TUBA1A tubulinopathies. Cytoskeleton 77:40. https://doi.org/10.1002/cm.21567
13. Hoff KJ, Aiken JE, Gutierrez MA, Franco SJ, Moore JK (2022) TUBA1A tubulinopathy mutants disrupt neuron morphogenesis and override XMAP215/Stu2 regulation of microtubule dynamics. elife 11:e76189
14. Park K et al (2021) Kinetically Stabilizing Mutations in Beta Tubulins Create Isotype-

Specific Brain Malformations. Front Cell Dev Biol 9:765992

15. Fothergill T et al (2014) Netrin-DCC signaling regulates corpus callosum formation through attraction of pioneering axons and by modulating Slit2-mediated repulsion. Cereb Cortex 1991(24):1138–1151
16. Howard LJ, Reichert MC, Evans TA (2021) The Slit-binding Ig1 domain is required for multiple axon guidance activities of Drosophila Robo2. Genes N. Y. N 2000 59:e23443
17. Silver J, Lorenz SE, Wahlsten D, Coughlin J (1982) Axonal guidance during development of the great cerebral commissures: descriptive and experimental studies, in vivo, on the role of preformed glial pathways. J Comp Neurol 210: 10–29
18. Bush JO, Soriano P (2009) Ephrin-B1 regulates axon guidance by reverse signaling through a PDZ-dependent mechanism. Genes Dev 23:1586–1599
19. Croteau L-P, Kao T-J, Kania A (2019) Ephrin-A5 potentiates netrin-1 axon guidance by enhancing Neogenin availability. Sci Rep 9: 12009
20. Ueno M et al (2020) Olig2-induced semaphorin expression drives corticospinal axon retraction after spinal cord injury. Cereb Cortex 1991(30):5702–5716
21. Tessier-Lavigne M, Goodman CS (1996) The molecular biology of axon guidance. Science 274:1123–1133
22. Rasband MN, Shrager P (2000) Ion channel sequestration in central nervous system axons. J Physiol 525(Pt 1):63–73
23. Hedstrom KL, Ogawa Y, Rasband MN (2008) AnkyrinG is required for maintenance of the axon initial segment and neuronal polarity. J Cell Biol 183:635–640
24. Ogawa Y, Rasband MN (2008) The functional organization and assembly of the axon initial segment. Curr Opin Neurobiol 18:307–313
25. Schafer DP et al (2009) Disruption of the axon initial segment cytoskeleton is a new mechanism for neuronal injury. J Neurosci 29: 13242–13254

Chapter 2

Electron Microscopy of Neurons on Biomimetic Substrates

Claudia Latte Bovio, Valentina Mollo, Anna Mariano, and Francesca Santoro

Abstract

Recent advancements in nano- and microfabrication techniques have led to the development of highly biomimetic patterned substrates able to guide neuronal sprouting, routing, elongation, and branching. Such substrates, recapitulating shapes and geometries found in the native brain, may pave the way toward the development of cell instructive paradigms able to guide morphogenesis at the neuron–material interface. In this scenario, high-resolution electron microscopy approaches, owing to their ability of discerning the details of neural morphogenesis at a nanoscale resolution, may play a crucial role in unravelling the fine ultrastructure of neurons interfacing with biomimetic structured substrates.

Key words Neuron ultrastructure, Biomimetic substrates, Neural interface engineering, Neuron, material interface, Electron microscopy

1 Introduction

The morphogenesis of neuronal processes (i.e., dendrites and axons) is an essential step in the establishment of the neural circuitry [1]. During morphogenesis, neurons respond to a plethora of biochemical, mechanical, and topographical cues provided by the neural extracellular matrix (ECM) and neighboring cells [2, 3]. These niche-specific stimuli, by guiding neural processes navigation, outgrowth, branching, and contributing to synapses formation, will ensure the correct wiring of the brain [4–7]. Among informative cues found in the neural ECM, topography has been found to strongly influence neuronal function and behavior [8, 9], making structured substrates of great interest in neural tissue engineering. In fact, considerable effort has been deployed in developing highly biomimetic substrates able to recapitulate the neural ECM topographical architecture in order to guide morphogenesis, paving the way toward the development of engineered substrates that could be effectively used to dictate a desired

Kazuhito Toyooka (ed.), *Neuronal Morphogenesis: Methods and Protocols*, Methods in Molecular Biology, vol. 2831, https://doi.org/10.1007/978-1-0716-3969-6_2,

neuronal behavior at the cell–material interface [10–12]. For instance, topographical isotropic cues recapitulating shapes and geometries of dendritic spines have been used to promote axonal sprouting and elongation [13–16] and neurite branching [14, 15, 17, 18]. Similarly, anisotropic features, recapitulating both ECM fibrils and axonal tracts have been employed to guide neurites directionality and outgrowth [19–25]. Given the pivotal role that such engineered substrates may play in the tissue engineering field and for neural repair strategies, the possibility of visualizing morphogenesis at the cell–material interface becomes critical. In this scenario, electron microscopy, owing to its ability to reveal high-resolution ultrastructural information, has been instrumental in revealing neuron–substrate interactions [26]. Among electron microscopy methods, scanning electron microscopy (SEM) and transmission electron microscopy (TEM) have been widely used to visualize neurons supported with biomimetic substrates. Conventional SEM, relying on the emission of secondary electrons from the surface of a specimen, provides detailed top-view micrographs of neurons cultured on structured materials (Fig. 1a) [26–28]. Such approach has been used to visualize neurons tightly adhering on a structured substrates and sending their projections between protrusions (Fig. 1b) [17, 27, 29–31]. Direct visualization of neuron–substrate interfacial interactions, such as plasma membrane ruffling, can be characterized at the nanoscale by investigating cross sections at sites of topography-mediated impingement by TEM and FIB-SEM [28, 32]. Both these approaches have been instrumental in the bioelectronics field to measure the extent of micro- and "nanoprotrusions" engulfment by neurons and to measure the cleft formed between the neural plasma membrane and the substrate (Fig. 1c) [19, 29, 31, 33, 34].

Here, we will describe TEM, SEM, and FIB-SEM specimen preparation and material–neuron interface characterization. For all three approaches, the process starts with hydrated-living cells and ends either with a water-free sample (SEM) or with a plastic embedded sample (TEM; FIB-SEM), which will be used for further mechanical sectioning or focused ion beam milling.

2 Materials

Prepare all the solutions in distilled water and under the chemical hood. Follow all waste disposal regulations when disposing waste materials.

1. Primary Fixative: 4% paraformaldehyde (PFA) and/or 2.5–3% glutaraldehyde (GA) in sodium cacodylate buffer (*see* **Note 1**).
2. Sodium cacodylate buffer: 0.1 M sodium cacodylate in distilled water (*see* **Note 2**).

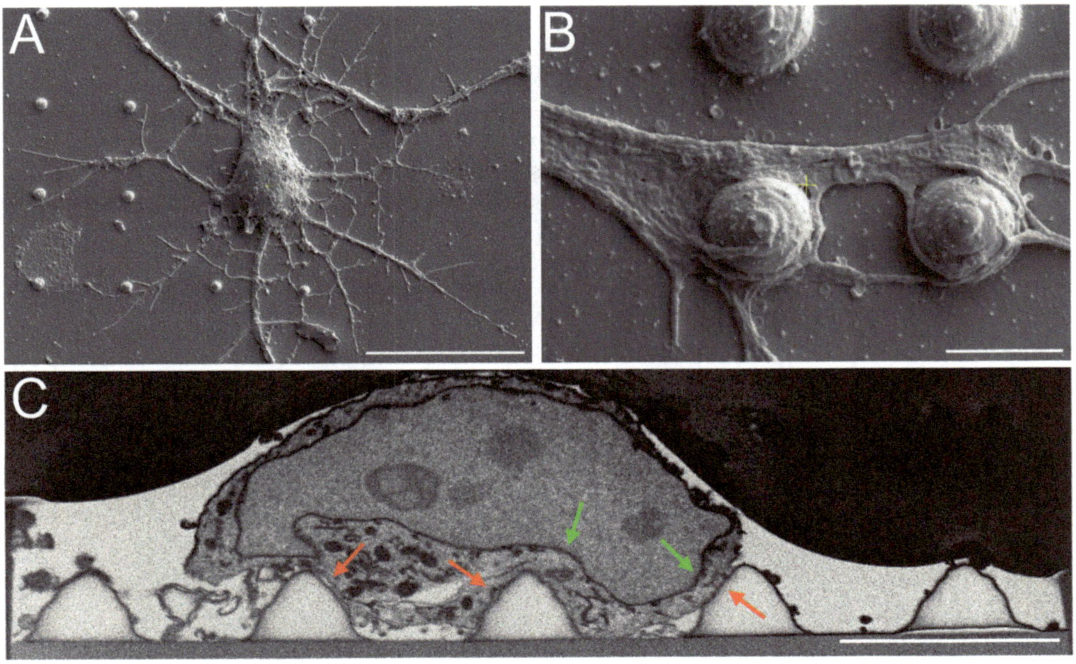

Fig. 1 High-resolution electron microscopy micrographs of neurons supported by biomimetic structured substrates. (**a**) Top-view scanning electron micrograph showing a neuron extending its neurites on a structured substrate. Scale bar: 20 μm. (**b**) Close-up view of a neurite adhering onto the topographical features of the substrate. Scale bar: 2 μm. (**c**) Cross section obtained by FIB milling and polishing showing neuron–substrate interactions. The topography-induced nucleus and cell membrane deformation are indicated by the green and red arrows, respectively. Scale bar: 4 μm

3. Glycine: 20 mM of glycine in 0.1 M sodium cacodylate.
4. Potassium ferrocyanide/osmium tetroxide: 1% (v/v) potassium ferrocyanide in 4% (v/v) osmium tetroxide in a 1:1 ratio, with a final concentration of 0.5% and 2%, respectively, in 0.2 M sodium cacodylate (*see* **Note 3**).
5. Thiocarbohydrazide (TCH): 1% (w/v) TCH in distilled water (*see* **Note 4**).
6. Osmium tetroxide: 2% (v/v) osmium tetroxide in distilled water.
7. En bloc staining solution: 0.5–2% (v/v) uranyl acetate in distilled water.
8. Tannic acid: 0.15% (v/v) tannic acid in in chilled distilled water.
9. Ethanol: 30–50–75–95% (v/v) ethanol in distilled chilled water. Isopropanol may also be used during sample dehydration.
10. Low-viscosity resin for embedding—Spurr's kit: prepare the stock of Spurr's resin following the instruction on the technical data sheet.
11. Embedding resin: 1:2, 1:1, 2:1 resin/ethanol (or isopropanol) (*see* **Note 5**).

3 Methods

TEM, SEM, and FIB-SEM sample preparation consists of three main steps as shown in Fig. 2: fixation, postfixation, and dehydration. Additional steps, such as contrast enhancement, embedding, and sectioning, are specific for TEM and FIB-SEM sample preparations. Similarly, post-dehydration by chemicals or critical point drying (CPD) are specific for SEM sample preparation.

Fixation (SEM, TEM, FIB-SEM) Fix the biological sample with 4% PFA or 2.5–3% GA. Depending on the thickness of the specimen, it is possible to use a mixture of GA and PFA. Incubate at 4 °C for 0.5–2 h if incubating with PFA, or overnight if incubating with GA (*see* **Note 6**). After incubation, rinse sample three times with sodium cacodylate buffer. Keep the samples at 4 °C (*see* **Note 7**).

Postfixation and Contrast Enhancement with Heavy Metal Staining (RO-T-O Staining) (SEM, TEM, FIB-SEM) Incubate the sample with 1% osmium tetroxide for 1 h at 4 °C. Avoid exposure to light. Osmium tetroxide, acting as a secondary fixative,

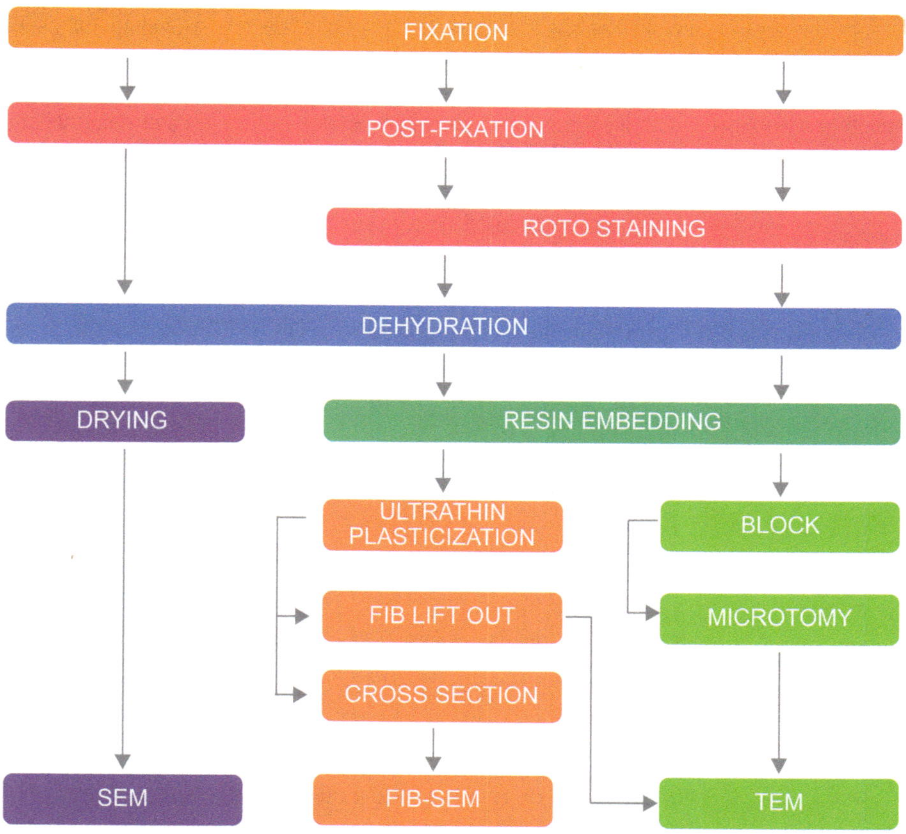

Fig. 2 Schematic representation of specimen preparation for SEM, FIB-SEM, and TEM

will (1) stabilize lipids and membranes, (2) provide contrast enhancement of cellular ultrastructure, and (3) preserve the sample from heating during the electron beam scanning. In the case of cross-sectioned specimens, additional contrast enhancement may be achieved by implementing the reduced osmium–thiocarbohydrazide–osmium (RO-T-O) procedure. Incubate samples with potassium ferrocyanide/osmium tetroxide (RO step). Then add 1% thiocarbohydrazide for 20–30 min at room temperature (T step). Incubate the sample with 1–2% osmium tetroxide (O step) for at least 20 min to 1 h at room temperature. After incubation, rinse sample three times with distilled water. Incubate the sample with uranyl acetate solution for an en bloc staining overnight at 4 °C and rinse with 0.15% tannic acid for 3 min at 4 °C.

Protein-Specific Contrast Enhancement Through Immunohistochemistry (Optional) (TEM, FIB-SEM) The immunogold labeling can be performed both pre- and post-embedding in resin for TEM sample preparation and imaging, and pre-embedding for FIB-SEM milling and imaging. Specific proteins can be additionally labeled through immunohistochemistry labeling, a process that binds gold nanoparticles to distinct cellular ultrastructures at the neuron–substrate interface providing contrast enhancement [35]. This procedure requires an incubation with a primary antibody binding to a target protein and a following step with a secondary antibody bearing both a fluorescent probe and gold nanoparticles. With this approach, the specimen can be first visualized with fluorescence microscopy and then a region of interest can be correlated with a subsequent electron micrograph. The nanogold particles are typically enhanced in size (gold enhancement technique) before the postfixation in osmium and uranyl acetate and tannic acid. The duration of nanogold particles growth depends on their size: 1–5 min are usually enough to enlarge 1.4 nm nanoparticles to 3–20 nm nanoparticles.

Dehydration (SEM, TEM, FIB-SEM) Incubate the sample with an ascendant series of ethanol or isopropanol dilutions (30%, 50%, 75%, 95%, 100%). Each step is carried out for 5–15 min at 4 °C. Rinse the sample twice with absolute ethanol at room temperature (*see* **Note 8**).

Samples used for SEM are further dried via air-drying with chemicals such as hexamethyldisilazane (HMDS) or via critical point drying (CPD). For air-dried samples, immerse the sample in increasing ratios of HMDS/ethanol (or isopropanol) until the solvent used for dehydration is completely replaced with HMDS.

For CPD, the ethanol (or isopropanol) used for dehydration is replaced with liquid CO_2 by using a critical point dryer (temperature, number of cycles, and cycle speed will vary depending on specimen's thickness and composition) (*see* **Note 9**). After drying, the sample can be directly coated with a thin layer of metal and used for SEM imaging. This step is necessary to make the sample conductive for SEM image acquisition. The coating is obtained via sputter coater machine. The thickness of the coating must be tuned depending on the samples.

Resin Embedding and Polymerization (TEM, FIB-SEM) After dehydration, the sample can be further processed for sectioning and TEM or FIB-SEM imaging. Here samples are infiltrated with liquid plastic resin and then the resin is polymerized. Incubate the sample in a resin/ethanol (or isopropanol) mixture in 1:2 ratio for 2–3 h at room temperature (RT). Move the sample in a resin/ethanol mixture in 1:1 ratio and incubate overnight at room temperature. Transfer the sample in a resin/ethanol in a 2:1 ratio for 3–4 h. To complete infiltration, move the sample in absolute resin (*see* **Note 10**). Fresh resin solution can be replaced up to three times every 4 h before leaving the samples for overnight infiltration at RT. Once samples are infiltrated, the resin can be cured.

Resin block embedding. *Transfer resin-infiltrated samples in silicon molds (or plastic capsules) for polymerization. Cure the resin in an oven at 60–70 °C for 48–72 h. Once polymerization is complete, the resin block containing the specimen can be further processed for mechanical or FIB sectioning.*

Resin flat-embedding. *Place the resin-infiltrated samples face down in resin-filled molds and cure the resin in an oven at 60–70 °C for 48–72 h.*

UTP. *Place the resin-infiltrated sample vertically for 3 h to remove the excess resin [36]. To facilitate this process, a low-viscosity resin (Spurr's resin) can be employed for the infiltration. Wash the sample with ethanol (or isopropanol) and cure the resin in an oven at 60–70 °C overnight.*

Sectioning (TEM, FIB-SEM) The sectioning can be performed both mechanically or by FIB-assisted milling. Before mechanical sectioning, the support substrate may be removed via immersion in liquid nitrogen (at −196 °C) or 49% hydrofluoric acid. After substrate removal the specimen might be re-embedded in resin (*see* **Note 11**).

Mechanical sectioning. *After resin block embedding, slice the sample using an ultramicrotome (equipped with a glass or diamond knife) to obtain a series of sections of about 50–100 nm of thickness for TEM imaging* (see ***Note 12***). *Leave the slices floating in a water bath to facilitate collection and transfer to TEM grids* (see ***Note 13***).

TEM lamellae preparation by focused ion beam. *Mill the sample surface with the focused ion beam (1–2 μm of thickness), then transfer the lamella to a customized grid for TEM imaging* (see **Note 14**). *Successful TEM imaging requires the lamella to be electron-transparent, which requires a thickness of ~100 nm. Once on the grid, reduce the size of the lamella down to 100–200 nm. Lift-out and transfer the lamella using an in situ micromanipulator* (see **Note 15**).

Focused ion beam milling. *After resin block embedding or UTP, the sample is further processed for focused ion beam milling. To protect the region of interest (ROI) from ion beam damage, deposit a Pt protective patch via electron-beam- and/or ion-beam-assisted chemical vapor deposition by tuning the size of the ROI and the thickness. Electron-beam-assisted deposition is less destructive to the sample and produces a smoother patch than the ion-assisted deposition; however, electron-beam-assisted deposition rate is very slow. A double deposition (electron-beam-assisted followed by ion-beam-assisted) may yield a more durable patch that protects the cell from beam damage. Start milling a wide rectangular trench few micrometers (μm) away from the Pt-protected region. The depth of the trench needs to be about three times the depth of the region of interest, and it should be high enough to avoid shadowing the region. At this stage, milling can be performed using relatively high currents such as 9.3 nA, or even more than 20 nA depending on the robustness of the resin to ion-induced local heat. Following the initial cross section, apply a lower current to polish the section and produce a cleaner and crisper image of the cross section. Draw a rectangular pattern to detect the area of interest for Pt deposition to protect the cell. Start the milling procedure in proximity of the Pt-protected region using a current of* $\leq$*2.8 nA.*

4 Notes

1. Both PFA and GA are highly toxic and must be handled under chemical hood. Wear protective equipment when handling.
2. Cacodylate buffer is toxic and must be handled under chemical hood. Wear protective equipment when handling.
3. Osmium tetroxide is toxic and must be handled under chemical hood. Wear protective equipment when handling.
4. TCH solution should be freshly prepared. To avoid TCH precipitation, both the sample and the TCH solution must be at room temperature.

5. The resin/solvent stock must be freshly prepared. Resin-based materials are highly toxic. Wear protective equipment when handling.
6. The sample can be stored for a maximum of 7 days at 4 °C.
7. Sample fixation with GA is mandatory for the EM sample preparation due to its capacity of irreversibly cross-linking proteins. PFA is not strictly needed.
8. The dehydration is a crucial step for EM sample preparation. Avoid to leave the sample dry during the final incubations with 70–95–100% ethanol (or isopropanol).
9. Despite CPD preserves the turgidity of the biological sample by preventing the evaporation of the solvent used for dehydration, this step will make the sample shrink of 20–30% of its original size.
10. Avoid the formation of bubbles before polymerization to prevent the formation of artefacts. Wear lab coat and gloves.
11. Substrate removal may fracture the resin and thus alter the cell–substrate interface. To overcome this issue, nanostructured substrates may be treated with a layer of octafluorocyclobutane (C4F8) before cell seeding. C4F8 exhibits anti-sticky properties, which is attributed to the extremely low surface energy of the fluorine-containing functional groups, and C4F8 coating facilitates the release of Si-based substrates upon resin polymerization.
12. Longitudinal sections are recommended to reduce sample damage.
13. The procedure can lead to some artefacts: chattering, compression, and knife marker. To facilitate slice distension water surface, chloroform vapors may be used.
14. The lateral dimensions (X, Y) of the lamella are sample-specific (cell size, density of the vertically aligned substrate). The bigger and heavier the lamella, the more difficult will be to transfer it in the grid. A lamella of 10 × 8 μm is usually ideal for TEM imaging (part of the lamella is sacrificed for its mounting on the grid via Pt welding).
15. The durability of the resin against ion damage is important to ensure the success of a lamella lift-out. A highly durable resin may expedite the milling process, allowing the use of higher currents as well as protecting the lamella from mechanical deformation.

References

1. Heuer K, Toro R (2019) Role of mechanical morphogenesis in the development and evolution of the neocortex. Phys Life Rev 31:233–239
2. Gattazzo F, Urciuolo A, Bonaldo P (2014) Extracellular matrix: a dynamic microenvironment for stem cell niche. Biochim Biophys Acta Gen Subj 1840:2506–2519
3. Mariano A, Bovio CL, Criscuolo V, Santoro F (2022) Bioinspired micro- and nano-structured neural interfaces. Nanotechnology. https://doi.org/10.1088/1361-6528/ac8881
4. Mason CA, Wang L-C (1997) Growth cone form is behavior-specific and, consequently, position-specific along the retinal axon pathway. J Neurosci 17:1086–1100
5. Ferrer-Ferrer M, Dityatev A (2018) Shaping synapses by the neural extracellular matrix. Front Neuroanat 12:40
6. Tamariz E, Varela-Echavarría A (2015) The discovery of the growth cone and its influence on the study of axon guidance. Front Neuroanat 9:51
7. Suter DM, Forscher P (2000) Substrate-cytoskeletal coupling as a mechanism for the regulation of growth cone motility and guidance. J Neurobiol 44:97–113
8. Janson IA, Putnam AJ (2016) Extracellular matrix elasticity and topography: material-based cues that affect cell function via conserved mechanisms. J Biomed Mater Res A 26. https://doi.org/10.1002/jbm.a.35254
9. Wang T, Nanda SS, Papaefthymiou GC, Yi DK (2020) Mechanophysical cues in extracellular matrix regulation of cell behavior. Chembiochem 21:1254–1264
10. Tonti OR, Larson H, Lipp SN, Luetkemeyer CM, Makam M, Vargas D, Wilcox SM, Calve S (2021) Tissue-specific parameters for the design of ECM-mimetic biomaterials. Acta Biomater 132:83–102
11. Jhala D, Vasita R (2015) A review on extracellular matrix mimicking strategies for an artificial stem cell niche. Polym Rev 55:561–595
12. Flemming RG, Murphy CJ, Abrams GA, Goodman SL, Nealey PF (1999) Effects of synthetic micro- and nano-structured surfaces on cell behavior. Biomaterials 20:573–588
13. Micholt L, Bartic C (2013) Substrate topography determines neuronal polarization and growth in vitro. PLoS One 8:14
14. Dowell-Mesfin NM, Abdul-Karim M-A, Turner AMP, Schanz S, Craighead HG, Roysam B, Turner JN, Shain W (2004) Topographically modified surfaces affect orientation and growth of hippocampal neurons. J Neural Eng 1:78–90
15. Belu A, Yilmaz M, Neumann E, Offenhäusser A, Demirel G, Mayer D (2018) Asymmetric, nano-textured surfaces influence neuron viability and polarity: asymmetric, nano-textured surfaces. J Biomed Mater Res 106:1634–1645
16. Bugnicourt G, Brocard J, Nicolas A, Villard C (2014) Nanoscale surface topography reshapes neuronal growth in culture. Langmuir 30: 4441–4449
17. Hoffman-Kim D, Mitchel JA, Bellamkonda RV (2010) Topography, cell response, and nerve regeneration. Annu Rev Biomed Eng 12:203–231
18. Hanson JN, Motala MJ, Heien ML, Gillette M, Sweedler J, Nuzzo RG (2009) Textural guidance cues for controlling process outgrowth of mammalian neurons. Lab Chip 9:122–131
19. Mariano A, Fasolino I, Dinger NB, Latte Bovio C, Bonadies I, Pezzella A, Ambrosio L, Raucci MG, Santoro F (2022) Eumelanin coated aligned PLA electrospun microfibers to guide SH-SY5YCells spreading, alignment and maturation. Adv Mater Interfaces. https://doi.org/10.1002/admi.202202022
20. Wong ST, Teo S-K, Park S, Chiam K-H, Yim EKF (2014) Anisotropic rigidity sensing on grating topography directs human mesenchymal stem cell elongation. Biomech Model Mechanobiol 13:27–39
21. Chua JS, Chng C-P, Moe AAK, Tann JY, Goh ELK, Chiam K-H, Yim EKF (2014) Extending neurites sense the depth of the underlying topography during neuronal differentiation and contact guidance. Biomaterials 35:7750–7761
22. Tonazzini I, Masciullo C, Savi E, Sonato A, Romanato F, Cecchini M (2020) Neuronal contact guidance and YAP signaling on ultra-small nanogratings. Sci Rep 10:3742
23. Slavík J, Čmiel V, Hubálek J, Yang Y, Ren T-L (2020) Hippocampal neurons' alignment on quartz grooves and Parylene cues on quartz substrate. Appl Sci 11:275
24. Chan LY, Birch WR, Yim EKF, Choo ABH (2013) Temporal application of topography to increase the rate of neural differentiation

from human pluripotent stem cells. Biomaterials 34:382–392

25. Song L, Wang K, Li Y, Yang Y (2016) Nanotopography promoted neuronal differentiation of human induced pluripotent stem cells. Colloids Surf B Biointerfaces 148:49–58

26. Gershenbaum MR, Roisen FJ (1981) Applications of scanning electron microscopy in neuroscience research. Trends Neurosci 4:109–114

27. Matino L, Mariano A, Ausilio C, Raghav G, Cohen-Karni T, Santoro F (2022) Modulation of early-stage neuronal outgrowth through out-of-plane graphene. Nano Lett. https://doi.org/10.1021/acs.nanolett.2c03171

28. Lunghi A, Mariano A, Bianchi M et al (2022) Flexible neural interfaces based on 3D PEDOT:PSS micropillar arrays. Adv Mater Interfaces 9:2200709

29. Ojovan SM, Rabieh N, Shmoel N, Erez H, Maydan E, Cohen A, Spira ME (2015) A feasibility study of multi-site, intracellular recordings from mammalian neurons by extracellular gold mushroom-shaped microelectrodes. Sci Rep 5:14100

30. Hällström W, Mårtensson T, Prinz C, Gustavsson P, Montelius L, Samuelson L, Kanje M (2007) Gallium phosphide nanowires as a substrate for cultured neurons. Nano Lett 7:2960–2965

31. Chiappini C, Chen Y, Aslanoglou S et al (2021) Tutorial: using nanoneedles for intracellular delivery. Nat Protoc 16:4539–4563

32. Milos F, Tullii G, Gobbo F, Lodola F, Galeotti F, Verpelli C, Mayer D, Maybeck V, Offenhäusser A, Antognazza MR (2021) High aspect ratio and light-sensitive micropillars based on a semiconducting polymer optically regulate neuronal growth. ACS Appl Mater Interfaces. https://doi.org/10.1021/acsami.1c03537

33. Santoro F, Panaitov G, Offenhäusser A (2014) Defined patterns of neuronal networks on 3D thiol-functionalized microstructures. Nano Lett 14:6906–6909

34. Li X, Matino L, Zhang W, Klausen L, McGuire AF, Lubrano C, Zhao W, Santoro F, Cui B (2019) A nanostructure platform for live-cell manipulation of membrane curvature. Nat Protoc. https://doi.org/10.1038/s41596-019-0161-7

35. Tao-Cheng J-H, Crocker V, Moreira SL, Azzam R (2021) Optimization of protocols for pre-embedding immunogold electron microscopy of neurons in cell cultures and brains. Mol Brain 14:86

36. Santoro F, Zhao W, Joubert L-M, Duan L, Schnitker J (2017) Revealing the cell–material interface with nanometer resolution by focused ion beam/scanning electron microscopy. ACS Nano 9. https://doi.org/10.1021/acsnano.7b03494

Chapter 3

Analysis of Neurite and Spine Formation in Neurons In Vitro

Jenny R. Diaz and Martin J. Sadowski

Abstract

Primary neuronal cultures are commonly used to study genetic and exogenous factors influencing neuronal development and maturation. During development, neurons undergo robust morphological changes involving expansion of dendritic arbor, formation of dendritic spines, and expression of synaptic proteins. In this chapter, we will cover methodological approaches allowing quantitative assessment of in vitro cultured neurons. Various quantitative characteristics of dendritic arbor can be derived based on immunostaining against anti-microtubule-associated protein 2 followed by dendrite tracing with the SNT plug-in of the FIJI software package. The number and subtypes of dendritic spines can be assessed by double labeling with DiI and Phalloidin iFluor448 followed by laser scanning confocal microscopy analysis. Finally, expression of presynaptic and postsynaptic proteins can be determined by immunohistochemistry and quantification using several available software packages including FIJI and Imaris, which also allows for 3D rendering and statistical displaying of the expression level of synaptic proteins.

Key words Dendritic arbor, Dendritic spines, FIJI, Fluorescent microscopy, Imaris, Laser scanning confocal microscopy, Neuronal cultures, Synaptic proteins

1 Introduction

Primary neuronal cultures, and most often cultures of primary hippocampal neurons are an in vitro experimental model, which is commonly used to characterize wide range of factors modulating neuronal development, synaptic plasticity, neurodegeneration, and to test neuroprotective agents in rescue experiments where neuronal toxicity is introduced beforehand [1–3]. Quantitative readouts of dendritic arbor morphology, dendritic spine density, frequency analysis of dendritic spine subtypes, and expression of various synaptic proteins are typical outcome measures used in such experiments. This chapter aims to familiarize the reader with detailed methodological approaches allowing for quantitative assessment of in vitro cultured neurons.

Neuronal morphology plays an essential role in neuronal communication. Establishment of effective neuronal connectivity relies

Kazuhito Toyooka (ed.), *Neuronal Morphogenesis: Methods and Protocols*, Methods in Molecular Biology, vol. 2831, https://doi.org/10.1007/978-1-0716-3969-6_3,

on a number of parameters including dendritic complexity, number and maturation of dendritic spines, and expression of pre- and postsynaptic proteins. Changes in any of these parameters are associated with different neurological diseases [4]. Dendrites receive and integrate signals from neighboring neurons and propagate them toward the perikaryon [5]. The number of all dendrites and complexity of their ramification is collectively named the dendritic arbor. The size of the dendritic arbor closely corelates with capacity to form neuronal networks [6]. Microtubule-associated protein 2 (MAP2) is involved in microtubule assembly and stabilization of dendrites [7]. Because MAP2 is localized only in dendrites, anti-MAP2 immunocytochemistry can be reliable and used to visualize morphology of the dendritic arbor. Several quantitative characteristics can be collected to analyze and compare the complexity of the dendritic arbor. For a given neuron, we routinely collect three different quantitative parameters: the combined length of all dendrites, the total arbor surface, and the number of intersections between dendrites and a series of concentric circles, which have a common center point in the center of the perikaryon [6, 8, 9]. The last analysis is also known as Sholl's analysis from the name of its inventor. It is a commonly used measure of the dendritic arbor complexity, while the first two measures characterize the actual arbor size.

Dendritic spines are miniscule dendritic protrusions, which have highly plastic nature. They are formed by sideways elongations of the dendritic cytoskeleton and are known to assume a spectrum of different shapes, which differ by organization of the actin cytoskeleton within them and represent various stages of maturation [10]. Immature forms of dendritic spines take shapes of filopodia or thin spines, while mature forms have stubby and mushroom appearance. Mature forms of dendritic spines form functional synaptic connections between dendrites and axons of other neurons or are pruned when they fail to form such connections [5]. The Golgi silver stain was originally developed as a method to visualize dendritic spines. Though very elegant, this silver staining techniques is arduous to perform and yields inconsistent and irreproducible results, hence it has limitation when used as a basis for quantitative analysis [11]. Instead, we recommend using the combination of DiI (1,1′-dioctadecyl-3,3,3′,3′-tetramethylindocarbocyanine perchlorate) and phalloidin labeling, as a convenient and reproducible method allowing to visualize the entire spectrum of dendritic spine forms. DiI is a lipophilic fluorescent dye, which diffuses through the cell membrane, including that of dendritic spines rendering the outline of the spine discernible [12]. DiI labels particularly well matures spine forms (stubby and mushroom) and can be visualized using a standard rhodamine filter set. Phalloidin is a naturally occurring bicyclic peptide, with inherent affinity toward actin filaments in many different species of animals and plants [13].

In contrast to DiI, phalloidin primarily labels immature spine forms including filipodia and thin spines. Phalloidin can be labeled with numerous fluorescent dyes including iFluor 488, which allows for its application in tandem with DiI [10]. Quantification of dendritic spine density can be done along secondary and tertiary dendrites, within test areas of predefined length. Main readouts from dendritic spine quantitative analysis include the total number of spines per unit of length, and relative incidence of various spine types.

There are a large number of functional proteins expressed along the dendrites, which contribute to establishing functional synaptic connections. Perhaps the best known are subunits of glutamate receptors such as α-amino-3-hydroxy-5-methyl-4-isoxazolepropionic acid (AMPAR) and *N*-methyl-D-aspartate (NMDAR), which are implicated in synaptic plasticity [14, 15]. The postsynaptic density protein 95 (PSD-95) is a major component of postsynaptic densities and is structurally and functionally associated with AMPAR and NMDAR [16, 17]. Another commonly used synaptic protein marker is synaptophysin (SYP), which is a vesicle protein involved in neurotransmitter release, synaptic connectivity, and synaptic plasticity [18]. Aforementioned proteins can be very selectively immunodetected and their expression along secondary and tertiary dendrites can be densitometrically quantified.

2 Materials

All solutions should be prepared using deionized water on the day of the experiments unless exemptions are otherwise stated. All tools and reagents used to culture neurons should be sterile.

2.1 Cell Culture

1. Sulfuric acid 70%.
2. Dulbecco's phosphate-buffered saline (DPBS).
3. Poly-D-lysine solution. Dilute poly-D-lysine (molecular weight 70,000–150,000) to the final concentration of 0.2 mg/mL using sterile, deionized water to obtain poly-D-lysine 2X stock. Sterilize the 2X stock by filtering it through a 0.2 μm syringe filter, makes aliquots, and store the aliquots at −20 °C until use. Thaw an aliquot immediately prior to use and mix it 1:1 with DPBS.
4. Hank's Balanced Salt Solution (HBSS) Ca^{2+}- and Mg^{2+}-free.
5. Papain stock solution. Reconstitute the entire content of a Papain vial (containing 100 units of Papain in 5 mL of Neurobasal medium. Incubate the vial in a 37 °C water bath until the Papain is entirely dissolved and the solution becomes clear. This can take up to 10 min. The resulting stock solution will contain 1 U of Papain in 50 μL volume (or 0.02 U/μL).

6. DNase I stock solution. Reconstitute the entire content of a DNAse I vial containing a 1000 units of DNAse I in 500 μL of Neurobasal medium. The concentration of DNAse in the resulting stock solution will be 2 U/μL.
7. Set of polished pipettes with gradually reducing aperture size. Reduce diameter tip of glass Pasteur pipettes by burning their tips in a burner flame while rapidly rotating the pipettes so the aperture will narrow evenly. For the first pipette, leave the pipette tip end in contact with the flame for around 2–3 s. For subsequent pipettes, gradually increase the time of contact with the flame to achieve gradual reduction in the aperture size, obtaining pipettes 2–4 from bigger to lower aperture. Leave the polished pipettes under UV or autoclave them being careful not to break the tips.
8. Neurobasal medium for neuronal plating. Supplement Neurobasal medium with 2% B-27 supplement, 2 mmol/L glutamine, 50 mg/mL of streptomycin, and 100 U/mL penicillin.
9. Neurobasal medium for neuronal growth. Supplement Neurobasal medium with 2% B-27 supplement, 50 mg/mL of streptomycin, and 100 U/mL penicillin.
10. Cytosine-β-D-arabinoside (AraC) stock solution. Dilute AraC to 5 mmol/L using sterile, deionized water. Aliquot AraC solution and store the aliquots at −20 °C protected from light.
11. Paraformaldehyde 2% and 4% solutions. Dilute paraformaldehyde in 0.1 M phosphate buffered pH 7.4. Aliquot and store the aliquots at −20 °C until use.

2.2 Immunocytochemistry

1. 0.01 M phosphate buffered saline (PBS) pH 7.4. Dilute 10X commercially obtained PBS stock in sterile deionized water.
2. PBS-Triton. Dilute Triton in 0.01 M PBS pH 7.4 to the final concertation of 0.25%.
3. PBS-Tween. Dilute Tween-20 in 0.01 M PBS pH 7.4 to the final concentration of 0.05%.
4. Avidin block (kit). Add 100 μL of Avidin solution to 2.5 mL of PBS-Tween.
5. Biotin block (kit). Add 100 μL of Biotin solution to 2.5 mL of PBS-Tween.
6. Bovine serum albumin (BSA) blocking solution. Dilute lyophilize BSA in PBS-Tween to obtain 10% (w/v) solution.

2.3 Antibodies and Dyes

1. Mouse monoclonal antibody against MAP2 diluted 4 μg/mL in PBS-Tween.
2. Mouse monoclonal antibody against PSD-95 diluted 10 μg/mL in PBS-Tween.

3. Rabbit monoclonal antibody against NR1 subunit of the NMDAR diluted 1 μg/mL in PBS-Tween.
4. Rabbit monoclonal antibody against SYP diluted 0.8 μg/mL in PBS-Tween.
5. Biotinylated secondary antibody against mouse or rabbit IgG diluted 2 μg/mL in PBS-Tween.
6. Cy3 conjugated Streptavidin. Prepare 1 μg/mL dilution of Streptavidin/Cy3 in PBS-Tween (*see* **Note 1**).
7. DAPI solution. Dilute DAPI in 80% methanol 1/1000. Store DAPI solution at 4 °C in a light-tight vial.
8. DiI working solution. Dissolve DiI in DMSO to the final concentration of 0.2 μg/mL to prepare the stock solution. Once the solution is homogeneous, dilute DiI stock in DPBS 1:500 to obtain working solution.
9. Phalloidin-iFluor 488 working solution. Dilute 1 μL of Phalloidin-iFluor 488 in 1 mL of PBS with 1% BSA. Mix well by pipetting.

2.4 Equipment and Software

AmScope Stereoscopic Microscope.

Fiji Software (NIH ImageJ2).

Fisherbrand™ Multi-Platform Shaker.

Imaris V 9.0 Software.

Micro-Bead Sterilizer with Glass Beads.

NU-4750 Water Jacketed CO_2 Incubator.

NU-540 Series F Class II, Type A2 Biosafety Cabinet.

Thermo IEC Centra CL2 Centrifuge.

Zeiss 880 Laser Scanning Confocal Microscope (LSCM) with ZEN Black 2.3 SP1 Acquisition Software (v. 14.0.18.201).

80i Nikon Epi-fluorescent Microscope with DS-Qi1Mc High-Sensitivity, Cooled, Monochrome Camera.

3 Methods

All procedures should be carried out in room temperature unless otherwise specified.

3.1 Cleaning of Coverslips for Neuronal Plating

1. Place 18 mm diameter glass coverslips in a closed glass container filled with 70% sulfuric acid. Incubate overnight while slowly shaking using orbital shaker.
2. Recover sulfuric acid the following day. The acid can be reused up to three times.

3. Wash the coverslips four times with distilled water, while gently shaking. Each wash should last approximately 2 h.
4. Let the coverslips dry on the filter paper.
5. Place the coverslips in a Petri dish for sterilization. If there is more than one layer of coverslips, separate individual layers with the filter paper to avoid coverslips sticking to each other. Sterilize coverslips through autoclaving. Once sterile, the coverslips can be stored for prolonged time. Open the Petri dish container with sterilized coverslips only under the hood.

3.2 Culturing of Primary Hippocampal Neurons

1. Coat the coverslips with poly-D-lysine by placing them in poly-D-lysine solution for 2 h at 37 °C. We suggest using a 12-well plate and placing one coverslip per well. Wash the poly-D-lysine-coated coverslips in sterile distilled water three times for 5 min each. Let the coverslip dry completely under the hood before plating the cells.
2. Sterilize all instruments before use by submerging them in 70% ethanol for 15 min, followed by posterior sterilization at 100 °C using a Micro-Bead Sterilizer for 30 s. Let the instruments cool down before use.
3. Extract brains from the skulls of instantaneously decapitated P0 mouse pups. Place the brains in a Petri dish and cover them well with ice-cold Ca^{2+}- and Mg^{2+}-free HBSS. Maintain the Petri dish on ice.
4. Remove meninges under a stereoscope, dissect the hemispheres, and isolate the hippocampi. Place the hippocampi in a 35 mm Petri dish filled with 1 mL of Ca^{2+}- and Mg^{2+}-free HBSS. Transfer the Petri dish under a cell culture hood and carry all subsequent steps there under aseptic conditions.
5. Transfer the hippocampi along with the entire volume of the HBSS solution into a 15 mL tube using a pipette.
6. Add 30 units of Papain (1500 μL of the Papain stock solution) and 200 units of DNAse I (100 μL of the DNAse I stock solution) and incubate for 30 min at 37 °C. During the incubation time, manually dissociate the hippocampal tissue by pipetting up and down ten times with three fire-polished glass pipettes as follows: after 5 min of incubation, pipette up and down with pipette 1; after 15 min of incubation, pipette up and down with pipette 2; and after 25 min of incubation, pipette up and down with pipette 3 (*see* **Note 2**).
7. Add 7 mL of plain Neurobasal medium to the triturated cell suspension. Transfer the entire volume to 15 mL tube and centrifuge for 6 min at 150× *g*.

8. Gently remove the supernatant from above the pellet and resuspend the pellet in 5 mL of Neurobasal media with 100 μL DNase I by pipetting up and down with the fire-polished glass pipette 4 (*see* **Note 3**).
9. Let the 15 mL tube stand undisturbed for 2–5 min to allow tissue debris to settle.
10. Transfer the supernatant to a 40 μm cell strainer presoaked with Neurobasal medium. Place the strainer in a 50 mL tube and centrifuge for 6 min at 150 × *g* to obtain the neuronal pellet.
11. Resuspend the neurons in 1 mL of Neurobasal plating medium (*see* **Note 4**). Count the cells using hemocytometer and carefully plate 50,000 cells as a single drop on a coverslip placed in the bottom of a 12-well plate. Place the 12-well plate in an incubator at 37 °C and 5% CO_2 for 1 and let the neurons attach to the coverslips. Add 1.5 mL of Neurobasal growth medium (*see* **Note 5**).
12. After 48 h, remove 500 μL of medium and replace it with the same amount a fresh Neurobasal growth medium containing 1.5 μL of AraC stock solution. AraC is added to inhibit growth of nonneural cells and its final concentration should be 5 μmol/L.
13. Change one-third of medium every 3 days with fresh Neurobasal growth medium (*see* **Note 6**).

3.3 Fixation of Neurons

1. At 17-day in vitro (DIV), carefully aspirate all medium and wash the neurons with DPBS one time for 5 min (*see* **Notes 7** and **8**).
2. Add 2% or 4% paraformaldehyde depending on the staining that will be performed (see below) and incubate at room temperature for 15 min.
3. Wash the neurons with PBS twice by adding 2 mL of PBS to each well for 5 min each time and then carefully aspirate PBS (*see* **Note 9**). Neurons, which are fixed and attached to the coverslips can be stored on the bottom of 12-well plate filled with PBS at 4 °C until further use. The 12-well plate should be sealed with parafilm to prevent evaporation of the PBS, which should be replenish if needed.

3.4 Dendritic Arbor Analysis

1. *Visualization of the Dendritic Arbor by MAP2 Immunocytochemistry*

   For MAP2 immunocytochemistry, we suggest using neurons fixed with 4% paraformaldehyde. At the beginning of the staining protocol, we place coverslips with affixed neuronal cultures cells up on the bottoms of the wells in a standard 12-well cell culture plate. All steps of the staining protocol are then carried out by exchanging reagents and washes in individual wells (*see* **Note 7**). We routinely use 500 μL of a reagent per well/step and 1000 μL of PBS or PBS-Tween wash per well/step unless otherwise stated. All reagents are prepared immediately before use. All steps of the protocol should be performed on a shaker rotating at low speed and at room temperature unless otherwise stated.

1. Wash three times with PBS for 15 min each time.
2. Permeabilize neuronal membranes using PBS-Triton for 15 min.
3. Block with the BSA blocking solution for 1 h.
4. Wash twice with PBS-Tween for 5 min each time.
5. Block with the Avidin block for 15 min.
6. Wash three times with PBS-Tween for 15 min each time.
7. Block with the Biotin block for 15 min.
8. Wash three times with PBS-Tween for 15 min each time.
9. Incubate with anti-MAP2 primary antibody for 1 h.
10. Wash three times with PBS-Tween for 15 min each time.
11. Incubate with anti-mouse IgG biotinylated secondary antibody for 30 min.
12. Wash three times with PBS-Tween for 15 min each time.
13. Incubate with Cy3/Streptavidin for 30 min.
14. Wash three times with PBS-Tween for 15 min each time.
15. Stain neuronal nuclei with DAPI solution for 40 s (*see* **Notes 10** and **11**).
16. Wash twice with cold Methanol 80% for 2 min each time (*see* **Note 12**).
17. Place the coverslips neurons down on acid-cleaned histological slides and affix them using a Fluoromount-G mounting medium.
18. Store the slides in a light-tight box at 4 °C.

2. *Quantitative Analysis of the Dendritic Arbor*

1. Take photographs of MAP2 immunostained neurons under 40x objective using a fluorescent microscope.
2. Take photographs of neurons immunostained with the omission of the primary antibody (anti-MAP2) to adjust background fluorescence (*see* **Note 13**).
3. Open FIJI [19].
4. Open the NeuronJ plugin for neurite tracing and analysis. Open the image in an 8-bit format and trace each dendrite. Point the cursor to the start of the dendrite right next to the soma. Using the "+" command start tracing and follow the path of the dendrite with the mouse. Trace all the dendrites of the neuron and click save tracing (Fig. 1b).

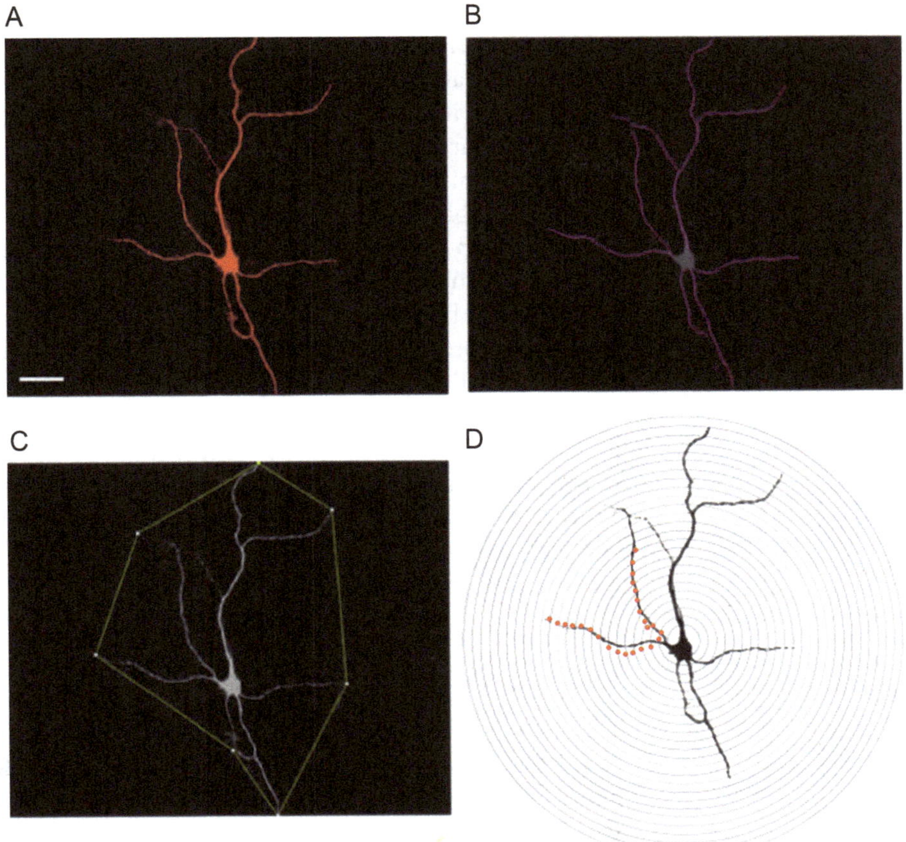

Fig. 1 Analysis of dendritic arbor analysis. (**a**) Representative microphotograph of a hippocampal neuron at 17-day in vitro, which was immunostained against MAP2. (**b**) Shown is tracing of the dendritic arbor using NeuronJ plugin. Combined dendrite length is calculated from the sum of the lengths of all traced dendritic branches for a given neuron. (**c**) Shown is measurement of the total arbor surface. The measurable surface is created by manually joining the tips of the most far out reaching dendrites using the FIJI freehand tool. (**d**) Sholl's analysis of dendrite branching. Shown is the schematic of the Sholl's grid superimposed on an analyzed neuron. The Sholl's grid represents a set of concentric circles, which are evenly spaced apart from the common center point. In this case the distance between each pair of circles is set at 3 μm. Red dots indicate the cross points made between selected dendritic branches and the concentric circles. Results of the Sholl's analysis can be reported as a function between the distance from the center point (the soma) and the number intersection at a given distance, the total number of intersections per neuron, or both. Scale bar 40 μm in (**a**)

5. To measure the total arbor surface area, draw the outline of the dendritic arbor by joining tips of the most far outreaching dendritic branches using the freehand draw tool (Fig. 1c). Use the measure tool to quantify the area occupied by the outlined dendritic arbor.
6. For Sholl's analysis, open the SNT plugin (*see* **Note 14**). Create the Sholl's grid of concentric circles spaced 3 μm (or as desired) over previously traced neuron. SNT enumerates the number of intersections between the dendrites and the Sholl grid.

3.5 Dendritic Spines Analysis

1. *Visualization of Dendritic Spines by Double Staining with DiI and Phalloidin-iFluor 488*

   For DiI/Phalloidin-iFluor 488 staining, we suggest using neurons fixed with 2% paraformaldehyde. Using higher concentrations of paraformaldehyde or methanol as the fixative can impair DiI staining (*see* **Note 15**). At the beginning of the staining protocol, we place coverslips with affixed neuronal cultures cells up on the bottoms of the wells in a standard 12-well cell culture plate. All steps of the staining protocol are then carried out by exchanging reagents and washes in individual wells (*see* **Note 7**). We routinely use 500 μL of a reagent per well/step and 1000 μL of PBS or PBS-Tween wash per well/step unless otherwise stated. All reagents are prepared immediately before use. All steps of the protocol should be performed on a shaker rotating at low speed and at room temperature unless otherwise stated. When executing the staining protocol, the 12-well plate should be covered using a light-tight material, e.g., aluminum foil.

   1. Wash three times with DPBS for 15 min each time.
   2. Stain with DiI working solution for 15 min.
   3. Wash twice with DPBS for 2 min each time.
   4. Leave the neurons in fresh DPBS overnight to allow for free-diffusion of DiI. This step is done without rotational shaking (*see* **Note 16**).
   5. Wash three times with PBS for 5 min each time.
   6. Stain with Phalloidin-iFluor 488 working solution for 90 min.
   7. Wash three times with PBS for 5 min each.
   8. Place the coverslips neurons down on acid-cleaned histological slides and affix them using a Fluoromount-G mounting medium.
   9. Store the slides in a light-tight box at 4 °C.
   10. Visualize the cells after 72 h to allow dyes to equilibrate across neuronal membranes and diffuse well within the spine structures.

2. *Image Acquisition and Analysis*

   1. Image DiI/Phalloidin-iFluor 488 stained neurons using a LSCM. Z-stack of images should be acquired through the entire thickness of the neuronal layer. Our verified imaging parameters include the 63 × 1.4 N.A. objective, a capture speed of 5, and a pinhole diameter of 1 Airy unit. In the *z*-axis, each two consecutively acquired slices have a 54% overlap between each other.

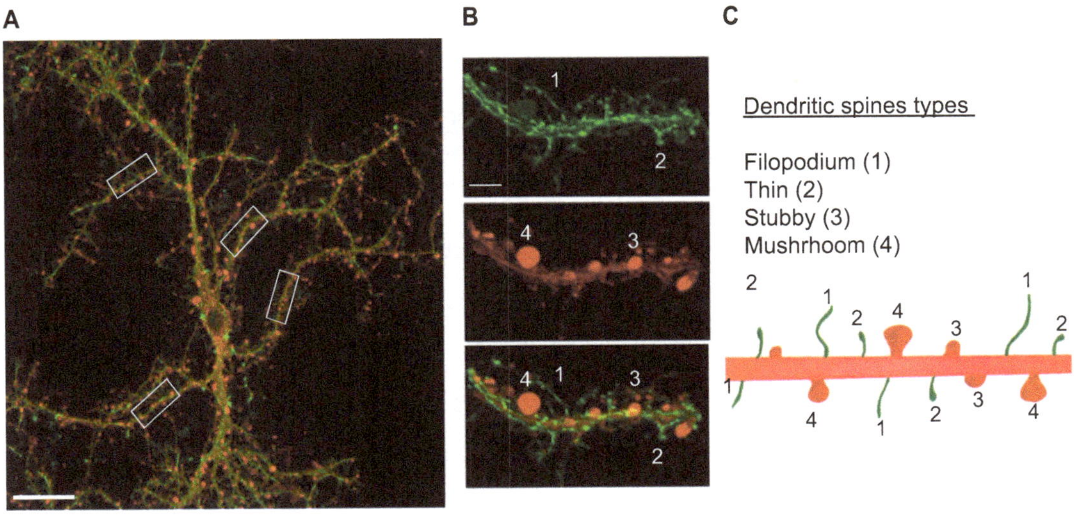

Fig. 2 Analysis of dendritic spines. (**a**) Shown is a micrograph of a representative hippocampal neuron at 17-day in vitro, which was co-labeled with DiI (red) and phalloidin-iFluor 488 (green). Rectangular test areas (20 μm × 10 μm) are superimposed on primary and secondary dendrites with the long axis of the test area oriented parallel to the long axis of the dendrite. (**b**) Shown are high magnification images of a representative test area captured under Fluorescein (top) and Rhodamine (center) filter sets and their merge (bottom). Particular types of dendritic spines are labeled with number, which correspond to the legend given in (**c**). Filopodia and thin spines represent immature spine forms and stain stronger with Phalloidin, while stubby and mushroom spines denote more mature spine forms and more intensely labeled with DiI. (**c**) Shown are morphological features used for classification of dendritic spine types. Scale bar 40 μm in (**a**) and 5 μm in (**b**)

2. Process acquired Z-stacks using the Fiji software. Collapse Z-stacks into a two-dimensional image using the maximum intensity projection function. Subtract the background with a rolling ball radius of 35 pixels and adjust gamma to 1.1. Proportionally increase the picture size by a factor of 5 in the *x*- and *y*-axis using the *Transform J* FIJI plugin in software.
3. Place four to five rectangular test area (4 μm × 20 μm) over secondary and tertiary dendrites aligning the long axis of the test area with the axis of the dendrite (Fig. 2a). Save the image of a neuron with superimposed test areas.
4. Identify dendritic spine subtype based on their distinct morphology: filopodium, thin, stubby, and mushroom-shaped and enumerate them [20] (Fig. 2b, c).
5. Average the total number of dendritic spines and their subtypes across all test areas taken from a single neuron.

3.6 Analysis of Synaptic Protein Expression in Primary and Secondary Dendrites

1. *PSD-95, NR1, and SYP Immunocytochemistry*

   For synaptic protein immunocytochemistry, we suggest using neurons fixed with 4% paraformaldehyde. At the beginning of the staining protocol, we place coverslips with affixed neuronal cultures cells up on the bottoms of the wells in a standard 12-well cell culture plate. All steps of the staining protocol are then carried out by exchanging reagents and washes in individual wells (*see* **Note 7**). We routinely use 500 μL of a reagent per well/step and 1000 μL of PBS or PBS-Tween wash per well/step unless otherwise stated. All reagents are prepared immediately before use. All steps of the protocol should be performed on a shaker rotating at low speed and at room temperature unless otherwise stated.

1. Wash three times with PBS for 15 min each time.
2. Permeabilize neuronal membranes using PBS-Triton for 15 min.
3. Block with the BSA blocking solution for 1 h.
4. Wash twice with PBS-Tween for 5 min each time.
5. Block with the Avidin block for 15 min.
6. Wash three times with PBS-Tween for 15 min each time.
7. Block with the Biotin block for 15 min.
8. Wash three times with PBS-Tween for 15 min each time.
9. Incubate with anti-PSD-95, anti-NR1, or anti-SYP primary antibodies for 1 h.
10. Wash three times with PBS-Tween for 15 min each time.
11. Incubate with anti-mouse IgG biotinylated secondary antibody for PSD-95 and NR1 stainings or with anti-rabbit IgG biotinylated secondary antibody for SYP staining for 30 min.
12. Wash three times with PBS-Tween for 15 min each time.
13. Incubate with Cy3/Streptavidin for 30 min.
14. Wash three times with PBS-Tween for 15 min each time.
15. Stain neuronal nuclei with DAPI solution for 40 s (*see* **Notes 10** and **11**).
16. Wash twice with cold Methanol 80% for 2 min each time (*see* **Note 12**).
17. Place the coverslips neurons down on acid-cleaned histological slides and affix them using Fluoromount-G mounting medium.

2. *Image Acquisition*

   1. Image immunostained neurons using a LSCM. Z-stack of images should be acquired through the entire thickness of the neuronal layer. Our verified imaging parameters include the 63 × 1.4 N.A. objective, a capture speed of 7, and a pinhole diameter of 1 Airy unit. In the *z*-axis, each two consecutively acquired slices have a 67% overlap between each other. The fluorescence level should be carefully adjusted to avoid saturation (*see* **Note 13**).

3. *Analysis of Synaptic Protein Expression Using FIJI*
   1. In FIJI, convert acquired Z-stacks into two-dimensional images using the maximum intensity projection function and save them into an 8-bit format.
   2. Process images by enhancing the contrast to a value of 0.3% normalized and adjust the threshold until single puncta representing synaptic proteins became visibly separated from each other (Fig. 3a, b). Then use the sharpen tool to better define the puncta shape (*see* **Note 13**).
   3. Place a rectangular test area (4 μm × 20 μm) over one of secondary or tertiary dendrites aligning the long axis of the test area with the axis of the dendrite (Fig. 3a, b).
   4. Use the command "analyze particles" to collect the integrated density value for the analyzed test area. The integrated density is defined as a product of the area and the mean intensity value for a positive immunostaining within the test area (Fig. 3c).
   5. Repeat **steps 3** and **4** four to five times for the same neuron and average obtained integrated density values.
4. *Analysis of Synaptic Protein Expression in Imaris*
   1. Open an image Z-stack using Imaris.
   2. Select the spot function for the channel and set diameter, quality, and threshold parameters needed to visualize small and large puncta. Maintain the same diameter, quality, and threshold parameters across all experimental groups.
   3. Generate dendritic surface to visualize dendritic outline.
   4. Statistical information function will automatically display the area of all puncta per neuron.
   5. For statistical visualization of synaptic protein expression, use the color tool to highlight differences across the experimental groups. Select statistical parameters such as area, or volume and assign a pseudocolor (Fig. 3d, e).
   6. Alternatively, to analyze five areas of interest per neuron, create a surface using the circle drawing mode, and copy and paste the same area for the first and the last slices of the stack. The first and last slices were selected such as they captured the entire thickness of each neuron.
   7. Create a mask with the surface selected into the channel of the puncta that you want to analyze (*see* **Note 17**).
   8. Create again spots for this new region of interest using the same parameters as before.
   9. The area, volume, and other characteristics of the puncta in the selected test areas will be displayed under statistical information. Repeat **steps 6–9** for up to five test areas and average obtained values for a given neuron.

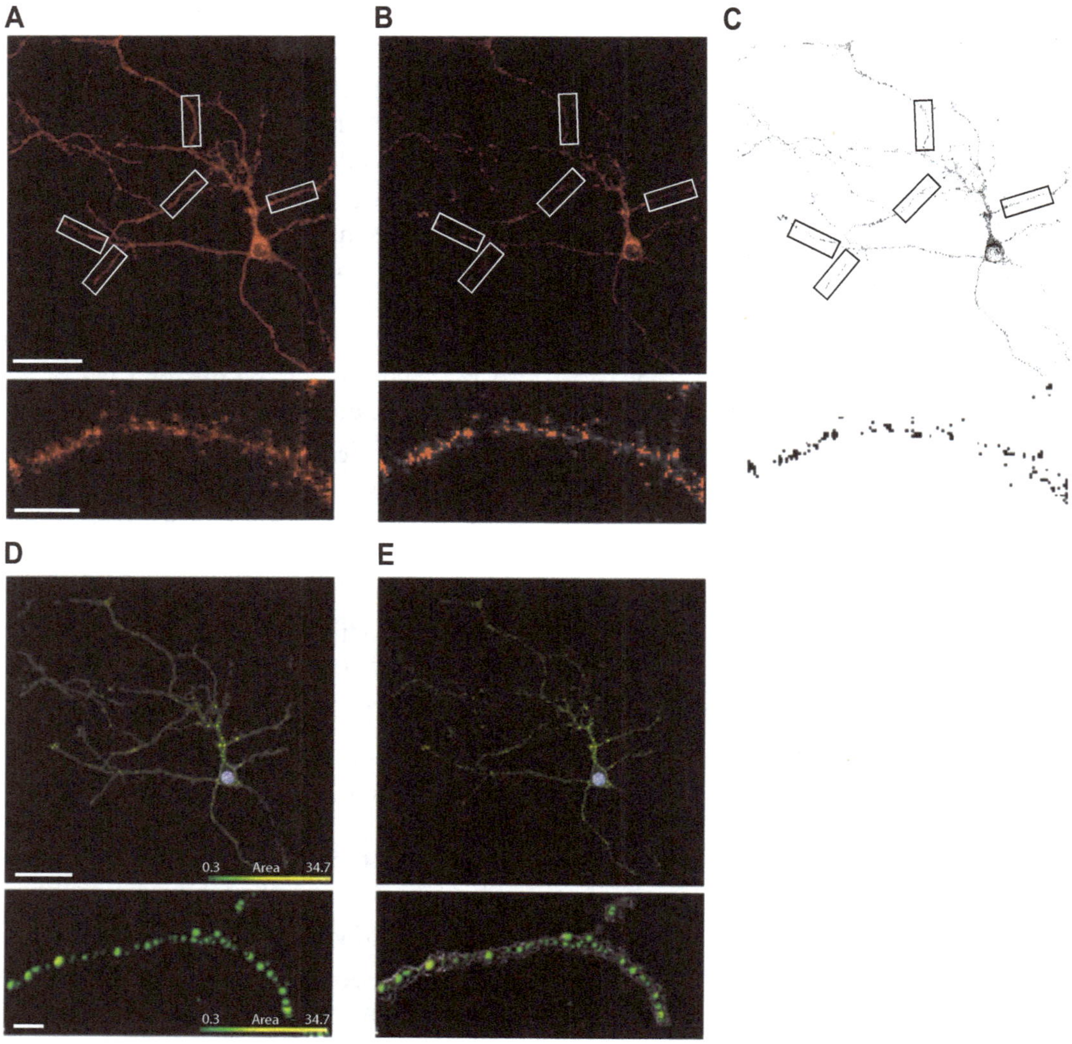

Fig. 3 Analysis of synaptic protein expression. (**a**) Shown is a collapsed Z-stack of confocal images through a representative hippocampal neuron immunostained against PSD−95. (**b**) Shown is a threshold correction of the neuron imaged in (**a**), which eliminates nonspecific staining and highlights PSD-95 specific particles (or puncta) distributed through soma and dendritic arbor along with the mask of thresholded particles. (**c**) Particles are routinely quantified within rectangular test areas (20 μm × 10 μm) superimposed over primary and secondary dendrites with the long axis of the test area oriented parallel to the long axis of the dendrite. Examples of test areas as shown underneath (**a–c**). (**d**) Shown is rendering of particle integrated density using Imaris's spot tool, where the integrated density of particles is represented as different size and color ball objects. The pseudocolor bar at the bottom of the image represents size range across particle. (**e**) Imaris's spot tool rendering of particle size with highlighted outline of the neuron. Selected sections of dendrites are enlarged underneath (**d**) and (**e**). In neurons depicted in (**d**) and (**e**), a DAPI stained nucleus is also visualized. Scale bar 40 μm in the main panel and 5 μm in the magnified section in (**a**) and (**d**)

4 Notes

1. To dilute Streptavidin/Cy3 in PBS-Tween, mix it vigorously in the shaker for 20 s, then mix the solution by inversion in a 3D shaker for at least 30 min before use.
2. Triturate hippocampal tissue very carefully, always putting the cell mixture back to the tube by directing the pipet tip toward the tube's wall and avoid making bubbles.
3. Remove the supernatant from above the cell pellet very carefully since the pellet can be easily disturbed.
4. Notice that only the Neurobasal plating medium but not the Neurobasal growth medium contains glutamine. Continuous use of glutamine can lead to neuronal death.
5. Always add fresh medium to the neurons very carefully to avoid detachment of the neurons from the coverslip. Our preferred technique is applying the medium slowly against the wall of the well.
6. Make sure to always maintain at least 1.5 mL of medium per well in the 12-well plate. If the medium accidentally evaporates, add more than one-third of fresh medium during the medium exchange.
7. Exchange of liquid reagents in the well must be done carefully but quickly to avoid leaving the neurons uncovered for prolonged time. Please pay attention when adding the reagents to the well. It should be done by gentle pipetting out the liquid against the wall of the well to prevent detachment of neurons from the coverslip.
8. Always check under the microscope that neurons are healthy before proceeding with the experiment.
9. Perform all steps under laminar flow biosafety hood. Minor contaminations with bacteria can produce background during immunocytochemistry.
10. DAPI working solution can be used twice.
11. If the DAPI staining is dim, consider increasing the incubation time.
12. Leave methanol at −20 °C for at least 1 h before its use.
13. Use the same parameters to image neurons across all experimental groups.
14. The SNT plugin is a toolbox for tracing and analyzing neuronal morphologies imaged using light or fluorescent microscopy. SNT is a recent replacement of the simple neurite tracer plugin. Like NeuronJ, SNT also can be used to quantify the combined dendritic length of the arbor. The choice of one plugin over another is a personal preference of the operator.

15. DiI staining is not compatible with any staining protocol that requires methanol or ethanol.
16. It is normal for the DPBS solution to become slightly pink on the second day, as the remaining crystals in the wells will solubilize into the PBS overnight. The solution with a very intense pink color likely means a high background due to the excess of crystals after the washes. On the other hand, if the DPBS solution appears almost transparent, some of the neurons may not be stained.
17. In the dialog window, check off the duplicate channel and set the voxels outside of the surface to 0 before applying the mask.

References

1. Diaz JR, Marta-Ariza M, Khodadadi-Jamayran A, Heguy A, Tsirigos A, Pankiewicz JE, Sullivan PM, Sadowski MJ (2022) Apolipoprotein E4 effects a distinct transcriptomic profile and dendritic arbor characteristics in hippocampal neurons cultured in vitro. Front Aging Neurosci 14:845291. https://doi.org/10.3389/fnagi.2022.845291
2. Kuszczyk MA, Sanchez S, Pankiewicz J, Kim J, Duszczyk M, Guridi M, Asuni AA, Sullivan PM, Holtzman DM, Sadowski MJ (2013) Blocking the interaction between apolipoprotein E and Abeta reduces intraneuronal accumulation of Abeta and inhibits synaptic degeneration. Am J Pathol 182(5): 1750–1768. https://doi.org/10.1016/j.ajpath.2013.01.034
3. Kuszczyk MA, Sadowski MJ, Antkiewicz-Michaluk L, Lazarewicz JW (2014) 1MeTIQ provides protection against Abeta-induced reduction of surface expression of synaptic proteins and inhibits H(2)O(2)-induced oxidative stress in primary hippocampal neurons. Neurotox Res 25(4):348–357. https://doi.org/10.1007/s12640-013-9440-1
4. Kulkarni VA, Firestein BL (2012) The dendritic tree and brain disorders. Mol Cell Neurosci 50(1):10–20. https://doi.org/10.1016/j.mcn.2012.03.005
5. Nguyen C, Thompson-Peer KL (2021) Comparing automated morphology quantification software on dendrites of uninjured and injured drosophila neurons. Neuroinformatics 19(4): 703–717. https://doi.org/10.1007/s12021-021-09532-9
6. Wang S, Tanzi RE, Li A (2019) Quantitative analysis of neuronal dendritic Arborization complexity in drosophila. J Vis Exp 143. https://doi.org/10.3791/57139
7. Harada A, Teng J, Takei Y, Oguchi K, Hirokawa N (2002) MAP2 is required for dendrite elongation, PKA anchoring in dendrites, and proper PKA signal transduction. J Cell Biol 158(3):541–549. https://doi.org/10.1083/jcb.200110134
8. Longair MH, Baker DA, Armstrong JD (2011) Simple Neurite Tracer: open source software for reconstruction, visualization and analysis of neuronal processes. Bioinformatics 27(17): 2453–2454. https://doi.org/10.1093/bioinformatics/btr390
9. Binley KE, Ng WS, Tribble JR, Song B, Morgan JE (2014) Sholl analysis: a quantitative comparison of semi-automated methods. J Neurosci Methods 225:65–70. https://doi.org/10.1016/j.jneumeth.2014.01.017
10. Bar J, Kobler O, van Bommel B, Mikhaylova M (2016) Periodic F-actin structures shape the neck of dendritic spines. Sci Rep 6:37136. https://doi.org/10.1038/srep37136
11. Staffend NA, Meisel RL (2011) DiOlistic labeling of neurons in tissue slices: a qualitative and quantitative analysis of methodological variations. Front Neuroanat 5:14. https://doi.org/10.3389/fnana.2011.00014
12. Sherazee N, Alvarez VA (2013) DiOlistics: delivery of fluorescent dyes into cells. Methods Mol Biol 940:391–400. https://doi.org/10.1007/978-1-62703-110-3_28
13. Fehrenbach T, Cui Y, Faulstich H, Keppler D (2003) Characterization of the transport of the bicyclic peptide phalloidin by human hepatic transport proteins. Naunyn Schmiedeberg's Arch Pharmacol 368(5):415–420. https://doi.org/10.1007/s00210-003-0814-4
14. Ehrlich I, Malinow R (2004) Postsynaptic density 95 controls AMPA receptor incorporation during long-term potentiation and experience-

driven synaptic plasticity. J Neurosci 24(4): 916–927. https://doi.org/10.1523/JNEUROSCI.4733-03.2004

15. Won S, Incontro S, Nicoll RA, Roche KW (2016) PSD-95 stabilizes NMDA receptors by inducing the degradation of STEP61. Proc Natl Acad Sci USA 113(32):E4736–E4744. https://doi.org/10.1073/pnas.1609702113
16. Yoo KS, Lee K, Oh JY, Lee H, Park H, Park YS, Kim HK (2019) Postsynaptic density protein 95 (PSD-95) is transported by KIF5 to dendritic regions. Mol Brain 12(1):97. https://doi.org/10.1186/s13041-019-0520-x
17. Zhang P, Lisman JE (2012) Activity-dependent regulation of synaptic strength by PSD-95 in CA1 neurons. J Neurophysiol 107(4):1058–1066. https://doi.org/10.1152/jn.00526.2011
18. Tarsa L, Goda Y (2002) Synaptophysin regulates activity-dependent synapse formation in cultured hippocampal neurons. Proc Natl Acad Sci USA 99(2):1012–1016. https://doi.org/10.1073/pnas.022575999
19. Schindelin J, Arganda-Carreras I, Frise E, Kaynig V, Longair M, Pietzsch T, Preibisch S, Rueden C, Saalfeld S, Schmid B, Tinevez JY, White DJ, Hartenstein V, Eliceiri K, Tomancak P, Cardona A (2012) Fiji: an open-source platform for biological-image analysis. Nat Methods 9(7):676–682. https://doi.org/10.1038/nmeth.2019
20. Cheng C, Trzcinski O, Doering LC (2014) Fluorescent labeling of dendritic spines in cell cultures with the carbocyanine dye "DiI". Front Neuroanat 8:30. https://doi.org/10.3389/fnana.2014.00030

Chapter 4

Quantifying Dendritic Arbors In Vitro and In Vivo in Rodent Models

Rebecca J. Wilson, Jessie R. Badley, and Pamela J. Lein

Abstract

Dendritic arborization is a critical determinant of neuronal connectivity. The structure of a neuron's dendritic arbor determines the number of synaptic inputs a neuron can receive and how it processes synaptic input from other neurons. Here, we describe methods for visualizing and quantifying the dendritic arbor in primary cell cultures and in the intact rodent brain. These techniques can be used to answer significant scientific questions, such as the effects of disease processes, drugs, growth factors, and diverse environmental stressors on dendritogenesis in both in vitro and in vivo rodent models.

Key words Dendritic arborization, Primary neuronal cell culture, Golgi stain, Adeno-associated viral vector, Fluorescent vector, In Vitro, In Vivo

1 Introduction

Most neurons in the vertebrate central and peripheral nervous systems extend two types of processes—axons and dendrites—which differ functionally, biochemically, and structurally. The primary function of axons is to convey electrical signals from the neuronal cell body to downstream cells in the neural circuit. In contrast, dendrites comprise the primary site of synaptic input to the neuron. Dendritic morphology is a critical determinant of neuronal function. Dendritic shape affects signal processing within the neuron, while their number, length, and branching patterns determine the pattern of synaptic connections, which in turn regulates the distribution of information within the nervous system. Experimental evidence indicates that even subtle perturbations of temporal or spatial aspects of dendritic growth can cause persistent changes in synaptic patterning in the developing brain. Clinical data confirm that altered dendritic structure is strongly associated with not only neurodevelopmental disorders but also neurodegenerative diseases. Based on such observations, chemical-induced changes in

Kazuhito Toyooka (ed.), *Neuronal Morphogenesis: Methods and Protocols*, Methods in Molecular Biology, vol. 2831,
https://doi.org/10.1007/978-1-0716-3969-6_4,

dendritic morphology are considered functionally relevant endpoints of neurotoxicity. Thus, there is considerable interest in using morphometric analysis of dendrites to screen genetic mutations and chemicals for effects on dendritic morphogenesis. Multiple approaches are used to measure dendritic morphology, but across all approaches, there are two key steps: (1) visualizing the dendritic arbor of neurons; and (2) quantifying dendritic morphology. Choosing which method to use depends upon the experimental model, e.g., in vitro or ex vivo. In this chapter, we present several methods for visualizing and analyzing dendritic growth of central neurons in vitro and ex vivo.

1.1 In Vitro Visualization of Dendritic Arbors

Primary hippocampal and cortical neurons have long been used to study neuronal morphogenesis and the effects of various endogenous and exogenous factors on dendritic growth. Traditionally, hippocampal and cortical neuronal primary cultures have been prepared from embryonic or early postnatal animals. In cell culture, these neurons undergo an orchestrated schedule of maturation. During stage 1, 0–6 h after cell plating, spherical neurons begin extending lamellipodial structures around the cell body and form neurites. During stage 2 (6–24 h), neurons typically have multiple minor neurites with growth cones at their tips. The growth cones allow the neurites to extend and/or retract short distances until stage 3 (24–72 h) when a single neurite begins to rapidly exceed the length of the other surrounding neurites by two- to three-fold, thus differentiating into an axon. In stage 4 (3–7 days), the other neurites begin to portray characteristics of dendrites. In stage 5 (7–28 days), as the culture ages, the complexity of the dendrites become more robust as synapses, dendritic spines, and neural circuits are formed [1]. This predictable schedule is what makes the primary hippocampal and cortical neuron culture an advantageous in vitro model for assessing chemical effects on dendritic morphology at varying stages of neuronal maturation.

In our lab, we utilize primary hippocampal and cortical neurons that are dissociated from rodent neocortices or hippocampi on postnatal day 0 or 1 (P0/P1). There are several reasons for this preference, including that neurons are less susceptible to damage during the dissociation process if prepared while somata are still small and before processes begin to extend. At this stage of development, the meninges and connective tissue are also easier to cleanly remove. The neurons can be cultured at a high cell density in the presence of endogenous glia, or at this same density or lower as cultures enriched for neurons. In culture, these neurons develop morphological features approximating those seen in vivo. At P0/P1, the neurons are postmitotic and committed in their differentiation. Individual neurons can be visualized in their entirety in the first days of culture before neural networks become complex. This characteristic allows for direct observation of growing axons and dendrites, their modes of branching, and the behavior of their growth patterns.

It is important to consider the substrate that the primary hippocampal and cortical neurons are plated on, especially when testing how exogenous factors affect neuron morphology. Plating on glass coverslips or tissue culture-treated glass-bottom well plates is the best practice because lipophilic compounds can adhere to plastic, which can prevent or minimize neuronal exposure. If glass coverslips are used, they must be cleaned and sterilized first. This can be done either with concentrated acid wash or a plasma cleaner. Not all glass coverslips are suitable for use. We recommend German glass coverslips and always test batches before buying big qualities of the same lot number.

1.2 In Vivo Visualization of Dendritic Arborization

Golgi staining is the most extensively published method for in vivo morphological assessment of neurons. Originally established in 1873 by Camillo Golgi, this technique utilizes metal precipitate to fill neuronal cells for visualization under a brightfield microscope [2]. Over the years it has been modified for visualizing neurons in different applications, including the Golgi-Cox, Rapid Golgi, and Golgi-Kopsch methods (reviewed in [3]). This chapter will describe the protocol we use in our lab for the Rapid Golgi method, adapted from the Golgi-Cox method to improve reliability and reduce time for labeling neurons and acquiring images.

In vivo neuronal visualization can also be achieved using adeno-associated viral (AAV) vectors [4]. Important considerations for this methodology include the following: (1) the serotype of the vector, because the AAV8 serotype has been shown to allow wide-spread delivery to rodent brain [5–7]; (2) the chosen promotor to enhance cell-specific expression, for which we use the neuron-specific promotor, the human Synapsin1 (hSyn1) [8, 9]; (3) the gene for the expressed fluorophore, in this case EGFP; and (4) the injection method. We and others have found that intracerebroventricular injections allow for dispersal throughout the brain via the ventricular system [10, 11]. The protocol for AAV vector injections and visualization as used in our lab is the final method described in this chapter.

2 Materials

2.1 For In Vitro Primary Neuron Culture and Transfection

1. German glass coverslips.
2. 24 well plates.
3. Plating media.
4. Growth media.

5. Feeding media.
6. Neurobasal Plus Medium.
7. Lipofectamine 2000.
8. MAP 2B plasmid.
9. Sterile Falcon tubes.
10. Treatment media.
11. 4% Paraformaldehyde (w/v) in 0.2 M phosphate buffer (0.2 M NaH_2PO_4, 0.2 M Na_2HPO_4, pH 7.2).
12. PBS (1x, w/o calcium and magnesium).
13. Coverslips.
14. ProLong Gold Mountant with DAPI (Invitrogen, Catalog #P36931, used as provided by the supplier).
15. 16-gauge needle.
16. Tweezers.
17. Disposable pipet.
18. Microscope slides.
19. Slide storage.

2.2 For Ex Vivo Tissue Preparation Prior to Golgi Staining

1. 0.1% Gelatin (w/v) in ddH_2O.
2. ddH_2O.
3. Microscope slides.
4. Glass coverslips.

2.3 For Ex Vivo Golgi Staining

1. FD Rapid GolgiStain Kit.
2. Aluminum foil.
3. 10% Sucrose (w/v) in PBS.
4. 30% Sucrose (w/v) in PBS.
5. Kimwipes.
6. Ethanol.
7. Xylenes.
8. ddH_2O.
9. Permount Mounting Medium.

2.4 For Viral Preparation

1. Sterile PBS.
2. pAAV-hSyn-EGFP (gift from Bryan Roth (Addgene viral prep # 50465-AAV8; http://n2t.net/addgene:50465; RRID: Addgene_50465)).
3. Microtubes.

2.5 For Viral Injection

1. 4 L Foam ice pan (Fisher, CAT: 03-395-153) with crushed ice.
2. Paper towel.
3. Stereotaxic apparatus, including dual-sided mouse ear bars (Stoelting CAT: 51617).
4. 10 μL Hamilton syringe Model 801 (Hamilton Company, Catalog # 7642-01).
5. Small box.
6. Heating pad.

2.6 For Tissue Collection Following AAV Injection

1. 24-well tissue culture plate (Sigma Aldrich, CAT: CLS3527) or polypropylene transport tubes (Fisher; CAT: 22-170-193).
2. 4% Paraformaldehyde(w/v) in 0.2 M phosphate buffer (0.2 M NaH_2PO_4, 0.2 M Na_2HPO_4, pH 7.2).
3. Razor blade or scalpel.
4. Permount Mounting Medium.

3 Methods

3.1 Lipid-Mediated Transfection of Cultured Neurons with Fluorescent Probes to Visualize Dendritic Growth

Neurons can be somewhat difficult to transfect because they are postmitotic, but there are physical methods to get cDNA into the cytoplasm. Lipid-mediated transfection is particularly useful for studying early stages of development. Primary central neurons can be transfected with low-efficiency lipid-mediated transfection, which is desirable for visualizing the dendritic arbor of individual neurons, particularly in high-density cultures. There are several commercially available cationic lipids that interact with cDNA to form complexes taken up by cells. The cDNA is not encapsulated by a liposome, but rather is coated with the lipids and neutralized, making fusion and endocytosis more efficient (we use Lipofectamine 2000, Invitrogen). For a low-efficiency transfection, we limit the incubation time for transfection to only 1–2 h, followed by total replacement of the transfection solution with cell culture media at the end of the incubation period. This also results in the desired low-efficiency transfection.

Expressing cDNA encoding dendrite-specific proteins linked to fluorescent tags or intracellular fluorescent proteins that can distribute throughout the whole cell is best practice when working with high-density cultures. Dendrites can be labeled by transfecting cultures with a plasmid encoding microtubule-associated protein-2B fused to either enhanced green fluorescent protein (MAP 2B-eGFP) or red fusion protein (MAP 2B-FusRed). Expression of these high molecular weight MAP 2B fusion proteins is localized to the somatodendritic compartment in cultured neurons and does not alter fundamental dendritic growth patterns [12]. We obtained these two stock plasmids, which are not commercially available, as a gift from Dr. Gary Wayman, and prepare glycerol stocks for use and long-term storage. A critical parameter of transfection is the quality of the plasmid DNA, and best results are achieved when plasmid DNA with high purity is used. Note: Regardless of transfection method used, it is important to perform quality control as recombination issues can occur. When the vector is expressed in a cell, it emits fluorescence, and recombination can diminish fluorescence.

DIV 0 Plating

1. After dissecting and dissociating primary cortical and/or hippocampal neurons, seed cells on a 24-well plate with German glass coverslips at a cell density of 150,000 cells/well (83,000 cells/cm^2).
2. Shake the plates lightly to ensure that cells are evenly distributed across the glass cover slips and there is no clumping.
3. Return the plates to the incubator and allow the cells to attach for 4 h.
4. After cells have attached, do a complete media change from plating media to growth media. Shake the plates to dislodge any cellular debris, aspirate all media from wells, and replace with 500 μL of growth media.
   (a) Apply the media slowly to avoid detaching cells. Do two plates at a time to prevent plates from drying out and becoming contaminated.

DIV 4 Media Replacement

1. Remove 250 μL of growth media with a multichannel pipette or aspirator.
2. Replace growth media (250 μL) in each well with feeding media. Feeding media only needs to be given once (on DIV 4).
   (a) This can be repeated every 4 days until the plates are used for experiments.

DIV 6 Transfection

1. Prewarm growth media in incubator at 37 °C. You will need 400 μL/well.
2. Transfer Neurobasal Plus medium *without* additives into a sterile Falcon tube. You will need 100 μL/well.
3. Add lipofectamine 2000 to Neurobasal Plus Medium and incubate at 37 °C for 5 min. You will need 2 μL/well.
4. Add plasmid to lipofectamine/Neurobasal Plus mixture and incubate at 37 °C for 30 min. You will need 0.8 μg cDNA/well of MAP 2B-pCAG-FusRed or MAP 2B-pCAG-eGFP plasmid.

5. Collect (*and SAVE*) conditioned media from wells into a sterile Falcon tube (pool together all the conditioned media per plate in the tube and label).
   (a) Sterile technique is extremely critical at this point. This is a frequent point of contamination as media is being transferred. Be mindful of this and your sterile technique.
   (b) Incubate pooled conditioned media in Falcon tubes at 37 °C.
6. Replace the conditioned media with 400 μL/well of pre-warmed growth media.
7. Add 100 μL/well of Neurobasal Plus/lipofectamine/cDNA mix.
8. Incubate plates at 37 °C for approximately 1–2 h.
9. Aspirate out all of transfection media.
10. Replace with 500 μL/well appropriate conditioned media saved earlier.
    (a) Evaporation may occur, so if you are a little short on conditioned media, add some growth media to make up the difference.
11. Return plates to the incubator at 37 °C.

DIV 7 Treatment

1. Prepare vehicle and test compound stocks in growth media the day of use.
   (a) When preparing test compound stock (treatment media), consider making twice the final concentration of your compound/treatment (i.e., if you want a treatment of 10 nM, make 20 nM treatment media so that the final concentration after a half-media change is 10 nM).
2. Remove 250 μL/well.
3. Replace with 250 μL/well of treatment media.
4. Return to incubate at 37 °C for 48 h.

DIV 9 Fixation and Coverslip Mounting

1. Remove all treatment media and dispose using proper chemical waste safety protocols.
2. Fix with 4% PFA (200 μL/well) for 45 min.
3. Wash 3× with 1× PBS (1 mL/well) for 5 min each.
4. Mount coverslips on slides using ProLong Gold mounting medium with DAPI.
   (a) We use a 16-gauge needle with the tip bent slightly to flip up the coverslip and a pair of tweezers to gently grab the

edge of the coverslip. The coverslip is positioned so that the cell side is down and it is placed atop a small amount of ProLong Gold mounting medium on a slide. We use a disposable pipet to place the mounting medium on the slides (a little goes a long way). Three coverslips fit on one slide. The weight of the coverslips usually spreads the mounting medium, but a small press can be given to the coverslip. This usually removes any bubbles as well.

5. Dry the newly mounted coverslips overnight at room temperature and protect from light (a slide book can be used).
6. Store at 4 °C until imaging. We use ImageXpress high-content imaging system.

3.2 High-Content Image Acquisition of Cultured Neurons

We use the ImageXpress high-content imaging system (Molecular Devices), which is interfaced to MetaXpress software (Molecular Devices), to acquire images from cell cultures that have been transfected with fluorescent cDNA [13].

Most of the basic steps described below are part of the plate acquisition setup. The Configure tab provides options for configuring a plate acquisition protocol and is organized in a top-to-bottom structure that helps guide and configure imaging acquisition.

1. *Objective and Camera tab*: Allows protocol choice of objective magnification, camera binning preferences, and gain. We use 10× objective magnification.
2. *Sites to Visit tab*: Configures the number of sites to be acquired on each coverslip. The number of sites acquired depends on the size of coverslip region and the objective magnification.
3. *Acquisition tab*: At a minimum, users must specify the autofocus options and acquisition wavelengths they wanted. In addition, users can choose specific journals to run during acquisition, like the slide acquisition journal we use for imaging sets of three slides.
4. *Autofocus*: Apply to multiple sites to ensure accuracy of the defined autofocus offset throughout the samples.
5. *Wavelengths*: One wavelength must be specified, but up to eight can be chosen. We use two for dendritic arborization: DAPI and Texas Red. Exposure settings can be chosen on individual wavelength tabs as well.

We acquire *25 images* per coverslip, with a total of 75 images per slide, and export them as 16-bit tiff stacks to be uploaded in MATLAB and analyzed using the custom Omnisphero algorithm. (See subheading 3.6 for endpoints and analysis.)

3.3 In Vivo Golgi Staining

Tissue Preparation for Golgi Staining

Carry out all procedures at room temperature unless otherwise specified.

1. Purchase or prepare gelatin-coated microscope slides.
   (a) Place clean microscope slides into staining racks and wash thoroughly in ddH_2O and dry in a dust-free area (37 °C incubator).
   (b) Make a 1% solution of gelatin (from porcine skin—Sigma Aldrich, Catalog # G2500) in ddH_2O and heat to 55 °C. Filter with filter paper into a clean histological staining box.
   (c) Add 1 g chromium potassium sulfate.
   (d) Place slides into the gelatin solution for 10 min before removing and drying in a dust-free area (37 °C incubator overnight).
   (e) Once dry, slides can be stored in a closed box at room temperature until use.
2. Deeply anesthetize the animal before tissue collection using IACUC approved methods. During tissue collection, the type of anesthetic being used must be considered for influences on the molecular profile of the tissue, which in turn alters the cellular pathways before tissue is even collected [14, 15].
3. Perfuse the animal with cold 1X PBS solution, remove brain and hemi-sect the brain into two hemispheres by cutting along the midsagittal plane with a glass coverslip.

Golgi Staining

1. Place brain tissue into impregnation solution (labeled Solution A/B in the kit; chromate solution consisting of potassium dichromate, mercuric chloride, and potassium chromate) using approximately 1.5× volume of solution to weight of tissue in a labeled plastic vial that is covered with aluminum foil. Leave at room temperature.
2. After 24 h, replace impregnation solution and leave tissue at room temperature for 2 weeks. Properly dispose of used solution in designated chemical waste.
3. Switch the brain tissue to Solution C for a total of 1 week and replace after the first 24 h. Use at least 7× volume of solution to weight of tissue to ensure proper clearing of background staining and nonspecific precipitate.
4. Place tissue in 10% sucrose solution in 4 °C fridge for 4 h and then switch to 30% sucrose solution and store in 4 °C fridge for at least 3 weeks before slicing.

5. Cut 100 uM sections using a vibratome and mount tissue onto the gelatin-coated slides using a paintbrush and a couple of drops of Solution C (3–6 drops) on the slide to assist with mounting.
6. After moving the section to an ideal spot on the slide with the toothbrush, remove excess Solution C from the slide with a Kimwipe. Be careful not to touch the sections with the Kimwipe.
7. Allow solutions to dry naturally at room temperature in a dust-free area (slide holder).
8. Follow FD RapidGolgi Kit manufacturer instructions for staining. You will need ddH_2O, Solution D and E (from kit), 50% EtOH, 75% EtOH, 95% EtOH, absolute EtOH, and xylene. Be careful to ensure that the samples do not dry out during the staining process.
9. Coverslip slides with Permount Mounting Medium once staining is completed. Using a transfer pipet, place 5–6 drops of Permount down the middle of the slide and slowly drop a coverslip onto the slide. Place one end of the coverslip down next to the label and use one hand to hold that in place while the other hand slowly guides the other end of the coverslip down, starting at about a 45° angle until the entire coverslip has been laid down over the samples. Ensure that there are no bubbles touching the tissue and remove any by adding light pressure to the top of the coverslip with a cotton swab.
10. Slides can be stored at room temperature and should be protected from light.
11. Image sections under the microscope as soon as possible, but they can be stored at room temperature in the dark for up to 1 year without a decrease in the staining intensity. See imaging section below for more information.

3.4 In Vivo AAV Vector Injections

Viral Preparation

1. Store the AAV vector at −80 °C when it is received. Thaw just before use and keep on ice.
2. Titer the vector to the desired concentration. The virus comes tittered at $\geq 7 \times 10^{12}$ vector genomes (vg)/mL. To make the stock titer, dilute in sterile PBS to 1×10^{12} vg/mL in a microtube.
3. Make small (10–20 uL) aliquots to prevent freeze–thaw cycles affecting viral efficacy.
4. Make working titers (between 2×10^{11} and 5×10^{11} vg/mL) by further diluting aliquots in sterile PBS before use.

Set-Up

1. Fill a foam ice pan with crushed ice and place one to two paper towels on top of the ice.
2. Set up the stereotactic apparatus so that the Hamilton syringe can be held securely on the Y-arm, and if you have an electronic stereotactic device, ensure that the cords are all plugged in correctly.
3. Place two ear bars with rubber ends facing inwards into the stereotactic apparatus and tighten one of the ear bars in place. If the ear bars are too high to reach a P0-P1 mouse pup, add a small metal stage underneath the ear bars to raise the pup to an appropriate height.
4. Place the Hamilton syringe into the clamp on the Y-arm.
5. Aspirate a minimum of 3 uL of the viral working stock and push out any air bubbles at the end of the needle with the plunger until you visibly see the viral stock coming out of the needle.
6. Set up two small, covered boxes (e.g., microtube boxes) with a heating pad inside each box set to 30 °C.
7. Right before beginning injections, remove the pups and their nest from the home cage and place on top of the heating pad inside one of the covered boxes. This will prevent pup temperature from dropping during isolation from the dam.

Viral Injection

1. For cryoanesthesia, follow all IACUC approved protocols and mandates.
2. Weigh the first pup and place it onto the paper towel for protection from direct contact with the ice in the foam ice pan and cover with another paper towel. Cover the top paper towel with small amount of crushed ice.
3. Leave pup for 5–6 min until the pup does not showing signs of movement, has changed color from pink to gray, and feels cool to the touch.
4. Place the anesthetized pup on the prepared stereotaxic apparatus.
5. Utilize ear bars with rubber stoppers to rest on either side of the head, gently holding the head in place. Do not push these tight against the head.
6. Once the pup is positioned, center the end of the needle at bregma, leaving some space above the head and the needle.
7. Move the needle (0.8 mm, 2.0 mm) and set it right at the level of the head so that it is touching the skin but not puncturing the skin. If you are using an electronic measurement, tare the Z measurement.

8. Using a quick but gentle movement, move the needle down on the Z plane by 2 mm to puncture the skull and enter the brain.
9. Slowly bring the needle to a Z position of 1.5 mm.
10. Inject 1 uL of the virus at a rate of 8 uL/min.
11. Wait 10 s following injection to ensure that the injected material begins distributing and then slowly remove the needle from the brain.
12. Remove the pup from the stereotax and inspect for any visible signs of bleeding or damage.
13. Place the pup into the second heated nest to allow for easy distinction between pups that have been injected and those that have not, and wait for the pup to wake up and begin moving before setting up the next injection.
14. Between injections, clean the needle with 70% ethanol and ensure that there is no air bubble at the end of the needle.

Tissue Collection

1. Before tissue collection, order or prepare gelatin-coated slides (previously described).
2. Deeply anesthetize pup before tissue collection using IACUC approved methods.
3. Perfuse the pup with cold 1X PBS solution, remove brain and hemi-sect the brain into two hemispheres by cutting along the midsagittal plane with a razor or scalpel.
4. Place each brain (or hemisphere) into individual wells of a 24-well cell culture plate filled with 2 mL of 4% PFA (alternatively, use polystyrene vials with 3 mL of 4% PFA) and store at 4 °C. In cases where tissue is not required to be used fresh, fixative is required for prevention of tissue degradation and is most commonly done with formalin [16], which cross-links proteins to halt cellular activity [17].
5. After 12–24 h of fixation, place tissue in 10% sucrose solution in 4 °C fridge for 4 h and then switch to 30% sucrose solution and store in 4 °C fridge for at least 3 weeks before slicing.
6. Cut 100 uM sections using a vibratome, and mount tissue onto the gelatin-coated slides using a paintbrush and a couple of drops of PBS (3–6 drops) on the slide to assist with mounting and to ensure the slices do not dry out.
7. After moving the section to an ideal spot on the slide with the toothbrush, remove excess PBS from the slide with a Kimwipe. Be careful not to touch the section with the wipe.
8. Coverslip slides with Permount Mounting Medium or ProLong Gold without DAPI.

9. Once slides are dry, they can be stored in a 4 °C fridge and visualized for up to 6 months without losing fluorescence.
10. Image with a confocal microscope using the fluorescent laser scanning capabilities.

3.5 Visualization Utilizing Confocal Microscopy

Confocal microscopy can be used to analyze Golgi-stained neurons and AAV-infected neurons. Golgi visualization is done with a brightfield setting. The AAV-infected neurons are imaged using the fluorescent capabilities of the microscope.

1. Using a Leica confocal microscope and 20× magnification, scan the slides for tissues with neurons having optimal qualities for imaging, such as:
   (a) A soma that is located near the middle of the Z-plane, allowing visualization of as much of the dendritic arbor as possible.
   (b) Minimal overlap of dendrites with neighboring neurons.
   (c) No obvious damage to the soma, dendritic arbor, or axon of the neuron.
   (d) Ability to fit the entire length of the dendritic arbor into the camera's field of view.
   (e) No impurities blocking the dendritic arbor.
2. For measurements of dendritic arborization, image at 40× magnification. For measurements of dendritic spines, image at 60× magnification.
3. Image the Z-stack for the neuron, including as much of the associated dendritic arbor as possible. Typically, Z-stacks will range between 65 and 85 um from top to bottom.
4. Program the microscope to take an image every 0.2 um through the stack to get a clear image that will be easy to trace during later analysis.
5. For each animal, image and average at least five neurons for every region of interest.
6. Tracing of each neuron can be done utilizing the NeuroLeucida program (MBF Bioscience). This program will output the endpoints you program in the interface and allow for 3D reconstruction of your neuron.

3.6 Common Endpoints of Analysis of Dendritic Arbors

All of the endpoints described below can be used for analysis of both in vitro and in vivo dendritic arbors unless otherwise specified.

1. Sholl analysis: Sholl analysis was created in 1953 and has since been a gold-standard tool for assessing dendritic morphology. The Sholl profile is obtained by counting the number of times a dendrite intersects an imaginary sphere at a fixed distance from the soma, therefore giving a one-dimensional representation of the dendritic complexity of a three-dimensional dendrite and allowing for comparison of neuronal structures [18].

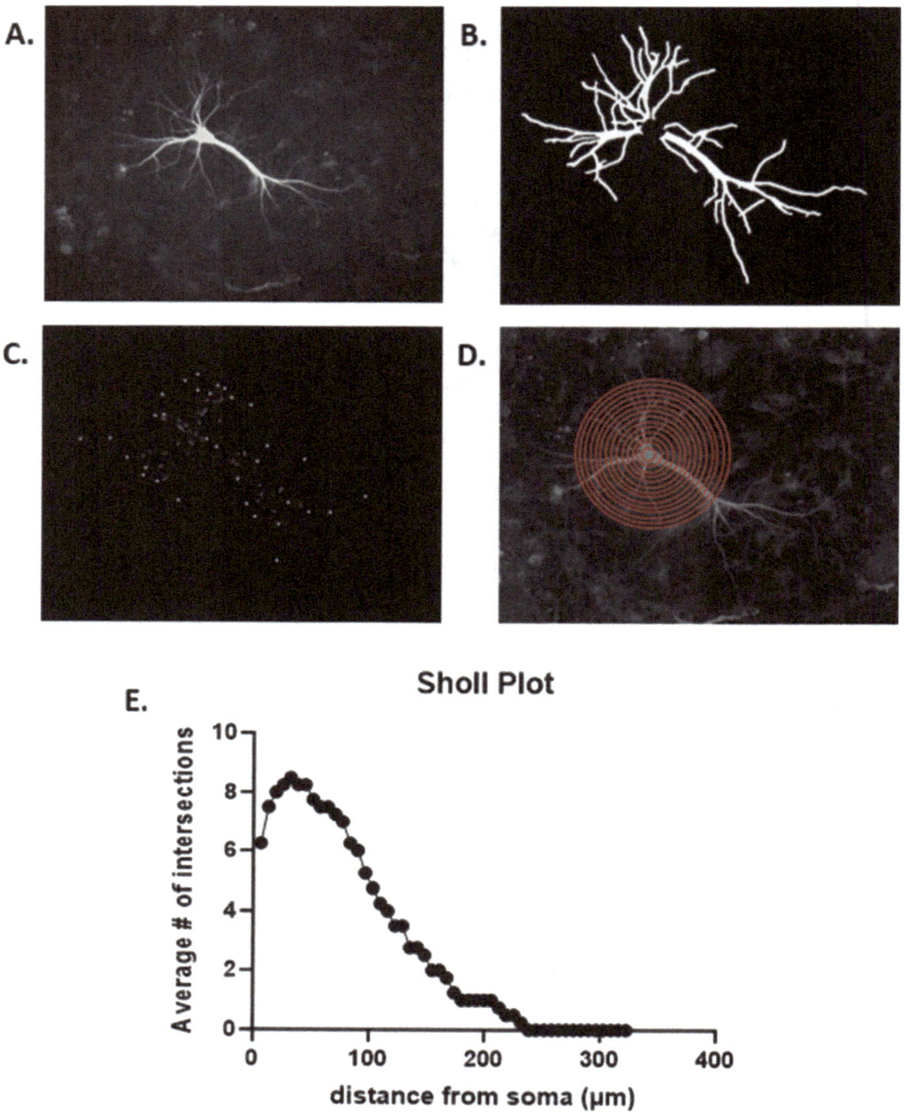

Fig. 1 Sholl analysis of dendritic arborization in primary rat cortical neuron-glia co-cultures. (**a**) Dendrites of individual rat cortical neurons transfected with MAP2B-eGFP are imaged using the MetaXpress software interfaced with the ImageXpress High Content imaging system. (**b**) Image of the skeletonization of the dendritic arbor automatically generated by the Omnisphero MATLAB algorithm. (**c**) The Omnisphero software also identifies points of intersection and terminal tips of dendrites. (**d**) To conduct a Sholl analysis, concentric circles are centered on the neuronal somata. (**e**) A representative Sholl plot illustrating the average number of dendritic intersections with each circle as a function of the distance of the circle from the soma

(a) Sholl analysis with Omnisphero for analysis of dendritic growth in vitro (Fig. 1): After images are acquired and processed from the high-content imaging system, Omnisphero, a custom -MATLAB algorithm, can be used for fully automated analysis of dendritic arborization data. Omnisphero can assess variable cell densities, inconsistent

Z-layers, and heterogeneous cell populations. It can identify classical endpoints, including average neuronal mass and neurite length, as well as the number of primary processes, branching points, end points, and neurons. It also gives individual Sholl curves of each neuron identified, the length gained per Sholl segment of each neuron, and total neurite length, number of branching points, and endpoints for reach individual neuron. Additionally, it has inclusions to assess endpoints like radial migration and neuronal density distribution within the migration area [19].

(b) Sholl analysis for in vivo images with ImageJ plugin (Fig. 2): The user can manually trace neurite structures by placing seeds that are eventually combined into lines, but because this method is time consuming and subjective, it is prone to biased analysis and is not ideal for screening assays or large datasets. The plugin automatically creates a Sholl mask by superimposing the binary image with rings places at fixed distances from each other centered around the neuronal soma and counting the number of dendritic intersections on each ring and then generating a distribution plot [13].

2. Total number of primary dendrites: A summation of the number of branches that originate from the soma of the neuron.
3. Length of primary dendrites: An average of the total length of each primary dendrite, including the secondary and tertiary branches for each individual neuron.
4. Dendritic tips: A measure of the total amount of ends/tips per primary dendrite wherein the primary dendrites are averaged, or the total amount of dendritic ends for the neuron wherein the number of tips on the primary dendrites are summed.
5. Dendritic spines: A measure of the number of spines on the secondary or tertiary branches of dendrites. Additionally, the morphological features of the spines can be assessed to differentiate between classes of spines (i.e., mushroom, thin, stubby, and filopodia) to get a more detailed analysis.

4 Notes

4.1. *Notes Regarding In Vitro Culturing*

1. Recombination issues: Recombination can occur and affect the fluorescence by diminishing it, causing downstream analysis issues.

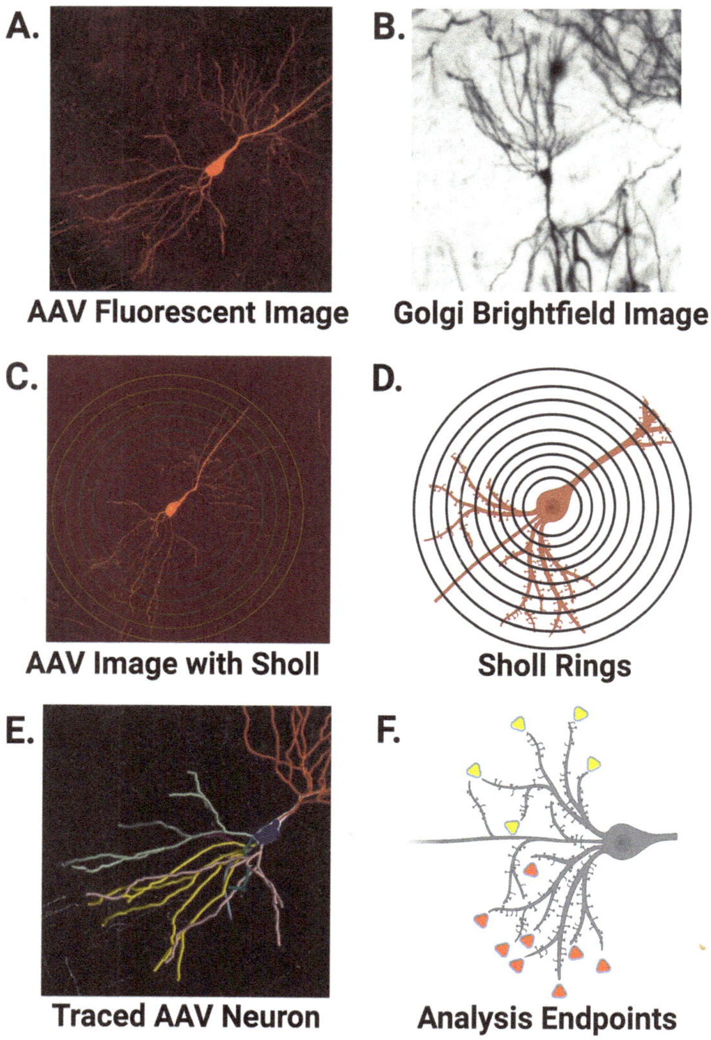

Fig. 2 Morphometric analysis of dendritic arborization in vivo. (**a**) Representative image of a hippocampal pyramidal neuron infected with AAV encoding a fluorescent protein. Image captured at 400X magnification using a STED microscope. (**b**) Representative brightfield image of a Golgi stained neuron. Image captured at 400X using a confocal microscope. (**c**) Sholl rings from the ImageJ plugin superimposed on the image of a neuron. (**d**) Diagram illustrating the Sholl rings centered on the soma of a neuron. (**e**) Image generated by NeuroLeucida software illustrating primary dendrites identified by different colors. (**f**) Illustrating identifying the dendritic tips (*yellow markers*) and dendritic intersections (*red markers*)

2. Contamination during transfection: This is a key concern when doing culture work. It is extremely important to use sterile technique to reduce the risk of contamination when removing and replacing media.
3. Immunocytochemical methods for visualizing dendrites: Labeling dendrites in primary neurons can be done with antibodies specific for MAP 2; however, this method is not ideal for high-density cultures because all dendritic

processes will be labeled. Rather, this method is best for use in low-density cell cultures or at very early time points. Materials needed for immunocytochemistry include: a fixative, usually 4% PFA; a permeabilization buffer, like Triton X-100, for antibodies to gain access to their intracellular antigens; a blocking buffer, such as PBS with 1–10% serum of the same species as the host species of secondary antibodies, to decrease nonspecific binding; a primary antibody that interacts with the antigen of interest; a fluorophore-tagged secondary antibody that cross-reacts with the primary antibody; and mounting medium, preferably with DAPI.

4.2. *Notes Regarding Golgi Staining*

1. Avoid metal instrumentation at all steps of Golgi staining. Plastic-coated magnetic stirring rods should be used to dissolve chemicals.
2. Store all solutions in the dark using aluminum foil or covered, lightproof boxes.
3. The time that the samples are kept in the impregnation solution can vary between species. For rat brain samples that are generally larger than similarly aged mouse brain samples, the brains may need to be kept in the impregnation solution longer for adequate staining, e.g., up to 3 weeks.

4.3. *Notes Regarding AAV Vector Utilization*

1. Additional serotypes that have been used for visualization of neurons in the central nervous system of rodents include AAV2 [20, 21], AAV9 [22], and AAV-PhP.eB [23–25].
2. Additional promotors that have been used for neuronal visualization are CAG promotor [26] and TTA/syn/tre.
3. Additional injection methods include intracisterna magna [25], intrathecal [27], and retro-orbital.

References

1. Flynn KC (2013) The cytoskeleton and neurite initiation. Bioarchitecture 3(4):86–109. https://doi.org/10.4161/bioa.26259
2. Valverde F (1970) The Golgi method. A tool for comparative structural analyses. In: Contemporary research methods in neuroanatomy. Springer, pp 12–31
3. Kang HW, Kim HK, Moon BH, Lee SJ, Lee SJ, Rhyu IJ (2017) Comprehensive review of Golgi staining methods for nervous tissue. Appl Microsc 47(2):63–69
4. Hastie E, Samulski RJ (2015) Adeno-associated virus at 50: a golden anniversary of discovery, research, and gene therapy success–a personal perspective. Hum Gene Ther 26(5): 257–265. https://doi.org/10.1089/hum.2015.025
5. Pignataro D, Sucunza D, Vanrell L, Lopez-Franco E, Dopeso-Reyes IG, Vales A, Hommel M, Rico AJ, Lanciego JL, Gonzalez-Aseguinolaza G (2017) Adeno-associated viral vectors serotype 8 for cell-specific delivery of

therapeutic genes in the central nervous system. Front Neuroanat 11(2). https://doi.org/10.3389/fnana.2017.00002

6. Howard DB, Powers K, Wang Y, Harvey BK (2008) Tropism and toxicity of adeno-associated viral vector serotypes 1, 2, 5, 6, 7, 8, and 9 in rat neurons and glia in vitro. Virology 372(1):24–34. https://doi.org/10.1016/j.virol.2007.10.007
7. Broekman ML, Comer LA, Hyman BT, Sena-Esteves M (2006) Adeno-associated virus vectors serotyped with AAV8 capsid are more efficient than AAV-1 or -2 serotypes for widespread gene delivery to the neonatal mouse brain. Neuroscience 138(2):501–510. https://doi.org/10.1016/j.neuroscience.2005.11.057
8. Chan KY, Jang MJ, Yoo BB, Greenbaum A, Ravi N, Wu WL, Sánchez-Guardado L, Lois C, Mazmanian SK, Deverman BE, Gradinaru V (2017) Engineered AAVs for efficient noninvasive gene delivery to the central and peripheral nervous systems. Nat Neurosci 20(8):1172–1179. https://doi.org/10.1038/nn.4593
9. Kügler S, Kilic E, Bähr M (2003) Human synapsin 1 gene promoter confers highly neuron-specific long-term transgene expression from an adenoviral vector in the adult rat brain depending on the transduced area. Gene Ther 10(4):337–347. https://doi.org/10.1038/sj.gt.3301905
10. Kim JY, Grunke SD, Levites Y, Golde TE, Jankowsky JL (2014) Intracerebroventricular viral injection of the neonatal mouse brain for persistent and widespread neuronal transduction. J Vis Exp 91:51863. https://doi.org/10.3791/51863
11. Hammond SL, Leek AN, Richman EH, Tjalkens RB (2017) Cellular selectivity of AAV serotypes for gene delivery in neurons and astrocytes by neonatal intracerebroventricular injection. PLoS One 12(12):e0188830. https://doi.org/10.1371/journal.pone.0188830
12. Wayman GA, Impey S, Marks D, Saneyoshi T, Grant WF, Derkach V, Soderling TR (2006) Activity-dependent dendritic arborization mediated by CaM-kinase I activation and enhanced CREB-dependent transcription of Wnt-2. Neuron 50(6):897–909. https://doi.org/10.1016/j.neuron.2006.05.008
13. Morgan RK, Schmuck M, Grodzki AC, Bruun DA, Matelski LE, Lein PJ (2021) Morphometric analysis of axons and dendrites as a tool for assessing neurotoxicity. In: Llorens J, Barenys M (eds) Experimental neurotoxicology methods. Springer US, New York, pp 51–87. https://doi.org/10.1007/978-1-0716-1637-6_4
14. Franks JJ, Wamil AW, Janicki PK, Horn J-L, Franks WT, Janson VE, Vanaman TC, Brandt PC (1998) Anesthetic-induced alteration of Ca2+ homeostasis in neural cells: a temperature-sensitive process that is enhanced by blockade of plasma membrane Ca2+-ATPase isoforms. J Am Soc Anesthesiol 89(1):149–164
15. Hollmann MW, Wieczorek KS, Berger A, Durieux ME (2001) Local anesthetic inhibition of G protein-coupled receptor signaling by interference with Gαq protein function. Mol Pharmacol 59(2):294–301
16. Dyrby TB, Baaré WF, Alexander DC, Jelsing J, Garde E, Søgaard LV (2011) An ex vivo imaging pipeline for producing high-quality and high-resolution diffusion-weighted imaging datasets. Hum Brain Mapp 32(4):544–563
17. Werner M, Chott A, Fabiano A, Battifora H (2000) Effect of formalin tissue fixation and processing on immunohistochemistry. Am J Surg Pathol 24(7):1016–1019. https://doi.org/10.1097/00000478-200007000-00014
18. Bird AD, Cuntz H (2019) Dissecting Sholl analysis into its functional components. Cell Reports 27:3081–3096. https://doi.org/10.1016/j.celrep.2019.04.097
19. Schmuck MR, Keil KP, Sethi S, Morgan RK, Lein PJ (2020) Automated high content image analysis of dendritic arborization in primary mouse hippocampal and rat cortical neurons in culture. J Neurosci Methods 341:108793. https://doi.org/10.1016/j.jneumeth.2020.108793
20. Passini MA, Wolfe JH (2001) Widespread gene delivery and structure-specific patterns of expression in the brain after intraventricular injections of neonatal mice with an adeno-associated virus vector. J Virol 75(24):12382–12392. https://doi.org/10.1128/jvi.75.24.12382-12392.2001
21. Kim JY, Ash RT, Ceballos-Diaz C, Levites Y, Golde TE, Smirnakis SM, Jankowsky JL (2013) Viral transduction of the neonatal brain delivers controllable genetic mosaicism for visualising and manipulating neuronal circuits in vivo. Eur J Neurosci 37(8):1203–1220. https://doi.org/10.1111/ejn.12126
22. Gholizadeh S, Tharmalingam S, Macaldaz ME, Hampson DR (2013) Transduction of the central nervous system after intracerebroventricular injection of adeno-associated viral vectors in

neonatal and juvenile mice. Hum Gene Ther Methods 24(4):205–213. https://doi.org/10.1089/hgtb.2013.076

23. Mathiesen SN, Lock JL, Schoderboeck L, Abraham WC, Hughes SM (2020) CNS transduction benefits of AAV-PHP.eB over AAV9 are dependent on administration route and mouse strain. Mol Ther Methods Clin Dev 19:447–458. https://doi.org/10.1016/j.omtm.2020.10.011
24. Brown D, Altermatt M, Dobreva T, Chen S, Wang A, Thomson M, Gradinaru V (2021) Deep parallel characterization of AAV tropism and AAV-mediated transcriptional changes via single-cell RNA sequencing. Front Immunol 12:730825. https://doi.org/10.3389/fimmu.2021.730825
25. Chatterjee D, Marmion DJ, McBride JL, Manfredsson FP, Butler D, Messer A, Kordower JH (2022) Enhanced CNS transduction from AAV.PHP.eB infusion into the cisterna magna of older adult rats compared to AAV9. Gene Ther 29(6):390–397. https://doi.org/10.1038/s41434-021-00244-y
26. Niwa H, Yamamura K, Miyazaki J (1991) Efficient selection for high-expression transfectants with a novel eukaryotic vector. Gene 108(2):193–199. https://doi.org/10.1016/0378-1119(91)90434-d
27. Castle MJ, Cheng Y, Asokan A, Tuszynski MH (2018) Physical positioning markedly enhances brain transduction after intrathecal AAV9 infusion. Sci Adv 4(11):eaau9859. https://doi.org/10.1126/sciadv.aau9859

Chapter 5

Visualizing and Measuring Dendrite Arborization in *Drosophila* Somatosensory Neurons

Yineng Xu, Inle Bush, and Chun Han

Abstract

Dendrites of neurons receive synaptic or sensory inputs and are important sites of neuronal computation. The morphological features of dendrites not only are hallmarks of the neuronal type but also largely determine a neuron's function. Thus, dendrite morphogenesis has been a subject of intensive study in neuroscience. Quantification of dendritic morphology, which is required for accurate assessment of phenotypes, can often be a challenging task, especially for complex neurons. Because manual tracing of dendritic branches is labor-intensive and time-consuming, automated or semiautomated methods are required for efficient analysis of a large number of samples. A popular in vivo model system for studying the mechanisms of dendrite morphogenesis is dendritic arborization (da) neurons in the *Drosophila* larval peripheral nervous system. In this chapter, we introduce methods for visualizing and measuring the dendritic arbors of these neurons. We begin with an introduction of da neurons and an overview of the methods that have been used for measuring da neuron dendrites. We then discuss the techniques and detailed steps of neuron visualization and image acquisition. Finally, we provide example steps for dendrite tracing and measurement.

Key words Dendritic arborization, Da neurons, Dendrite, Neuronal marker, *Drosophila*, Branch number, Branch order, Strahler analysis, Sholl analysis

1 Introduction

1.1 Dendritic Arborization (Da) Neurons

The *Drosophila* larval dendritic arborization (da) neurons are segmentally repeated peripheral sensory neurons that can be categorized into four classes (I–IV) with increasing dendritic arbor size and complexity (Fig. 1) [7]. Dendrites of class I (C1da) and II (C2da) neurons cover limited territories on the body wall, while class III (C3da) and IV (C4da) neurons exhibit complete and nonoverlapping dendritic coverage of the entire larval skin. In addition, C3da and C4da neurons have many dynamic high-order branches that constantly extend and retract [11, 30], while branches of C1da and C2da are much more static after the establishment of dendritic arbors at early larval stages. Lastly, C4da neurons are unique in

Kazuhito Toyooka (ed.), *Neuronal Morphogenesis: Methods and Protocols*, Methods in Molecular Biology, vol. 2831, https://doi.org/10.1007/978-1-0716-3969-6_5,

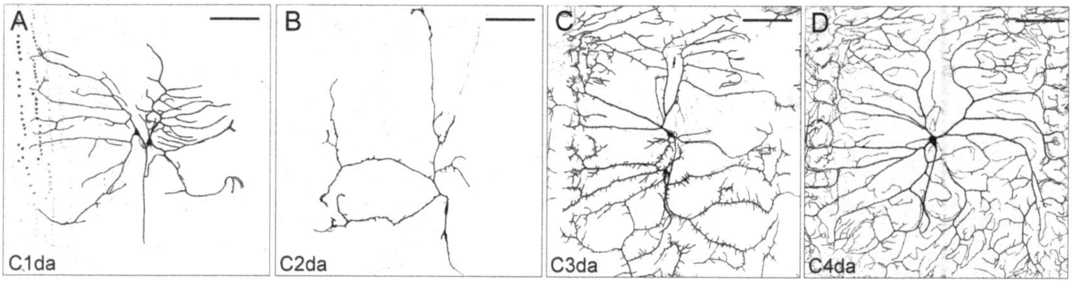

Fig. 1 Four classes of *Drosophila* da neurons. (**a**) A C1da neuron labeled by *Gal4*$^{2-21}$*>CD4-tdTom*. (**b**) A C2da neuron labeled by *Gal4*^{R20C11}*>CD4-tdTom*. (**c**) A C3da neuron labeled by *LexA*^{nompC}*>CD4-IFP2.0-T2A-HO1*. (D) A C4da neuron labeled by *ppk-CD4-tdGFP*. Scale bar, 100 μm

that they are space-filing, with a tendency to occupy all empty space on the skin with high-order branches [8, 9].

Da neurons have been a favorite in vivo model system for studying biology of dendrites, including branching, arbor-scaling, tiling, and remodeling [12]. This model offers multiple advantages. First, da neurons lie beneath the transparent larval body wall, whose dendrites can be imaged in live animals without dissection when a proper fluorescent marker is expressed in the neurons. Second, the neurons can be easily and reliably identified in stereotypical locations across segments. Third, this system is genetically tractable: The neurons and their interacting cells are easy to access using genetic drivers; genes expressed by these cells can be manipulated or tagged with powerful genetic tools available in *Drosophila*. Finally, robust genetic markers are available to specifically visualize single neuron types.

1.2 Genetic Labeling and Visualization of da Neurons

Various methods have been developed for labeling neurons in *Drosophila*. A method that was discovered early on but is still popular now is to use antibodies that recognize horseradish peroxidase (HRP). Anti-HRP antibodies specifically recognize neuronal tissues of insects at almost all developmental stages [31]. However, a limitation of HRP staining is that it lacks cell-type specificity, making it ineffective for tracing neuronal branches of single neurons. A more common method is to use the Gal4/upstream activator sequence (UAS) binary expression system [4]. This system allows type-specific neuron labeling, given that broad choices for tissue-specific Gal4 drivers (Table 1) and UAS-controlled florescence proteins are available. Despite the flexibility of the Gal4/UAS system, additional methods are necessary for simultaneously labeling two cell populations independently or labeling one cell population while manipulating another. One option to achieve this is to introduce another orthogonal binary system, such as LexA/LexAop [15] or Q system [25]. Alternatively, a Gal4-independent marker can be expressed in specific neurons under the control of a neuron-type-specific promotor or enhancer [8, 9]. For using the

Table 1
Neuron-type-specific Gal4 drivers identified for each class of da neurons

Cell type	Gal4	References
C1da neuron	2–21 R10D05	Grueber et al. [8, 9] Poe et al. [23]
C2da neuron	R20C11	Poe et al. [23]
C3da neuron	nompC ss 19–12	Cheng et al. [5] Parrish et al. [22] Rumpf et al. [26]
C4da neuron	4–77 ppk	Grueber et al. [8, 9] Kuo et al. [14]

second approach, special vectors that offer strong enhancer-driven transgene expression in *Drosophila* have been developed [10].

Because the brightness of cytosolic fluorescent proteins (FPs) is proportional to the volume of the cytoplasm, fine dendrites are difficult to visualize by these proteins due to the small cytoplasmic volume. In comparison, membrane-targeted fluorescent markers are more suitable for labeling axonal and dendritic processes of neurons. To anchor fluorescent proteins on the cell membrane, researchers have fused them with membrane-targeting signals. Two cell-surface glycoproteins, CD4 and CD8, can serve as membrane-targeting proteins, with CD4 being more efficient than CD8 for membrane delivery [10, 17]. Another option is a lipid-anchor signal, such as the myristylation signal (Myr) derived from Src64B [20]. Myristylation at the N-terminus of a fluorescent protein results in protein distribution evenly on the plasma membrane.

1.3 Methods for Tracing and Measuring Dendritic Arbors of da Neurons

Dendrite morphology of neurons is sensitive to perturbations of the cell state. Accurately measuring morphological features of the dendritic arbor is necessary for understanding cellular pathologies resulting from various insults and genetic manipulations. Parameters such as total dendrite length, territory area, and the number of dendrite branches can be used to quantify dendritic morphology using either manual or automated image analysis approaches. Here, we list some representative methods for tracing and quantifying dendrites.

Semiautomated dendrite tracing can be conducted in ImageJ/Fiji [27] (https://fiji.sc/) with the Simple Neurite Tracer (SNT) plug-in [19] (https://imagej.net/plugins/snt/). This plug-in enables users to quantify dendrite branch number and length by tracing and registering individual branches with respect to the cell body. During tracing, the beginning and the end of each dendrite are manually defined; the dendrite fragment in between is then automatically traced. This process is repeated for all dendrites in each neuron. Newer versions of this plug-in can also automatedly trace dendrites with segmented images. The total dendrite branch number is extracted from the ImageJ path for data output, specifically as the number of terminal branch tips. The path lengths of individual dendrites are summed to reveal the total dendrite length of the neuron. An advantage of this method is the ability to export the tracing into file formats compatible with neuronal reconstructions. However, the manual tracing steps are tedious and time-consuming. To address the low throughput of manual tracing, several automated algorithms have been developed.

DeTerm is a software developed by Kanaoka et al. [13]. It uses an artificial neural network to automatically recognize and quantify dendrite branch terminals. Once an image file of a neuronal dendritic arbor is imported, DeTerm can accurately label neuron terminals and provide positional data for each terminal. This automated approach provides high precision in identifying and quantifying dendrite branch terminals.

Imaris software (https://imaris.oxinst.com) is a software package for 2D/3D image analysis, with a powerful filament tracing and analysis module compatible with dendrite analysis. Image processing involves adjusting threshold levels to remove background noise for each image. Neuron reconstruction is performed using the Filament Tracer tool's automated detection. The generated filaments are then manually edited in the creation wizard window to correct misdetected and undetected branches. Imaris can count the total number of dendrite segments and the number of primary, secondary, and other branches.

ImagePro (https://mediacy.com/image-pro/) is another automatic method for morphology analysis. The "smart" segmentation tool is used to identify the areas occupied by the background and object, creating a mask of the neuron. By selecting the type of measurement as "area," the selected regions of interest are counted for the area of each neuron. Alternatively, by selecting the type of measurement as "region: number of dendrites," the selected regions of interest are counted for the total number of dendrites per neuron. The type of measurement can also be selected as "length," with regions of interest counted for length measurements and summed for each neuron.

1.4 Parameters to Measure

Da neurons exhibit segmental repetition and consistent morphology across segments and animals [8, 9]; thus, consistent defects in dendrite morphology can be distinguished reliably. By characterizing neuron morphology with meaningful metrics, it is possible to quantitatively describe the effects of genetic manipulation or environmental impacts. The results can provide deeper insights into the mechanisms underlying dendrite development and reveal potential therapeutic targets for neurological disorders.

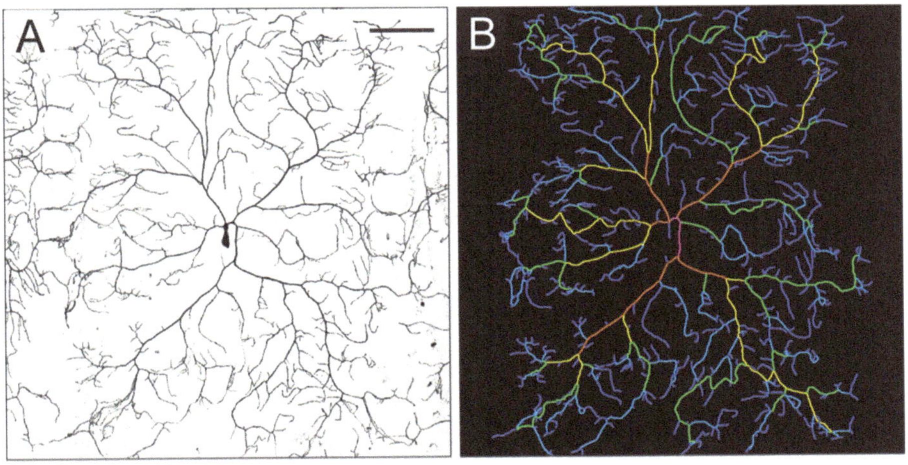

Fig. 2 Tracing of a C4da neuron with branch orders specified. (**a**) An image of a C4da neuron generated by 2D projection of a 3D stack. (**b**) A skeleton of the same neuron with branches belonging to the same order shown in the same color. Magenta: 1st order; orange: 2nd order; yellow: 3rd order; green: 4th order; light blue: 5th order; dark blue: 6th order. Scale bar, 100 μm

Among the basic parameters of dendrite morphology are the size of the dendritic arbor, total dendrite length, and total branch number, which can reveal dendrite reduction or overgrowth. However, it is worth noting that these parameters are closely related to the developmental stage and thus the animal size. Therefore, when analyzing data from different animals, it is important to control animal size or to normalize these parameters by the segment width.

Neuronal branches can be categorized into different orders using the Strahler analysis (Fig. 2) [29]. Metrics such as dendrite length, branch number, and the distance of branching points from the soma can be analyzed for each order to determine whether order-specific changes exist. Furthermore, this analysis can reveal whether changes in the total dendrite length or the branch number involve alterations of the number of branch order, and thus the arbor complexity. Finally, branches classified as the same order can grow from parent branches belonging to very different orders. The distribution of parent branch orders reflects fundamental branching logic of an arbor.

Sholl analysis [28] is a widely used method to assess the spatial distribution of dendritic branches. This technique involves counting the number of intersections between dendrites and concentric circles as a function of the distance from the soma. The Sholl profile is a product of dendrite spatial coverage, total dendrite length, and the root angle distribution [3], and thus is an effective way of measuring the arbor complexity.

In this protocol, we introduce our methods of animal preparation for imaging, image acquisition, image processing, and analysis.

2 Material and Equipment

1. *Drosophila* strains (e.g., da-specific Gal4 lines and UAS-CD4-FP markers).
2. Plain Microscope slides, Globe Scientific, Catalog# 1301.
3. Glass coverslip, Azer Scientific, Catalog# 1152250.
4. Vacuum grease, Dow Corning, Item# 1597418.
5. Glycerol, Avantor, Catalog# 56-81-5.
6. Isoflurane, Midwest Veterinary Supply, Catalog# 193.33161.3.
7. Immersion oil, Electron Microscopy Sciences, Catalog# 16915-01.
8. Kimwipe, Kimberly-Clark.
9. Leica SP8 microscope (or other confocal systems).
10. *Drosophila* medium, standard yeast-glucose medium (https://doi.org/10.1101/pdb.rec10907).

3 Methods

In this section, we first describe our methods for acquiring confocal images of da neuron using the Leica SP8 system. We then provide a step-by-step guide for processing the images and measuring a number of metrics of a dendritic arbor based on methods we have previously published [23, 24].

3.1 Larva Preparation

1. Cross da-specific Gal4 flies with UAS-CD4-FP marker flies in a vial with yeast paste. Leave the vial in 25 °C incubator and transfer the flies into a new vial every 2–3 days. A desirable egg density is 80–150 eggs per vial.
2. Grow larvae in 25 °C incubator until the desired stage.
3. Use forceps to pick larvae out of the vial and rinse larvae thoroughly with tap water in a Petri dish to get rid of food particles on larvae. Dry larvae gently on a piece of Kimwipe (*see* **Notes 1–3**).
4. Apply a small amount of vacuum grease (a 2 mm × 2 mm × 2 mm cube) on each of the four comers of a slide as spacers. Add a drop of glycerol in the center of the slide (Fig. 3).
5. (Optional) For larvae before the third instar, anesthetize them in a covered Petri dish with 8–12 isoflurane drops (160–240 μL) (Fig. 4a) (*see* **Note 4**).
6. Place a dried larva in the glycerol drop. Orient the larva vertically on the slide.

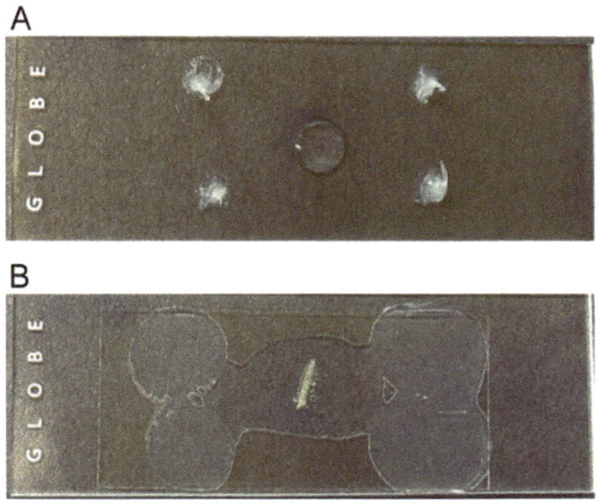

Fig. 3 Larvae mounting. (**a**) A glass slide with vacuum grease and glycerol. (**b**) A larva mounted on the slide

Fig. 4 larvae anesthetization. A small piece of moistened tissue paper is in a small (35 mm) Petri dish without a lid. The small Petri dish is placed in a large (60 mm) Petri dish containing a piece of dry tissue paper. Larvae are put in the small Petri dish. Isoflurane is added onto the dry tissue paper

7. Put a coverslip on the slide. Press down the coverslip gently until it touches the larva. Without applying much pressure, slide the coverslip laterally to roll and straighten the larva. Position the larva dorsal up.
8. Use fingers to press firmly the four corners of the coverslip until vacuum grease cannot spread any more. Be careful not to apply too much force onto the center of the coverslip where the larva is (Fig. 4b).

3.2 Imaging with a Leica SP8 Confocal Microscope

1. Mount the slide on the microscope stage.
2. Choose the 20X/NA0.75 multi-immersion objective. Apply immersion oil and focus on the larval body wall.
3. Choose appropriate lasers and detection windows. Locate the neuronal cell body and dendrites using live mode. Usually, neurons in the A2 and A3 segments are imaged due to their more consistent morphology. Position the cell body to the center of the frame. Rotate the field of view so that the lateral segment borders are vertical.
4. Define start/end positions of the Z stack and acquire the image stack. We use the following imaging settings: format 1024 × 1024 pixel; scanning speed 600 Hz; zoom 0.75; 2× line averaging with a HyD detector; pinhole 2.5 AU; Z step-size 3.5 μm.
5. Save image stacks in the .lif format.

3.3 Image Processing

All steps are completed in Fiji/ImageJ. The *italicized texts* are commands that can be accessed through the menus or the Search Bar. Tools available on the Toolbar are indicated by underlines. The names of plugins are indicated by font. The steps may be semiautomated to be applied in batch using ImageJ scripting languages (*see* **Note 5**).

3.3.1 Preprocessing

1. Export raw images in .lif files into individual .tif files using the Bio-Formats plug-in [18].
2. Open a 3D image stack. Go through each Z slice to inspect if there are interfering signals (e.g., autofluorescent internal organs and cuticles) that will overlap with neuronal signals after 2D projection.
3. If interfering signals are identified, manually outline the signals using the Polygon selections or Freehand selections tool in individual slices. Avoid selecting signals from the neuron to be analyzed. Fill the selection with black or a color sampled from a background region (*Edit>Fill*). Alternatively, interfering signals can be painted over using the Paintbrush tool with black as the foreground color.
4. Project the 3D stack into a 2D image using maximum intensity Z-projection (*Image>Stacks>Z Project...*).
5. Manually remove the axon from the projected image using the Paintbrush tool similarly to **step 3** (*see* **Note 6**).
6. Manually outline the dendrite arbor using Polygon selections tool by clicking the most distal tips of the arbor. This outlined region should be saved as a region of interest (ROI) or an

overlay (*Image>Overlay>Add Selection...*) along with the image for later retrieval.

7. If necessary, connect discontinuous branches using the Paintbrush tool.

3.3.2 Segmentation

Before one can analyze dendrite morphology, dendrites need to be separated from other areas in the image. Segmentation is the process of creating a binary mask containing only those pixels that belong to the dendrites. We describe two segmentation methods: auto-local thresholding and Trainable Weka Segmentation (TWS) [1]. Although TWS involves more initial manual steps to train a machine learning model, it is generally more accurate in separating neurons from background signals.

ImageJ Auto-local Thresholding:

1. If dendrite signals are low, the image can be brightened by a Gamma correction (*Process>Math>Gamma...*), with a gamma value between 0.5 and 0.9.
2. Remove noise by blurring the image slightly by applying a Gaussian Blur (*Process>Filters>Gaussian Blur...*) with a sigma of 0.8–1.5 pixels.
3. Apply "Auto Local Threshold" (*Image>Adjust>Auto Local Threshold*) with the Phansalkar method, a radius of 50, parameter 1 = 0 and parameter 2 = 0. Alternative methods are also available (*see* **Notes** 7 and **8**).

Trainable Weka Segmentation:

1. Train and save a segmentation model using the Trainable Weka Segmentation plug-in (*Plugins>Segmentation>Trainable Weka Segmentation*). Instructions are available at https://imagej.net/plugins/tws/.
2. Apply the model to all preprocessed 2D images individually or in-batch using a script. An example script is available at https://imagej.net/plugins/tws/scripting.

3.3.3 Skeletonization

1. Remove isolated particles using Particles8 in the Morphology plug-in (*Plugins>Morphology>Particles8*) [16], with the Filter by size option and Minimum size of 500 pixels. Examine mask as necessary (*see* **Note 9**).
2. From a segmented mask, generate a single pixel-wide skeleton using the Skeletonize (2D/3D) plug-in (*Plugins>Skeleton>Skeletonize (2D/3D)*).
3. To remove skeletons outside of dendritic field, select the dendrite arbor ROI, run Make Inverse (*Edit>Selection>Make Inverse*) and Fill (*Edit>Fill*) with black as the foreground color.

3.3.4 Quantification (See Note 10)

Dendrite Territory:

1. After selecting the outline of the dendrite arbor, run the Convex Hull method (*Edit> Selection> Convex Hull*).
2. Click the convex hull selection and record the area of the convex hull using the "Measure" command (*Analyze> Measure*).

Strahler Branch Ordering:

1. Run the Strahler Analysis plug-in [2]. The plug-in generates basic statistics on branch orders as well as an image in which branches of different orders are in different pixel values.
2. To generate an image containing branches of a single order for further analysis, identify the value of the pixels in branches of the desired order and keep only pixels of the determined value using the "Threshold" command (*Image> Adjust> Threshold*. . .).

Branch Length Distribution and Branch Number:

1. On the skeleton image of the neuron, run the Analyze Skeleton plug-in (*Analyze> Skeleton> Analyze Skeleton (2D/3D)*) [21].
2. The length and spatial position of each branch will be calculated and provided in a table.

Sholl Analysis:

1. On the skeleton image of the neuron, mark the center of the soma using the Point Selection Tool.
2. With the skeleton image open, run the Sholl Analysis plug-in (*Plugins> Neuroanatomy> Sholl> Sholl Analysis (From Image)*. . .) command [6] with default parameters.

4 Notes

1. Thorough washing of larvae with tap water is necessary for removing food particles on their body surface, as these particles can interfere with imaging due to their autofluorescence.
2. Prolonged suffocation and physical damage by forceps can alter the morphology of neurons; so it is important to handle larvae gently when using forceps and to avoid leaving them in water for extended periods. Additionally, it is recommended not to image larvae that have been mounted on slide for a long time (20 min).
3. To ensure consistency in results during quantification, it is crucial to select larvae that are at a similar stage of development.

4. Imaging animals with a small step size can increase the time needed for acquiring a complete 3D stack, during which the larva could move and cause "ghost" images after 2D projection. The use of isoflurane anesthesia can help to keep the animals still during imaging.
5. Once a suitable image processing and quantification protocol has been established and tested for your data, the protocol can be semiautomated by scripting the commands using the ImageJ macro language, BeanShell, Java, or Python. Semiautomation can save time in the application of the protocol and make the procedure more easily reproducible. However, one should weigh the time taken to code the macro and the loss of human input checking many steps of the process into the decision to semiautomate.
6. As axons may be falsely identified as primary order branches later in quantification, it is necessary to remove them.
7. The ideal auto-local thresholding method may differ among images to survey which method is better for your data, run the Auto Local Threshold plug-in with "Try All" selected in place of Phansalkar. This will produce an image with each of the possible thresholding methods displayed and labeled. Choose the best thresholding method from these.
8. The parameters for Gamma correction, Gaussian Blur, and Auto Local Threshold may differ among images depending on the image quality. These parameters should be optimized by preliminary testing.
9. It is important to manually examine the quality of automatically generated masks and skeletons. To do this, use *Merge Channels* command to combine the raw image and the generated mask, then ensure the white of the mask is only present at pixel locations where the neuron is also present. Often the mask will not initially be perfect, and one will need to tune parameters to fit the procedure to the data.
10. When presenting quantification results, it is recommended to normalize them by segment width.

Acknowledgments

This work was supported by NIH grants (R24OD031953 and R21NS127052) awarded to C.H.

References

1. Arganda-Carreras I, Kaynig V, Rueden C, Eliceiri KW, Schindelin J, Cardona A, Sebastian Seung H (2017) Trainable Weka Segmentation: a machine learning tool for microscopy pixel classification. Bioinformatics 33(15): 2424–2426
2. Arshadi C, Gunther U, Eddison M, Harrington KIS, Ferreira TA (2021) SNT: a unifying toolbox for quantification of neuronal anatomy. Nat Methods 18(4):374–377
3. Bird AD, Cuntz H (2019) Dissecting Sholl analysis into its functional components. Cell Rep 27(10):3081–3096 e3085
4. Brand AH, Perrimon N (1993) Targeted gene expression as a means of altering cell fates and generating dominant phenotypes. Development 118(2):401–415
5. Cheng LE, Song W, Looger LL, Jan LY, Jan YN (2010) The role of the TRP channel NompC in Drosophila larval and adult locomotion. Neuron 67(3):373–380
6. Ferreira TA, Blackman AV, Oyrer J, Jayabal S, Chung AJ, Watt AJ, Sjostrom PJ, van Meyel DJ (2014) Neuronal morphometry directly from bitmap images. Nat Methods 11(10):982–984
7. Grueber WB, Jan LY, Jan YN (2002) Tiling of the Drosophila epidermis by multidendritic sensory neurons. Development 129(12): 2867–2878
8. Grueber WB, Jan LY, Jan YN (2003a) Different levels of the homeodomain protein cut regulate distinct dendrite branching patterns of Drosophila multidendritic neurons. Cell 112(6):805–818
9. Grueber WB, Ye B, Moore AW, Jan LY, Jan YN (2003b) Dendrites of distinct classes of Drosophila sensory neurons show different capacities for homotypic repulsion. Curr Biol 13(8): 618–626
10. Han C, Jan LY, Jan Y-N (2011) Enhancer-driven membrane markers for analysis of non-autonomous mechanisms reveal neuron–glia interactions in *Drosophila*. Proc Natl Acad Sci 108(23):9673–9678
11. Han C, Wang D, Soba P, Zhu S, Lin X, Jan LY, Jan YN (2012) Integrins regulate repulsion-mediated dendritic patterning of Drosophila sensory neurons by restricting dendrites in a 2D space. Neuron 73(1):64–78
12. Jan Y-N, Jan LY (2010) Branching out: mechanisms of dendritic arborization. Nat Rev Neurosci 11(5):316–328
13. Kanaoka Y, Skibbe H, Hayashi Y, Uemura T, Hattori Y (2019) DeTerm: software for automatic detection of neuronal dendritic branch terminals via an artificial neural network. Genes Cells 24(7):464–472
14. Kuo CT, Jan LY, Jan YN (2005) Dendrite-specific remodeling of *Drosophila* sensory neurons requires matrix metalloproteases, ubiquitin-proteasome, and ecdysone signaling. Proc Natl Acad Sci 102(42):15230–15235
15. Lai S-L, Lee T (2006) Genetic mosaic with dual binary transcriptional systems in Drosophila. Nat Neurosci 9(5):703–709
16. Landini G (2008) Advanced shape analysis with ImageJ. In: Proceedings of the second ImageJ user and developer conference. Luxembourg, pp 116–121
17. Leahy DJ (1995) A structural view of CD4 and CD8. FASEB J 9(1):17–25
18. Linkert M, Rueden CT, Allan C, Burel JM, Moore W, Patterson A, Loranger B, Moore J, Neves C, Macdonald D, Tarkowska A, Sticco C, Hill E, Rossner M, Eliceiri KW, Swedlow JR (2010) Metadata matters: access to image data in the real world. J Cell Biol 189(5):777–782
19. Longair MH, Baker DA, Armstrong JD (2011) Simple Neurite Tracer: open source software for reconstruction, visualization and analysis of neuronal processes. Bioinformatics 27(17): 2453–2454
20. Mauss A, Tripodi M, Evers JF, Landgraf M (2009) Midline signalling systems direct the formation of a neural map by dendritic targeting in the Drosophila motor system. PLoS Biol 7(9):e1000200
21. Mertzanidou T, Hipwell JH, Reis S, Hawkes DJ, Ehteshami Bejnordi B, Dalmis M, Vreemann S, Platel B, van der Laak J, Karssemeijer N, Hermsen M, Bult P, Mann R (2017) 3D volume reconstruction from serial breast specimen radiographs for mapping between histology and 3D whole specimen imaging. Med Phys 44(3):935–948
22. Parrish JZ, Xu P, Kim CC, Jan LY, Jan YN (2009) The microRNA bantam functions in epithelial cells to regulate scaling growth of dendrite arbors in Drosophila sensory neurons. Neuron 63(6):788–802
23. Poe AR, Tang L, Wang B, Li Y, Sapar ML, Han C (2017) Dendritic space-filling requires a neuronal type-specific extracellular permissive

signal in *Drosophila*. Proc Natl Acad Sci 114(38):E8062–E8071

24. Poe AR, Xu Y, Zhang C, Lei J, Li K, Labib D, Han C (2020) Low FoxO expression in Drosophila somatosensory neurons protects dendrite growth under nutrient restriction. elife 9:e53351
25. Potter CJ, Tasic B, Russler EV, Liang L, Luo L (2010) The Q system: a repressible binary system for transgene expression, lineage tracing, and mosaic analysis. Cell 141(3):536–548
26. Rumpf S, Lee SB, Jan LY, Jan YN (2011) Neuronal remodeling and apoptosis require VCP-dependent degradation of the apoptosis inhibitor DIAP1. Development 138(6): 1153–1160
27. Schindelin J, Arganda-Carreras I, Frise E, Kaynig V, Longair M, Pietzsch T, Preibisch S, Rueden C, Saalfeld S, Schmid B, Tinevez JY, White DJ, Hartenstein V, Eliceiri K, Tomancak P, Cardona A (2012) Fiji: an open-source platform for biological-image analysis. Nat Methods 9(7):676–682
28. Sholl DA (1953) Dendritic organization in the neurons of the visual and motor cortices of the cat. J Anat 87(4):387–406
29. Strahler AN (1957) Quantitative analysis of watershed geomorphology. EOS Trans Am Geophys Union 38(6):913–920
30. Sturner T, Tatarnikova A, Mueller J, Schaffran B, Cuntz H, Zhang Y, Nemethova M, Bogdan S, Small V, Tavosanis G (2019) Transient localization of the Arp2/3 complex initiates neuronal dendrite branching in vivo. Development 146(7):dev171397
31. Wang X, Sun B, Yasuyama K, Salvaterra PM (1994) Biochemical analysis of proteins recognized by anti-HRP antibodies in Drosophila melanogaster: identification and characterization of neuron specific and male specific glycoproteins. Insect Biochem Mol Biol 24(3): 233–242

Chapter 6

Analyses of Neurite and Spine Formation by In Utero Electroporation in Mice

Hidenori Tabata, Nanako Hamada, and Koh-ichi Nagata

Abstract

Dendrite morphology and dendritic spines are key features of the neuronal networks in the brain. Abnormalities in these features have been observed in patients with psychiatric disorders and mouse models of these diseases. In utero electroporation is an easy and efficient gene transfer system for developing mouse embryos in the uterus. By combining with the Cre–loxP system, the morphology of individual neurons can be clearly and sparsely visualized. Here, we describe how this labeling system can be applied to visualize and evaluate the dendrites and dendritic spines of cortical neurons.

Key words In utero electroporation, Dendrite, Spine, Synapse, Cerebral cortex, Neurodevelopmental disorder

1 Introduction

Disturbances in dendritic arbors and spines are hallmarks of neurodevelopmental and other psychiatric disorders [1]. Golgi staining and virus vectors expressing fluorescent proteins are widely used to assess these phenotypes in disease model mice; however, these methods require special skills and are time-consuming. In utero electroporation (IUE) is a simple and efficient in vivo gene transfer method in which plasmid DNA can be introduced into neural stem cells (radial glia) in developing mouse embryos in the uterus [2–4]. Because neurons do not proliferate after they differentiate, the introduced plasmid persists without being diluted out, and the expression can be seen even after 6 months [2, 5]. The copy number of plasmids introduced by IUE is high (hundreds of copies in general), which is helpful for the visualization of neurons by introducing expression vectors of fluorescent proteins such as green fluorescent protein (GFP). Moreover, the application of the Cre–loxP system to IUE greatly enhances the signal-to-noise ratio of fluorescent labeling and reduces the overlapping of labeled

Kazuhito Toyooka (ed.), *Neuronal Morphogenesis: Methods and Protocols*, Methods in Molecular Biology, vol. 2831,
https://doi.org/10.1007/978-1-0716-3969-6_6,

neurons in the objective field [6–8]. In this chapter, we describe the methods for visualizing individual neurons using IUE and the Cre–loxP system and evaluating their morphology. This strategy can be used to easily and quickly assess the effects of knockdown or overexpression of desired genes on neuronal morphology.

2 Materials

2.1 In Utero Electroporation (IUE)

1. Electric square pulse generator for in vivo electroporation (*see* **Note 1**).
2. Tweezers-type electrode (5 mm diameter) (*see* **Note 2**).
3. Aspirator tube assembly composed of a mouthpiece, a silicon tube, and a capillary holder (*see* **Note 3**).
4. Optical fiber light.
5. Plate warmer: 23 × 16 cm or larger. Set the temperature to 38 °C.
6. Surgical instruments: fine forceps × 2, surgery scissors × 2, ring forceps (8 × 10 mm rings), and needle holder (for suturing).
7. Surgical suture and needle (e.g., 5–0 silk).
8. Autoclip.
9. Sterile gauze.
10. Micropipettes for injection: prepare 1 mm diameter glass capillary tubes with a puller (*see* **Note 4**). Cut the tip of the micropipette obliquely using fine forceps under a dissecting microscope and attach the micropipettes to the holder of an aspirator tube assembly (Drummond).
11. HEPES-buffered saline (HBS): 5 g/L of HEPES, 8 g/L of NaCl, 0.37 g/L of KCl, 0.251 g/L of $Na_2HPO_4{\cdot}12H_2O$, 1 g/L of Glucose in water. Alternatively, purchase 2× concentrate solution (Sigma) and dilute with autoclaved water at a 1:1 ratio.
12. Autoclaved phosphate-buffered saline (PBS).
13. MMB anesthetic: a mixture of medetomidine (30 μg/mL), midazolam (0.4 mg/mL), and butorphanol (0.5 mg/mL) in autoclaved water (*see* **Note 5**).
14. Atipamezole solution: Use as an antagonist for MMB anesthetic. Prepare a 150 μg/mL solution in autoclaved water.
15. Plasmid solution: Prepare using conventional methods, such as with the Qiagen kit. Plasmids are dissolved in HBS (*see* **Note 6**). The plasmid mix used to sparsely label neurons comprises pCAG-turboRFP (a red fluorescent protein expression vector

for visualization of electroporated area, 0.5 μg/μL), pCAG-LNL-EGFP (a Cre-dependent EGFP expression vector, 0.5 μg/μL), and pCAG-M-Cre (a Cre recombinase expression vector, 0.5 ng/μL) [9]. An expression vector or knockdown vector of the desired gene can be added to this plasmid mix. To monitor the location of the injected DNA solution in the brain, add 1/10 volume of 0.2% Fast Green solution (w/v in distilled water) to the DNA solution.

2.2 Immunohistochemistry

1. Isoflurane.
2. 4% (w/v) paraformaldehyde (PFA) in PBS (*see* **Note** 7).
3. Vibrating blade microtome.
4. Silan-coated slide or equivalent (*see* **Note 8**).
5. PBST: 0.05% Triton X-100 (v/v) in PBS.
6. Blocking solution: 0.1% bovine serum albumin (w/v) and 0.5% Triton X-100 (v/v) in PBS.
7. Anti-GFP antibody (purified from rabbit serum) and anti-rabbit IgG Alexa Fluor 488 antibody (*see* **Note 9**).
8. Laser scanning microscope.

3 Methods

3.1 In Utero Electroporation (IUE)

1. Anesthetize a pregnant mouse at E14 by intraperitoneally injecting MMB anesthetic at a dose of 10 μL/g body weight.
2. Ten minutes later, place the mouse on its back on a plate warmer that maintains the temperature at 38 °C (*see* **Note 10**).
3. After disinfecting the abdominal skin with 70% ethanol, make an approximately 2 cm skin incision along the midline, extending anteriorly from the midpoint between the two most posterior pairs of nipples to the level of the umbilicus, which is visible on the peritoneum, using a set of forceps and scissors (Fig. 1a, b).
4. Then, make a 2 cm midline incision in the abdominal wall along the linea alba using another set of forceps and scissors. Make a hole in the center of the sterile gauze, place over the incision, and gently draw one uterine horn out through the hole in the gauze using ring forceps (Fig. 1c).
5. After observing the orientation of the embryos through the wall of the uterine horn, insert a micropipette into one side of the lateral ventricle, and inject 1 μL of plasmid solution via expiratory pressure using the aspirator tube assembly (*see* **Note 11**).

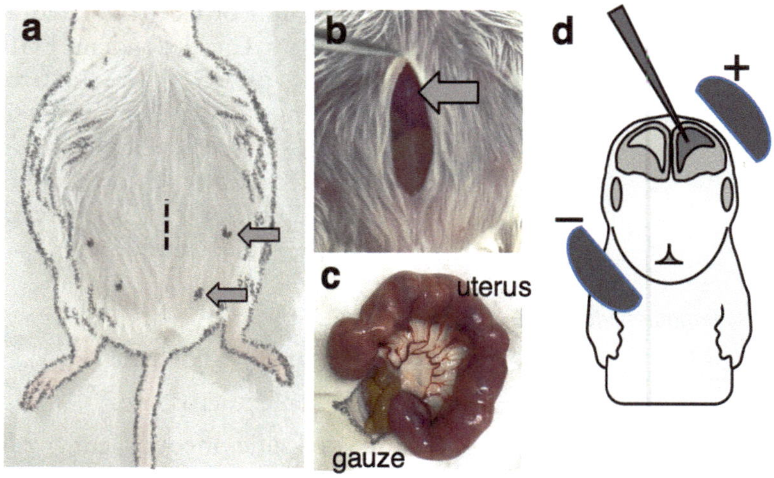

Fig. 1 In utero electroporation. Panel (**a**) shows position of the skin incision (broken line). The most posterior two nipples are pointed by arrows. (**b**) Through the skin incision, abdominal wall and the umbilicus (arrow) on it are observed. (**c**) The uterine horn (E17) was exposed and placed on gauze. (**d**) The schematic diagram shows how the electrodes are placed on the head of embryos. The anode should cover the injected hemisphere obliquely

6. Pinch the head of the embryo with a tweezer-type electrode (CUY650P5), and apply electronic pulses using an electroporator (35 V, 50 ms, 5 times with 500 ms intervals). In this step, the electrode should be placed obliquely so that the anode covers the injected hemisphere. The actual current should be 40–60 mA (Fig. 1d) (*see* **Note 12**).
7. After completing the procedure on the first uterine horn, place it back in the abdominal cavity, and expose the other uterine horn and repeat the same procedure.
8. After the procedure has been completed on the second uterine horn, pour 1–2 mL of sterile PBS into the abdominal cavity (*see* **Note 13**).
9. Apply atipamezole solution into the abdominal cavity at a dose of 10 μL/g body weight (*see* **Note 14**).
10. Close the abdominal wall and skin with surgical sutures. Alternatively, the skin can be closed using an autoclip.
11. Warm the mice on a plate warmer for approximately 30 min after surgery (*see* **Note 10**).

3.2 Tissue Preparation and Immunohistochemistry

1. After surgery, the mice should give birth to electroporated pups normally. At postnatal day 7 (P7, to analyze dendritic arbor) or P30 (to analyze dendritic spines), anesthetize the pups with isoflurane and perfuse with 4% PFA in PBS. Then, dissect out the brains and immerse them in 4% PFA for 2–16 h at 4 °C.

2. After washing the brains with PBS, embed the brains in 3% agarose in PBS and section them coronally with a vibrating blade microtome at 100 μm thickness (*see* **Note 15**).
3. Mount the slices on Silan-coated glass slides and dry them for 5–10 min (*see* **Note 16**).
4. Wash the slides with PBST.
5. Incubate the slides in antigen retrieval solution in a glass jar at 70 °C for 20 min, and then place the jar at room temperature for 20 min to allow it to cool down gradually (*see* **Note 17**).
6. Wash the slides with PBST.
7. Incubate the slides with blocking solution at room temperature for 30 min.
8. Remove the blocking solution and incubate with the primary antibody at 4 °C overnight. The primary antibody should be diluted in PBST.
9. Wash the slides with PBST thrice.
10. Apply appropriate secondary antibody and incubate at 4 °C overnight.
11. Wash the slides with PBST thrice, and mount the cover glass with a medium for fluorescence microscopy (such as PermaFluor).

3.3 Analysis of Dendritic Arbor

1. Acquire Z stack images (0.8–1.0 μm-pitch) of individual neurons visualized by GFP using a 20× objective lens of a laser scanning microscope (Fig. 2a) (*see* **Note 18**).
2. The data can be analyzed using a simple neurite tracer (SNT), a plug-in of ImageJ. Using this software, dendrites can be traced semiautomatically and quantified by Sholl analysis (the number of dendrite intersections against the radial distance from the soma center) (*see* **Note 19**).

3.4 Analysis of Dendritic Spines

1. Acquire Z stack images (0.2–0.3 μm-pitch) of individual dendrites visualized by GFP using a 63× objective lens with 2× digital zoom of a laser scan microscope (Fig. 2b) (*see* **Note 20**).
2. The data can be analyzed using the Imaris software. Using the "filament tracer" function, dendrites can be traced semiautomatically, and the spines on the dendrite are automatically detected. Imaris computes spine density, length, head diameter, and so on. This software also classifies spines into four types, namely stubby, mushroom, long thin, and filopodia (Fig. 2c, d), which represent different maturities (*see* **Note 21**).

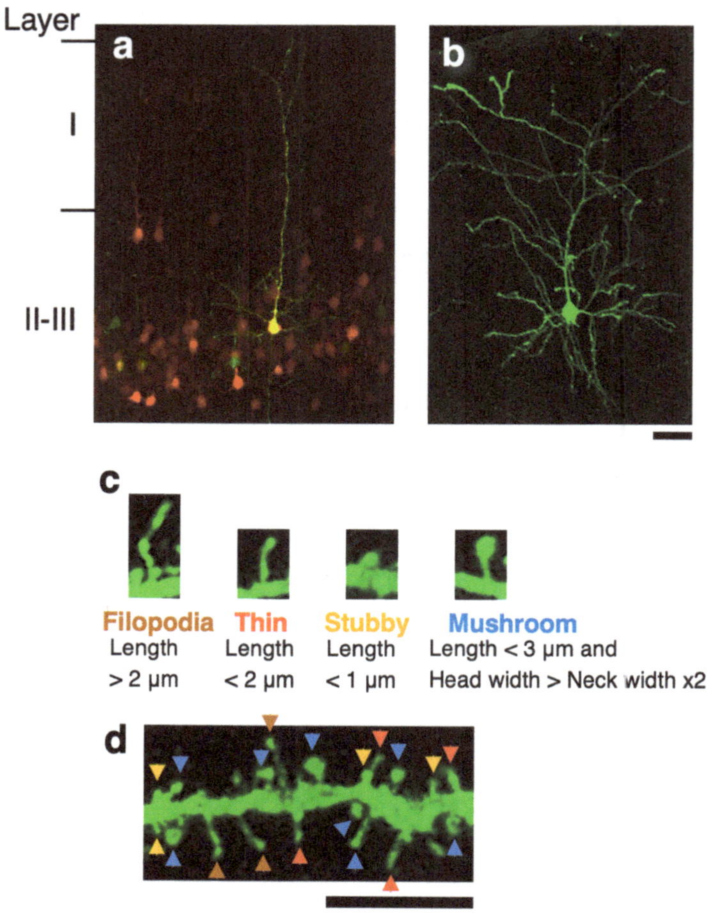

Fig. 2 Analysis of dendritic arbor and spine. (**a**, **b**) pCAG-LNL-EGFP (0.5 μg) was co-electroporated with pCAG-M-Cre (0.5 ng) for sparse together with pCAG-RFP (0.5 μg) at E14.5. Coronal sections were prepared with brains collected at P7 (**a**) and P30 (**b**), then immunostained for GFP. Scale bar, 20 μm. (**c**) Common dendritic spine types found in cortical pyramidal neurons. (**d**) Dendritic spines on the first-order basal dendrites in (**b**). Different spine types are indicated by arrows, color-coded to match (**c**). Scale bar, 5 μm

4 Notes

1. We are using CUY21E and NEPA21 (NEPA gene).
2. For example, CUY650P5 (NEPA gene). Electrodes with 5 mm diameter pads are used for E14 or later embryos, which label layers II/III neurons. For IUE in younger embryos, 3 mm diameter electrodes are used.
3. We are using a product from Drummond.
4. We are using a product from Narishige (GD-1).

5. Purchase three components individually and make a mixture in an appropriate amount (25 mL or so). The mixture can be stored at 4 °C in the dark for a few months.
6. If precipitates form in the plasmid solution, further purification with phenol/chloroform followed by chloroform extraction is required because such precipitates may cause hydrocephalus.
7. 4% PFA can be stored at 4 °C for up to 2 weeks. If longer storage is desired, make aliquots and store them at −20 °C.
8. We are using MAS-coated glass slides (Matsunami).
9. Other applicable anti-GFP antibodies derived from other species (chickens and rats) are commercially available.
10. The MMB anesthetic significantly lowered the body temperature of mice. Warming during and after surgery is necessary.
11. Alternatively, a plasmid solution can be injected using a microinjector. Make sure the injectant fills the lateral ventricle, which emerges as the whole shape of the lateral ventricle with the color of Fast Green.
12. If the electrodes are placed horizontally, the labeling efficiency is reduced. By placing the electrodes obliquely, the direction of the electronic pulse against the ventricular surface becomes perpendicular, which maximizes the gene transfer efficiency.
13. This is to avoid dehydration.
14. Sometimes the mice wake up immediately after applying atipamezole. Thus, we apply it just before placing the last suture on the abdominal wall.
15. If the brain is removed from the gel during sectioning, the slices are shattered. To avoid this, PBS around the brains was carefully removed using tissue paper, and after embedding the brains, the gel was solidified slowly on a sheet of Styrofoam. Sectioning the brains into 100 μm thickness may cut the dendrite arbor. However, this did not become a problem unless the cell bodies were located near the surface of the slices. The SNT plug-in performs Sholl analysis on 2D projection images.
16. Do not dry the slices completely, as it may reduce the immunohistochemical signals. When the slices become transparent to a certain extent, proceed to the next step. We used HistoVT One (Nacalai Tesque) as the antigen-retrieval solution.
17. Quick cool-down can reduce the nuclear staining with DAPI.
18. Neurons to be analyzed should be selected based on the following criteria: the cell body is not located near the surface or bottom of the slice, the dendrites of which may be seriously cut off, and the GFP signal is high enough for later analysis.
19. The SNT plug-in is not a preinstalled function. Install the Neuroanatomy plug-in, which includes SNT.

20. Upon acquisition of microscopic images for spine analysis, select the dendrites that extend around the same depth in the slice, do not intersect with other neurites, and have a strong GFP signal.
21. The Imaris software constructs a 3D rendering model of the dendrites and spines on them. The parameters are modified such that the reconstructed model fits well with the original fluorescent image.

Acknowledgments

This work was supported by JSPS KAKENHI (Grant Number JP21K07309) to H. Tabata.

References

1. Martínez-Cerdeño V (2017) Dendrite and spine modifications in autism and related neurodevelopmental disorders in patients and animal models. Dev Neurobiol 77:393–404
2. Tabata H, Nakajima K (2001) Efficient in utero gene transfer system to the developing mouse brain using electroporation: visualization of neuronal migration in the developing cortex. Neuroscience 103:865–872
3. Fukuchi-Shimogori T (2001) Neocortex patterning by the secreted signaling molecule FGF8. Science 294:1071–1074
4. Saito T, Nakatsuji N (2001) Efficient gene transfer into the embryonic mouse brain using in vivo electroporation. Dev Biol 240:237–246
5. Ito H, Morishita R, Iwamoto I et al (2014) Establishment of an in vivo electroporation method into postnatal newborn neurons in the dentate gyrus. Hippocampus 24:1449–1457
6. Hamada N, Ogaya S, Nakashima M et al (2018) De novo PHACTR1 mutations in West syndrome and their pathophysiological effects. Brain 141:3098–3114
7. Hamada N, Ito H, Iwamoto I et al (2015) Role of the cytoplasmic isoform of RBFOX1/A2BP1 in establishing the architecture of the developing cerebral cortex. Mol Autism 6:56
8. Nishikawa M, Ito H, Tabata H et al (2022) Impaired function of PLEKHG2, a Rho-guanine nucleotide-exchange factor, disrupts corticogenesis in neurodevelopmental phenotypes. Cells 11:696
9. Koresawa Y, Miyagawa S, Ikawa M et al (2000) Synthesis of a new Cre recombinase gene based on optimal codon usage for mammalian systems. J Biochem 127:367–372

Chapter 7

Analysis of Neuronal Morphology by Two-Photon Microscopy

Chia-Wei Huang, Yen-Lin Su, and Jin-Wu Tsai

Abstract

During the development of mammalian brains, pyramidal neurons in the cerebral cortex form highly organized six layers with different functions. These neurons undergo developmental processes such as axon extension, dendrite outgrowth, and synapse formation. A proper integration of the neuronal connectivity through dynamic changes of dendritic branches and spines is required for learning and memory. Disruption of these crucial developmental processes is associated with many neurodevelopmental and neurodegenerative disorders. To investigate the complex dendritic architecture, several useful staining tools and genetic methods to label neurons have been well established. Monitoring the dynamics of dendritic spine in a single neuron is still a challenging task. Here, we provide a methodology that combines in vivo two-photon brain imaging and in utero electroporation, which sparsely labels cortical neurons with fluorescent proteins. This protocol may help elucidate the dynamics of microstructure and neural complexity in living rodents under normal and disease conditions.

Key words Dendritic spine, Dendrite branches, Confocal microscopy, Two-photon microscopy

1 Introduction

During the development of cerebral cortex in the mammalian brain, excitatory pyramidal neurons are generated from neuroepithelial cells (NECs) and radial glial cells (RGCs) [1–3]. These neurons migrate to the outer region of the cortex and eventually form six specific cortical layers under a series of molecular control [4]. Once these cortical neurons reach the destinations, they extend numerous dendrites and link to neighboring cells and incoming axons from other brain regions to establish a highly complex and functional network [5, 6]. Along the dendrites of pyramidal neurons, dendritic spines, which are short protrusions of 0.5–3 μm long [7, 8], form and function as a postsynaptic structure, receiving signals from other neurons [9]. The formation, maturation, and elimination of dendritic spines are crucial to the plasticity of learning and memory [10–12].

Kazuhito Toyooka (ed.), *Neuronal Morphogenesis: Methods and Protocols*, Methods in Molecular Biology, vol. 2831,
https://doi.org/10.1007/978-1-0716-3969-6_7,

Based on their morphology, dendritic spines can be categorized into four major types: filopodia, stubby, thin, and mushroom [13]. Filopodia are headless and hair-like protrusions from dendrites. Filopodia are constantly formed and eliminated and considered as a key player in fast learning processes [14]. Thin spines have a small head and a long neck and is abundant in the developing brain, particularly during adolescence and young adult. Stubby spines are bulbous head protrusions without obvious neck and are considered as a relatively immature type compared to mushroom [7]. Some thin spines develop into mushroom spines over time [15]. Mushroom spines have a large and bulbous head, and a thin and small neck. They are considered as the most mature, memory-related, and stable type of spines [15]. The types of spines strongly correlate with neuronal functions through the difference in the property of receptor and the stability [7]. Clinically, changes in dendrite length, dendrite number, spine density, spine types, and spine dynamics are highly associated with many neurodegenerative [16] and neurodevelopmental [12, 17] disorders.

To investigate these macro- and microstructure of dendritic architecture, neurons were usually labeled with chemical dyes, antibodies, and fluorescent proteins. Since the first use of two-photon microscopy in biological samples, in vivo brain imaging has become a useful tool to unveil the mechanism of dendritic spine development in living animals [18, 19]. Neurons were made visible under two-photon microscopy by fluorescent protein expression through genetically knocking in enhanced green fluorescent protein (EGFP) or yellow fluorescent protein (YFP) driven by Thyl promoter [20, 21]. Recently, in utero electroporation of plasmids encoding a variety of fluorescent proteins were also used to label neurons of specific types [22–28].

This protocol provides a series of procedures, including in utero electroporation, in vivo two-photon imaging, and analysis of dendritic structure [25, 29]. By utilizing in utero electroporation of Cre-dependent GFP-expressing construct (pCALNL-GFP) and construct encodes cDNA of Cre-recombinase (pCAG-Cre), neurons can be labeled with variable density through adjusting the concentration of pCAG-Cre [30]. Overall, these methods provide us a powerful tool to monitor the morphological changes of dendrites and spines in individual neurons.

2 Materials

2.1 In Utero Electroporation (IUE)

1. Pregnant mice. Mice reported in this chapter are *Mbnl2*$^{+/+}$, *Mbnl2*$^{+/f}$, and *Mbnl2*$^{f/f}$ mice (mix background of 129S and B6 mice) (*see* **Note 1**).

2. Isoflurane, Isoflurane Vapomatic Anesthetic Vaporizer, pure oxygen (*see* **Note 2**).
3. PCR micropipette 1–10 μL, including bores (borosilicate glass) and plungers (stainless steel); Micropipette puller and microgrinder.
4. Maxiprep of pCALNL-GFP and pCAG-Cre plasmid DNA, Plasmid Maxi Kit (*see* **Note 3**).
5. Fast green FCF, Certifiable Grade.
6. Square Wave Electroporator, Foot Pedal, and Tweezertrodes™ Electrodes Kits (*see* **Note 4**).
7. Ring forceps, iris forceps, iris scissors, spay hook (blunt), needle holder, and monofilament surgical sutures. All surgical instruments must be sterilized before surgery.

2.2 Confocal Imaging of Brain Slices

1. Vibrating blade microtome, SUPER Gillette BLUE BLADES, low gelling temperature Agarose, delicate task wipers, 24-well cell culture plate.
2. Microslides, Microscope cover glasses, VECTASHIELD® Antifade Mounting Medium, and clear nail polish.
3. Confocal microscopy (Zeiss LSM 700) equipped with Plan-Apochromat 20×/0.8 DIC and Plan-Apochromat 40×/1.4 Oil DIC objective.

2.3 In Vivo Two-Photon Brain Imaging of Living Mice

1. P30 mice. Mice reported in this chapter are $Mbnl2^{+/+}$ and $Mbnl2^{f/f}$ mice (*see* **Note 5**).
2. Isoflurane, Isoflurane Vapomatic Anesthetic Vaporizer, pure oxygen.
3. Iris forceps, iris scissors, needle holder, and monofilament surgical sutures. All surgical instruments must be sterilized before surgery.
4. IDEAL Micro Drill, 1.0 mm/1.2 mm tips, Micro blade, SUPER Gillette BLUE BLADES, Instant glue, mouse head holder, compressed (canned) air.
5. Two-photon microscopy (Zeiss LSM 7MP) equipped with W Plan-Apochromat 20×/1.0 DIC III objective, mouse head holder, complete temperature control system for mice, artificial cerebral spinal fluid (ACSF), Axiovision software, Fluorescence stereomicroscope equipped with Plan Apo S 1.0× FWD 60 mm objective.

2.4 Analysis of Dendritic Structure

1. ZEN Microscopy Software (black/blue edition).
2. Customized user interface developed by using MATLAB environment [25, 31].

3 Methods

To access the images of dendritic structure from individual cortical neurons, in utero electroporation of plasmid DNA expressing fluorescent protein is used to label neurons with fluorescence.

3.1 In Utero Electroporation

1. Anesthetize pregnant mice at embryonic day (E) 14.5 with isoflurane (induction: 4% in plastic chamber; maintenance: 0.5–1.5% with mask) and inject Carprofen (5 mg/kg, subcutaneous (SC) injection) to relieve pain.
2. Shave the fur above abdominal area and clean whole surgical area with povidone-iodine and alcohol prep pad (professional/ hospital grade).
3. Make an incision (3–5 cm) through the skin and abdominal muscle along the linea alba to expose the underlying viscera.
4. Externalize the uterine gently and place it onto the moist gauze. Retain moisture of uterine and viscera with preheated warm (35–40 °C) sterile PBS (phosphate buffered saline) during the entire surgery.
5. Inject 0.5–1 μL of the mixture of plasmids and Fast Green FCF dye (2 μL dye/12 μL mixture) into the lateral ventricle of each embryo's brain. The amount of each injection is based on the brain size.
6. Hold the heads of the embryos and introduce mixture of plasmids into the neural progenitors located in the ventricular surface of lateral ventricle through BTX™ Tweezertrodes and Square Wave Electroporator (40 V × 5 pulses, duration: 50 ms, interval: 450 ms; *see* **Note 6**).
7. Place the uterine back into the abdominal cavity after electroporation and close the incision by utilizing surgical sutures.
8. Keep pregnant mice warm until recovery (be awake) and carefully monitor them following the Institutional Animal Care & Use Committee (IACUC) guideline for several days (at least first 3 days) after the surgery.

3.2 Confocal Imaging of Brain Slices

1. Anesthetize and perfuse the electroporated mice at P30 (*see* **Note 7**).
2. Harvest, dry, and embed the brains into low gelling temperature Agarose IITM (40–45 °C) pre-poured in a 35 mm Petri dish (*see* **Note 8**).
3. Cut out a trapezoidal or square column contained the brains from Agarose IITM.

4. Make coronal brain sections 100 μm in thickness by using Vibrating blade microtome. Collect these brain slices in the 24-well cell culture plates.
5. Use a preprocessed glass Pasteur pipette (L shape, processed by Bunsen burner) to move the brain slices. Gently move the brain slices onto microslides.
6. Mount the brain slices (six to eight brain slices for each mouse) with VECTASHIELD® Antifade Mounting Medium, cover with microscope cover glasses, and seal the edge of microscope cover glasses by using clear nail polish.
7. Put the sample on the confocal microscopy (LSM 700). (1) For spine imaging: Use Apochromat 40×/1.4 Oil DIC objective, zoom in two times, and randomly image 10–15 apical or basal dendrites (depends on the purpose) of cortical layer II/III neurons from each mouse. To reach certain quality for analysis, the parameters are as follows: frame size of each image, 1024 (X) × 1024 (Y) pixels; the size of each pixel (X and Y: <0.1 μm), 0.078 μm × 0.078 μm × 0.42 μm (X, Y, Z). (2) For dendrites imaging: Find the GFP+ neurons showing clear and complete structure without being cut. Use Plan-Apochromat 20×/0.8 DIC objective, zoom in one time, and image at least five neurons from each mouse. The parameters are as follows: frame size of each image, 1024 (X) × 1024 (Y) pixels; the size of each pixel, 0.31 μm × 0.31 μm × 1.00 μm (X, Y, Z).

3.3 *In Vivo* Two-Photon Brain Imaging of Living Mice

Day 0

1. Anesthetize P30 mice with isoflurane (induction: 4% in plastic chamber; maintenance: 0.5–1.5% with mask) and inject Carprofen (5 mg/kg, subcutaneous (SC) injection) to relieve pain on the first day (day 0) of experiment.
2. Use blade to shave the fur above the cortical region (*see* **Note 9**).
3. Cut the skin with scissors and expose the skull above the cortical region.
4. Use fluorescence stereomicroscope to check if IUE successfully labels the region of interest (ROI, here we use somatosensory cortex as example) (*see* **Note 10**). You may see obvious GFP signal through the skull.
5. Mark a dot on the skull above ROI (Fig. 4a).
6. Make a blade stack consist of three SUPER Gillette BLUE BLADES. Use instant glue to adhere them with each other tightly.
7. Dry the skull and remove the debris of skull by compressed air.
8. Align the central hollow of blade stack and the marked dot on skull and then glue them together (*see* **Note 11**).

9. Put the mouse head onto the mouse head holder (Fig. 4a).
10. Gently dig a shallow and round hole (1 mm in diameter) above the ROI (marked dot) with IDEAL Micro Drill (1.0 mm/ 1.2 mm tips) (*see* **Note 12**) under fluorescence stereomicroscope.
11. Gently shave the skull within center of ROI into 20–30 μm in thickness (*see* **Note 13**) fluorescence stereomicroscope. It is very important to retain moisture of the skull in ROI with ACSF (room temperature). Make sure no microbleeds or skull damages are generated during the entire procedure.
12. Place the mouse onto a LSM 7MP two-photon microscope and use flexible silicone rubber heating pad to maintain the body temperature of mice.
13. Positioning of the target cells: Choose a region that is abundant of GFP+ (or other fluorescent protein) neurons under the visible light. Images showing the target neurons and surrounding large/unique blood vessels are recorded as a reference for the imaging on the other day (Carl Zeiss AxioVision SE64 Rel. 4.9) (Fig. 4b). Please fix the region and do not change XY axis of objective before finishing the two-photon imaging in following few minutes to hours.
14. Use zoom-in tool to magnify four times (20X × 4 = 80X) into multiple subregions in the same field. Randomly select and monitor several apical dendrites extending from cortical layer II/III GFP+ pyramidal neurons (at least 15 apical dendrites from more than three mice in each group) by a two-photon microscopy. To scan fast and to assure enough quality of images for further analysis, we scan these dendrites with frame size 512 (X) × 512(Y) and make the size of each pixel 0.208 μm × 0.208 μm × 0.638 μm (X, Y, Z) (*see* **Note 14**).
15. Remove the blades on the skull, retain moisture of the skull, and close the incision on skin by using surgical sutures.

Day 2

16. Anesthetize P32 mice with isoflurane as previously described after 2 days (day 2).
17. Gently reopen the skin along with the sutures have been made and retain moisture of the skull in ROI with ACSF (room temperature).
18. Repeat **steps 6** and 7.
19. Gently shave the skull within center of ROI to remove connective tissue (*see* **Note 15**).
20. Repeat **step 11**.

21. Find the same region have been imaged on day 0 under the visible light by recognizing the features, such as the shape of blood vessels and GFP+ neurons (*see* **Note 16**).
22. Repeat **step 12**. Take second images on the same dendrites imaged on day 0, respectively. Remember to make the resolution equal to the previous one.

3.4 Analysis of Dendritic

Analysis for Dendritic Branches

For this analysis, images from both confocal (brain slice in microslide) and two-photon (in vivo mouse brain) microscopy are applicable.

1. Load the data (lsm5 format) acquired from Zeiss LSM 700 (confocal microscopy) or LSM 7MP (two-photon microscopy) into customized user interface [25, 31] by using MATLAB 2012b.
2. Define the nuclei and trace the dendrites of each neuron selected for further analysis.
3. Perform Sholl analysis [32] with proper radius depending on the size of neuron and interest of the investigators.
4. Count and analyze the "number of intersections" on each "shell" (concentric circles around the cell body) in each neuron (Fig. 1a).
5. Use statistics to compare the number of intersections between groups (Fig. 1b).

Analysis for Dendritic Spine Density and Types

1. Load the data (lsm5 format) acquired from Zeiss LSM 700 (confocal microscopy) or LSM 7MP (two-photon microscopy) into customized user interface [25, 31] by using MATLAB 2012b.
2. Trace the dendrites that are selected for further analysis.
3. Use spine add/editing tool to mark the spines around the traced dendrite (Fig. 2a, b). At the same time, label the types (mushroom, stubby, thin, filopodia; [7, 12]) of each spine (Fig. 3a, b).
4. Count spine density and percentage of each spine type of each dendrite. Thousands of dendritic spines from more than three mice in each group are analyzed. Spine density = total number of spine/total length of dendrite.
5. Analyze the average dendritic spine density and types of each mouse and compare the data between all groups (Figs. 2c and 3c).

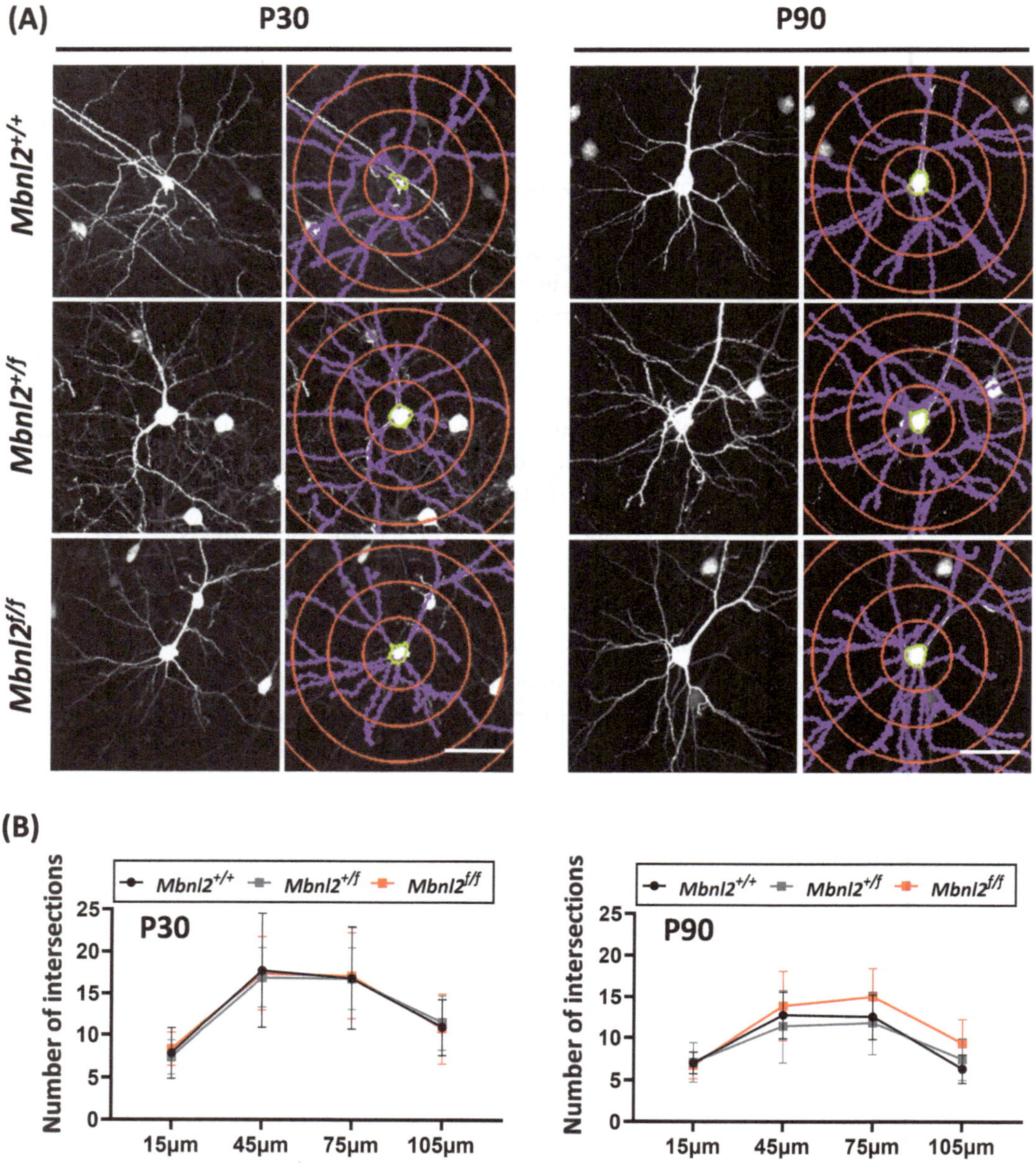

Fig. 1 Sholl analysis of Cre-electroporated cortical neurons at layer II/III cortices. (**a**) Representative images of cortical neurons generated from neural progenitors. *Mbnl2*$^{+/+}$, *Mbnl2*$^{+/f}$, and *Mbnl2*$^{f/f}$ embryos were electroporated by in utero electroporation at E14.5. Brain slices were imaged at P30 and P90. The number of intersections on the concentric circles (red circles) were counted. Bar = 50 μm. (**b**) Graph shows the number of dendritic branches in *Mbnl2*$^{+/+}$ (black), *Mbnl2*$^{+/f}$ (gray), and *Mbnl2*$^{f/f}$ (red) mice. Error bars represent SD. One-way ANOVA test. (Reproduced from Ref. [25] with permission)

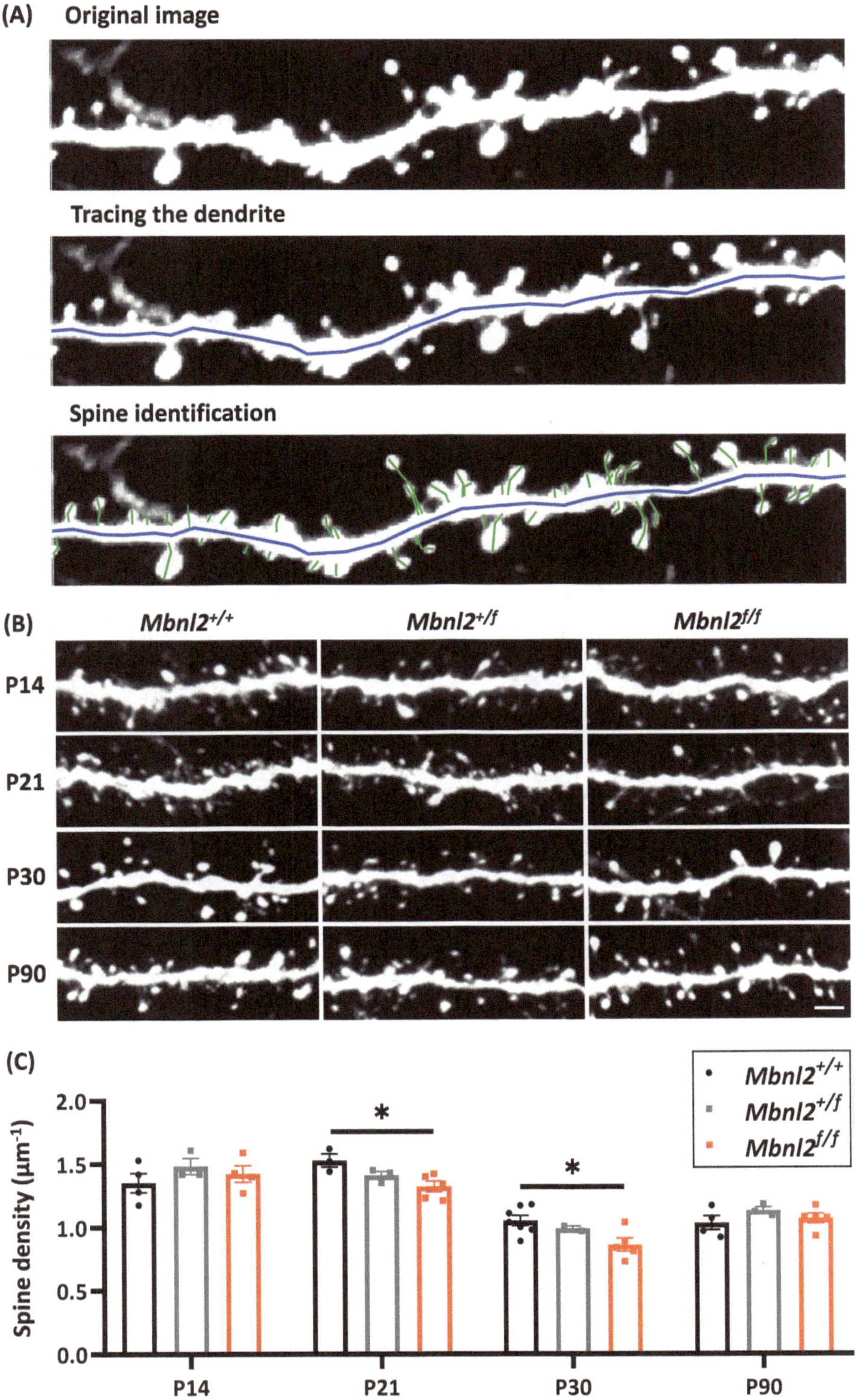

Fig. 2 Analysis of dendritic spine density. (**a**) Dendrites of original image were traced and the dendritic spines on the dendrite surface were then labeled manually. (**b**) Images of basal dendrites with spines at P14, P21, P30, and P90 in *Mbnl2*$^{+/+}$, *Mbnl2*$^{+/f}$, and *Mbnl2*$^{f/f}$ mouse cortices. Bar = 2 μm. (**c**) Statistical analysis of dendritic spine density at P14, P21, P30, and P90 showed a significant reduction in the dendritic spine density in Mbnl2 KO cortical neurons at both P21 and P30. Error bars represent SEM. One-way ANOVA test. *: $p < 0.05$. (Reproduced from Ref. [25] with permission)

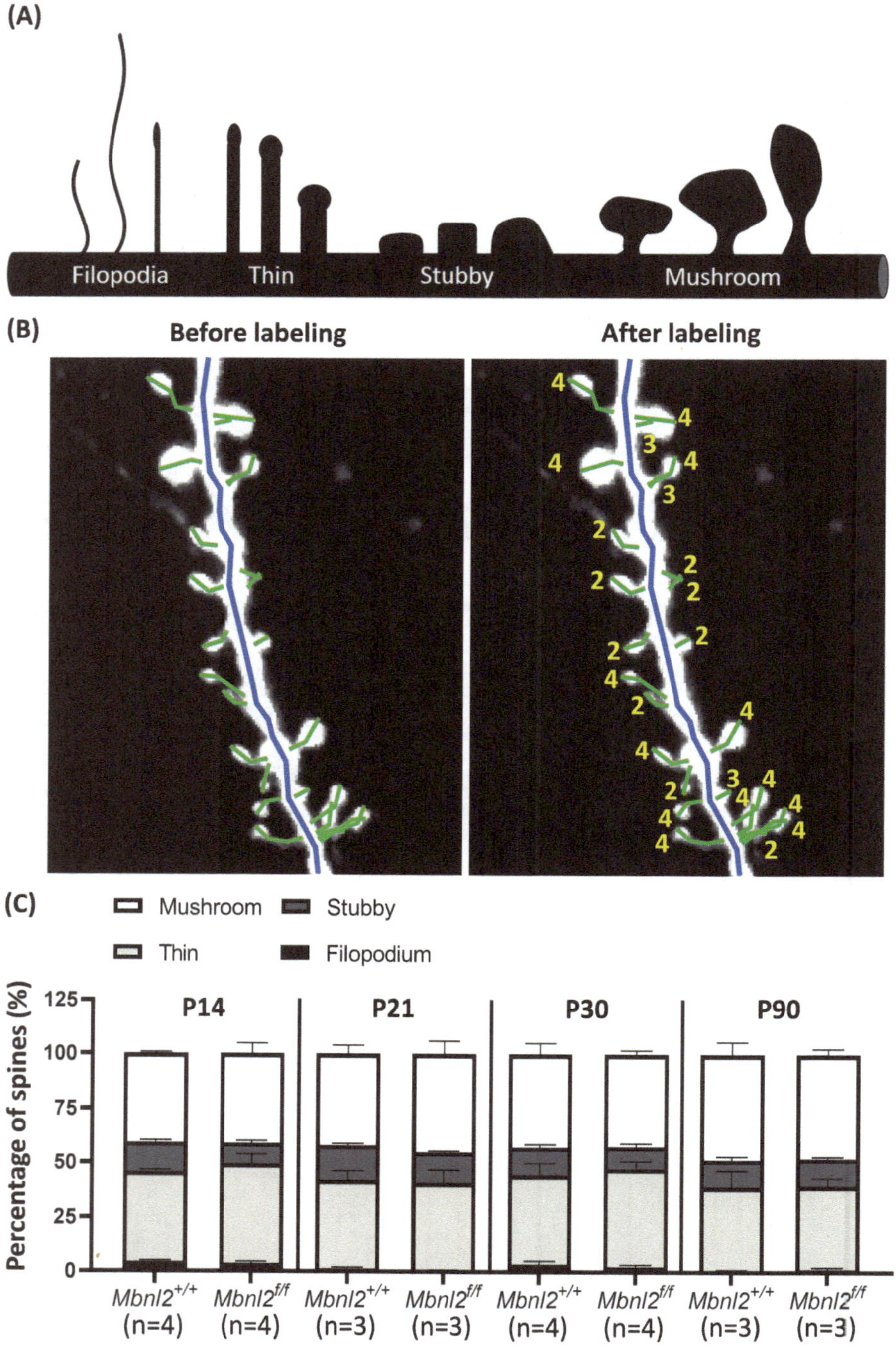

Fig. 3 Analysis of dendritic spine types. (**a**) Drawing of four typical types (mushroom, stubby, thin, filopodia) of dendritic spines. (**b**) Dendritic spines were labeled with different types (1: filopodia, 2: thin, 3: stubby, 4: mushroom) during identification of spines. (**c**) Bar graph shows similar distributions of each spine types in both *Mbnl2*$^{+/+}$ and *Mbnl2*$^{f/f}$ mice. Data was collected from P14, P21, P30, and P90 mice with more than three mice in each group. Error bars represent SEM. Student's t-test. (Reproduced from Ref. [25] with permission)

Analysis for Dendritic Spine Dynamics

1. Load both the day 0 and day 2 data (lsm5 format) acquired from LSM 7MP (two-photon microscopy) into customized user interface [25, 31] by using MATLAB.
2. Trace the dendrites that are selected for further analysis.
3. Use spine add/editing tool to mark the spines around the traced dendrite.
4. Mark the "stable" spines that are present in both images.
5. Mark the "new spines" that are newly formed in day 2 image.
6. Mark the "eliminated spines" that are eliminated in day 2 image.
7. Analyze the formation rate, elimination rate, and stable rate of dendritic spines (Fig. 4c). For the analysis of spine dynamics, statuses of dendritic spines were defined as follows: Formation rate (%) = number of new spines/preexisting spines; Elimination rate (%) = number of eliminated spines/preexisting spines; Stable rate (%) = number of stable spines/preexisting spines.
8. Use statistics to compare the formation, elimination, and stable rate between groups (Fig. 4d).

4 Notes

1. This in utero electroporation method is suitable for mouse and rat study. ICR, C57BL/6, and transgenic mice with C57BL/6 background are widely used in our lab and many other labs in the world. Here, we performed IUE at E14.5 pregnant mice and modulate cells that are going to differentiate into cortical layer II/III neurons. By changing timing of electroporation, neurons in deeper cortical layers can also be modulated: in brief, IUE at E12.5 labels layer V/VI neurons; IUE at E13.75 labels layer IV neurons; IUE at E14.5 labels layer II/III [33]. If you use rat as a model animal, the gestation of rat is slightly different from mouse.
2. Other anesthetics can also be used in this experiment. However, vaporized isoflurane is safer than intraperitoneal (i.p.) injection of controlled medication. Therefore, we highly recommend user chosen isoflurane for anesthetics. The use of anesthetics should follow the permission given by the Institutional Animal Care & Use Committee (IACUC). This is a crucial point during anesthesia of pregnant mice.

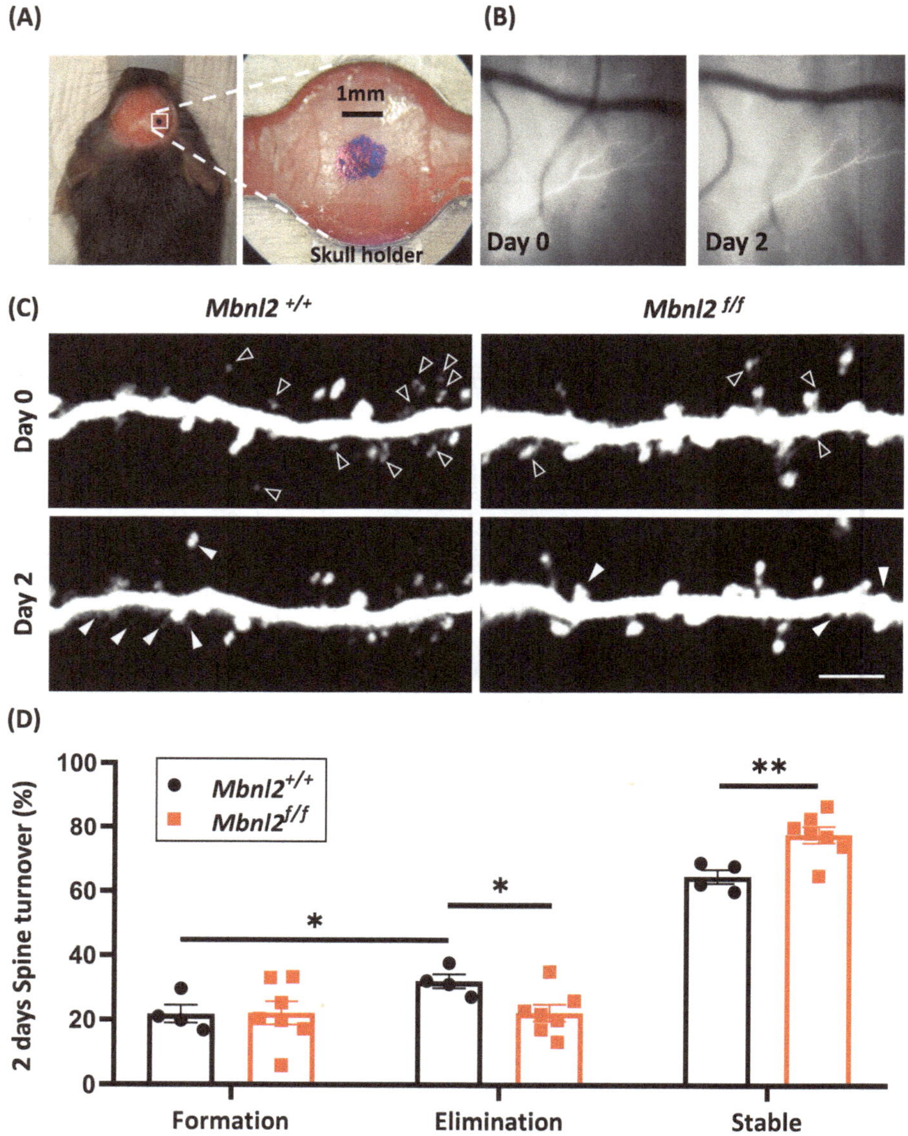

Fig. 4 Analysis of dendritic spines dynamics. (**a**) Thinned skull preparation. The mouse head was adhered to blades stack and fixed on the head holder. A small region of skull 1 mm in diameter above the somatosensory cortex was gently scraped. Bar = 1 mm. (**b**) Images taken under visible light at experimental day 0 (P30) and day 2 (P32). Blood vessels surrounds the GFP+ neurons were imaged and used for positioning neurons. (**c**) Representative images of dendritic spine dynamics from cortical neurons electroporated with Cre-recombinase in *Mbnl2*$^{+/+}$ and *Mbnl2*$^{f/f}$ mice. Newly formed spines (solid arrowheads) and eliminated spines (hollow arrowheads) were marked. Spine without labeling were stable spines. (**d**) Bar and dot graph showing the ratio of spine formation, elimination, and stability of *Mbnl2*$^{f/f}$ ($n = 7$ mice; 24 dendrites) compared to *Mbnl2*$^{+/+}$ mice ($n = 4$ mice; 17 dendrites). Error bars represent SEM, Student's t-test. *: $p < 0.05$, **: $p < 0.01$. (Reproduced from Ref. [25] with permission)

3. Quality of plasmid DNA is important. Maxiprep of plasmid DNA is essential for electroporation. pCAG-Cre is construct containing a sequence encoded Cre-recombinase. pCALNL-GFP is a Cre-dependent GFP expressing construct that GFP can only be expressed when Cre-recombinase is present.
4. Here, we use BTX™ Tweezertrodes™ Electrodes 5 mm in diameter to perform IUE on E14.5 pregnant mice. For performing IUE before E13.5, electrodes 3 mm in diameter should be considered [34].
5. Since the skull of mouse younger than P21 is small and fragile, monitor spine dynamics before P21 is challenging.
6. If you perform IUE on the other embryonic days, a revised protocol and parameters must be considered.
7. Mouse brains that are electroporated at embryonic stages with pCALNL-GFP (concentration: 1.5 μg/μL) and pCAG-Cre (0.5 μg/μL) can express fluorescent protein in neurons at least until 5–6 months old. Attention: intensity of GFP may decrease noticeably after 3 months old.
8. Ensure the brains are moderately dried by delicate task wipers (KIMTECH SCIENCE KIMWIPES, Kimberly-Clark, USA) gel contacts all surfaces of the brain by swirling the surrounding gel several times. Bury the Petri dish into ice for 5–10 min to solidify the agarose.
9. The whole procedure of thinned-skull preparation is suggested to be controlled within 30 min.
10. Incorrect labeling may happen due to technical issues during the IUE.
11. Avoid adhering blade stack to cranial sutures or connective/ muscle tissue. Adhering to cranial sutures may easily cause damage. Adhering to connective/muscle tissues will fail to fix the head-on-head holder.
12. IDEAL Micro Drill is used to roughly dig the outer and solid part of skull. Please do not dig too deep at this step (before reach the bone marrow).
13. After shaving, skull will become flexible and transparent so that we can see blood vessels clearly under fluorescence stereomicroscope.
14. Adjust the intensity of laser until the center of dendrite is overexposed but not dendritic spine.
15. If you work on spine dynamics after 7 days or later, additional shaving is needed.
16. Neurons have been damaged on the first day or during the preparation on the second day may cause neuroinflammation. Dendrites of degenerating neurons usually change the morphologies such as bending and swollen dendrite.

References

1. Malatesta P, Hartfuss E, Gotz M (2000) Isolation of radial glial cells by fluorescent-activated cell sorting reveals a neuronal lineage. Development 127(24):5253–5263
2. Miyata T et al (2001) Asymmetric inheritance of radial glial fibers by cortical neurons. Neuron 31(5):727–741
3. Noctor SC et al (2001) Neurons derived from radial glial cells establish radial units in neocortex. Nature 409(6821):714–720
4. Molyneaux BJ et al (2007) Neuronal subtype specification in the cerebral cortex. Nat Rev Neurosci 8(6):427–437
5. Radnikow G, Feldmeyer D (2018) Layer- and cell type-specific modulation of excitatory neuronal activity in the neocortex. Front Neuroanat 12:1
6. Spruston N (2008) Pyramidal neurons: dendritic structure and synaptic integration. Nat Rev Neurosci 9(3):206–221
7. Berry KP, Nedivi E (2017) Spine dynamics: are they all the same? Neuron 96(1):43–55
8. DeFelipe J (2015) The dendritic spine story: an intriguing process of discovery. Front Neuroanat 9:14
9. Fu AK, Ip NY (2017) Regulation of postsynaptic signaling in structural synaptic plasticity. Curr Opin Neurobiol 45:148–155
10. Kasai H et al (2010) Structural dynamics of dendritic spines in memory and cognition. Trends Neurosci 33(3):121–129
11. Ma S, Zuo Y (2022) Synaptic modifications in learning and memory – a dendritic spine story. Semin Cell Dev Biol 125:84–90
12. Runge K, Cardoso C, de Chevigny A (2020) Dendritic spine plasticity: function and mechanisms. Front Synaptic Neurosci 12:36
13. Peters A, Kaiserman-Abramof IR (1970) The small pyramidal neuron of the rat cerebral cortex. The perikaryon, dendrites and spines. Am J Anat 127(4):321–355
14. Ozcan AS (2017) Filopodia: a rapid structural plasticity substrate for fast learning. Front Synaptic Neurosci 9:12
15. Bourne J, Harris KM (2007) Do thin spines learn to be mushroom spines that remember? Curr Opin Neurobiol 17(3):381–386
16. Herms J, Dorostkar MM (2016) Dendritic spine pathology in neurodegenerative diseases. Annu Rev Pathol 11:221–250
17. Phillips M, Pozzo-Miller L (2015) Dendritic spine dysgenesis in autism related disorders. Neurosci Lett 601:30–40
18. Sahu P, Mazumder N (2020) Advances in adaptive optics-based two-photon fluorescence microscopy for brain imaging. Lasers Med Sci 35(2):317–328
19. Moyer CE, Zuo Y (2018) Cortical dendritic spine development and plasticity: insights from in vivo imaging. Curr Opin Neurobiol 53:76–82
20. Feng G et al (2000) Imaging neuronal subsets in transgenic mice expressing multiple spectral variants of GFP. Neuron 28(1):41–51
21. Porrero C et al (2010) Mapping of fluorescent protein-expressing neurons and axon pathways in adult and developing Thy1-eYFP-H transgenic mice. Brain Res 1345:59–72
22. Chang HY et al (2019) Multiple functions of KBP in neural development underlie brain anomalies in Goldberg-Shprintzen syndrome. Front Mol Neurosci 12:265
23. Chen JL, Chang CH, Tsai JW (2019) Gli2 rescues delays in brain development induced by Kif3a dysfunction. Cereb Cortex 29(2): 751–764
24. Chen YA, Lu IL, Tsai JW (2018) Contactin-1/F3 regulates neuronal migration and morphogenesis through modulating RhoA activity. Front Mol Neurosci 11:422
25. Huang CW et al (2023) Muscleblind-Like 2 knockout shifts adducin 1 isoform expression and alters dendritic spine dynamics of cortical neurons during brain development. Neuropathol Appl Neurobiol 49:e12890
26. Lu IL et al (2018) Identification of genes associated with cortical malformation using a transposon-mediated somatic mutagenesis screen in mice. Nat Commun 9(1):2498
27. Tsai MH et al (2020) Impairment in dynein-mediated nuclear translocation by BICD2 C-terminal truncation leads to neuronal migration defect and human brain malformation. Acta Neuropathol Commun 8(1):106
28. Tsai MH et al (2020) Pathogenic variants in CEP85L cause sporadic and familial posterior predominant Lissencephaly. Neuron 106(2): 237–245 e8
29. Ma L et al (2016) Experience-dependent plasticity of dendritic spines of layer 2/3 pyramidal neurons in the mouse cortex. Dev Neurobiol 76(3):277–286
30. Matsuda T, Cepko CL (2007) Controlled expression of transgenes introduced by in vivo electroporation. Proc Natl Acad Sci USA 104(3):1027–1032

31. Liu YT et al (2016) PRRT2 mutations lead to neuronal dysfunction and neurodevelopmental defects. Oncotarget 7(26):39184–39196
32. Sholl DA (1953) Dendritic organization in the neurons of the visual and motor cortices of the cat. J Anat 87(4):387–406
33. Yamashiro K, Ikegaya Y, Matsumoto N (2022) In utero electroporation for manipulation of specific neuronal populations. Membranes (Basel) 12(5):513
34. LoTurco J, Manent JB, Sidiqi F (2009) New and improved tools for in utero electroporation studies of developing cerebral cortex. Cereb Cortex 19(Suppl 1):i120–i125

Chapter 8

In Ovo Electroporation of Embryonic Chicken Spinal Cord in Combination with Ex Vivo Slice Culture for Live Imaging of Vertebrate Neuronal Differentiation

Victoria E. Higgs, Gabriela Toro-Tapia, Holly B. Burbidge, and Raman M. Das

Abstract

To investigate the cell behavior underlying neuronal differentiation in a physiologically relevant context, differentiating neurons must be studied in their native tissue environment. Here, we describe an accessible protocol for fluorescent live imaging of differentiating neurons within ex vivo embryonic chicken spinal cord slice cultures, which facilitates long-term observation of individual cells within developing tissue.

Keywords Neuronal differentiation, Chicken embryo, Development, Spinal cord, In ovo electroporation, Ex vivo slice culture, Live imaging, Fluorescence microscopy, Widefield microscopy, Axon outgrowth

1 Introduction

During vertebrate embryonic development, differentiation of neural progenitor cells into neurons is characterized by a series of morphological transformations that are finely regulated by signaling cues from the surrounding tissues [1]. These morphological transformations lead to newborn neurons acquiring the characteristic features required for them to become functional units of the nervous system, and therefore facilitate the formation of complex neural circuitry. Over the years, dissociated hippocampal neuron cultures [2] have remained the dominant experimental model to study the mechanisms directing early neuronal differentiation, revealing much about the localization and function of subcellular components and how these facilitate events like axon formation. However, neurons cultured in vitro lack their endogenous tissue microenvironment and the crucial extracellular cues that control their differentiation in vivo. Consequently, the morphological

Kazuhito Toyooka (ed.), *Neuronal Morphogenesis: Methods and Protocols*, Methods in Molecular Biology, vol. 2831,
https://doi.org/10.1007/978-1-0716-3969-6_8,

events underlying neuronal differentiation in vitro can widely differ from those occurring within the embryo [1, 3, 4]. Thus, a more physiologically relevant model system is essential to fully understand the complex dynamic transition from neural progenitor to neuron within a tissue context. However, observing neuronal differentiation live within vertebrate tissue, such as in an ex vivo slice culture system, has historically been challenging. For example, for rodent embryos, this requires in utero electroporation and advanced systems to maintain the tissue ex vivo in the long term [5]. In addition, there are significant challenges associated with imaging tissue slices for long periods at high resolution without causing damage.

Chicken embryos have long been used to explore the various morphogenetic events that characterize vertebrate embryonic development due to developmental similarities with the early human embryo, ease of accessibility, and experimental manipulation in comparison to other vertebrate embryos [6]. In the embryonic spinal cord, newborn neurons transit through well-characterized morphological stages as they differentiate [7–10], making it an ideal tissue for studying how neuronal differentiation is controlled within the embryo. Here, we describe a protocol to monitor the cell behavior that drives neuronal differentiation within a tissue context at high spatiotemporal resolution by combining embryonic chicken slice culture and a gentle widefield imaging system. Briefly, developing spinal cords of early chicken embryos are electroporated in ovo with constructs to express fluorescently tagged proteins. Following this, transverse embryo slices are cultured ex vivo within a collagen matrix and can be imaged over long periods to observe cell behaviors in the neural tube, from progenitor divisions to axon outgrowth. Using this technique, we have revealed novel cell biological processes such as apical abscission [9] and primary cilium remodeling [10], which have transformed our understanding of neuronal differentiation and cell-state transitions during development. In this protocol, we present detailed instructions for efficiently preparing, culturing, and imaging differentiating neurons in embryonic slices and provide the key information required for successful long-term image acquisition.

2 Materials

2.1 Electroporation

1. Fertilized chicken (*Gallus gallus domesticus*) eggs.
2. 37 °C incubator.
3. 1% Brilliant Blue solution: Add 10 mg Brilliant Blue G powder to 1 mL deionized water. Store at room temperature (*see* **Note 1**).

4. Electroporator (e.g., BTX) and gold or platinum-tip electrodes (e.g., BTX, 5 mm L-shape with gold tips).
5. Aspirator tube assembly.
6. Glass needles: Glass microcapillary tubes (1.0 mm OD × 0.58 mm ID × 100 mm L) pulled with a heated needle puller.
7. Fine forceps: Dumont no. 4 or 5.
8. Curved blunt forceps.
9. Egg stand (*see* **Note 2**).
10. Stereomicroscope with gooseneck LED illuminators (*see* **Note 3**).
11. Clear sticky tape: extra wide if possible.

2.2 Slice Culture and Imaging

1. CO_2 incubator.
2. 2× fine forceps: Dumont no. 4 or 5.
3. Small scissors: Vannas-Tübingen spring scissors.
4. Large scissors: Noyes spring scissors, straight.
5. Egg stand.
6. Stereomicroscope (as above).
7. Fluorescence stereomicroscope.
8. 90 mm plastic Petri dishes (*see* **Note 4**).
9. Dish to secure embryos for slicing (pin dish): Prepare in advance and reuse. Fill a 90-mm plastic Petri dish approximately a third of the way to the top with SYLGARD 184 silicone elastomer (prepared according to manufacturer's instructions) and allow to set at room temperature overnight. Place approximately 40 entomology pins into the set SYLGARD using fine forceps.
10. Ice-cold 1× L-15.
11. 3 mL plastic Pasteur pipettes. One with ~1 cm of the tip cut off to increase diameter.
12. 200 μL micropipette tips.
13. P2 and P20 micropipettes.
14. Collagen mixture: To 300 μL collagen (type I), add 100 μL 0.1% acetic acid solution and vortex well. Then add 100 μL 5× L-15 medium and vortex well again. The solution should turn yellow. Add 15–20 μL 7.5% sodium bicarbonate and vortex well. The solution should appear slightly pink. Keep collagen mixture on ice and use it on the day of preparation (*see* **Note 5**).
15. Slice culture medium: In a sterile hood, to one 500 mL bottle of neurobasal medium, add 125 mL Optiprep density gradient medium (Sigma; *see* **Note 6**). Store 10 mL aliquots at 4 °C. On

the day of slicing, to the 10 mL aliquot, add 200 μL 50× B-27 supplement, 100 μL GlutaMAX supplement (Gibco), and 10 μL 10 mg/mL gentamicin (*see* **Note 7**).

16. FluoroDish imaging dish (WPI FluoroDish tissue culture dish 35 mm with 23 mm cover glass bottom).
17. 50 mm Petri dishes.
18. Lens cleaning tissue.
19. Widefield inverted microscope with temperature and CO_2-controlled chamber (*see* **Note 8**).
20. For hand slicing only: Microknife (Mani Microknife Straight 15°: MST15; *see* **Note 9**).
21. For vibratome slicing only:
    (a) Vibratome (e.g., Leica VT 1200S).
    (b) 4% low-melting agarose in PBS.
    (c) Water bath.
    (d) Vibratome blades.
    (e) Plastic base molds, dimensions 7 × 7 × 5 mm.
    (f) Cyanoacrylate glue (i.e., super glue).
    (g) Ice.
    (h) Thin paintbrushes.

3 Methods

3.1 Electroporation

1. Incubate chicken eggs in their boxes, blunt-end upright, in a 37 °C humidified incubator to the desired stage (*see* **Note 10**).
2. Prepare DNA solution by diluting plasmids to approximately 25–200 ng/μL in deionized water plus 1 μL 1% Brilliant Blue solution up to a total volume of 15 μL (*see* **Note 11**).
3. Set up the electroporator with the following settings: 3 × 18–20 V pulses, 50 ms pulse length with a 950 ms interval between pulses.
4. Insert the blunt end of a glass needle into the silicone rubber nosepiece of an aspirator tube assembly. Using fine forceps, open the end of the needle by pinching off a few millimeters from the tip. The needle tip must remain as sharp as possible to cleanly pierce the embryo while being wide enough to draw the DNA solution.
5. Draw the DNA solution into the needle, filling it roughly halfway, by gentle aspiration through the mouthpiece at the other end of the aspirator tube assembly (*see* **Note 12**).

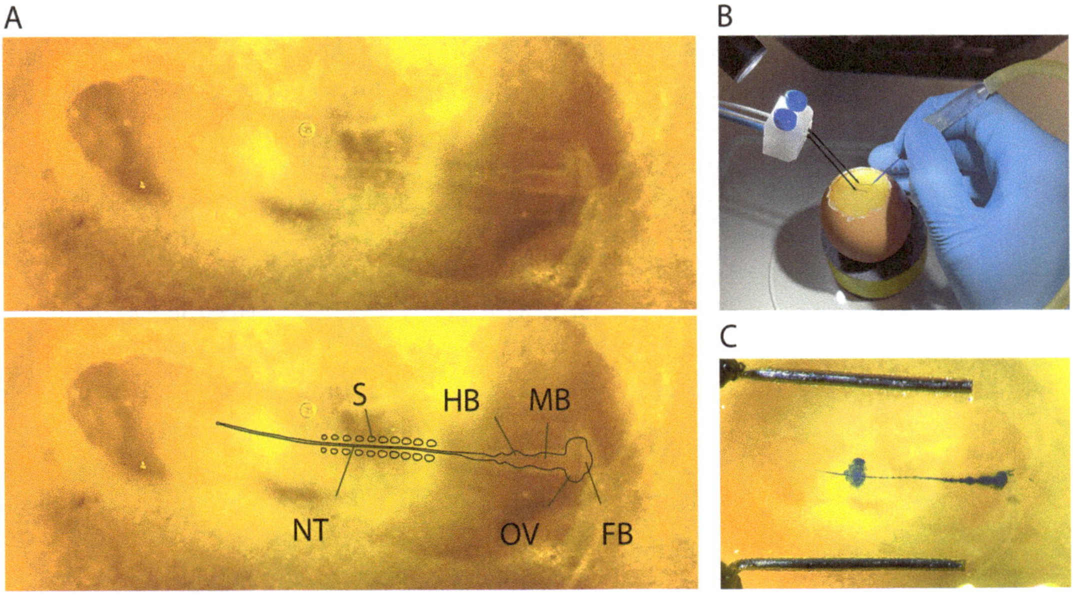

Fig. 1 In ovo electroporation. (**a**) Hamburger–Hamilton stage 10 embryo before electroporation with (below) and without (above) anatomic annotation. (**b**) Electroporation setup on stereomicroscope stage. (**c**) Embryo after injection of DNA solution into the neural tube with correctly positioned electrodes. FB forebrain, HB hindbrain, MB midbrain, NT neural tube, OV optic vesicle, S somite

6. Remove eggs from incubator (*see* **Note 13**) and place egg blunt-end up in the stand. Gently tap the top of the egg with the blunt end of curved forceps to create a small dent. Use the tips of the curved forceps to carefully create a window in the top of the egg approximately 3 cm in diameter.
7. Carefully peel away the inner shell membrane using fine forceps, taking care not to pierce or rip the underlying tissue.
8. Place the egg on the stereomicroscope stage and adjust gooseneck LED illuminators to illuminate the window. Carefully adjust the stereomicroscope zoom so that the window fills the field of view and focuses on the top of the yolk. The embryo should be in the region of the white disk on the top of the yolk (*see* **Note 14**).
9. Position the egg with the anterior end of the embryo on the right (Fig. 1a).
10. With the electrode holder coming from the left, carefully position the electrodes to gently rest on the yolk horizontally in parallel on either side of the embryonic spinal cord, approximately 5 mm apart.
11. Ensure the egg is stable, and viewing through the eyepiece, insert the needle at a shallow angle into the lumen of the neural tube, anterior of the region to be electroporated (Fig. 1a, b).

12. Slowly dispense a minimal amount of DNA solution to fill the developing spinal cord (Fig. 1c).
13. Apply current to the embryo to transfect cells of the neural tube (*see* **Note 15**).
14. Raise electrodes. Carefully insert the tip of a 3 mL plastic Pasteur pipette into the egg alongside the yolk and remove approximately 2–4 mL of albumen, being careful not to pierce the yolk. Clean electrodes with distilled water.
15. Seal the window with clear sticky tape, place the egg in an egg box, and transfer back to the 37 °C humidified incubator.
16. Repeat **steps 6–15** for the remaining eggs (*see* **Note 16**).
17. Incubate eggs to the desired stage (*see* **Note 17**).

3.2 Identifying Transfected Tissue

1. Prepare fresh collagen mixture (*see* Subheading 2.2) and place on ice.
2. Prepare fresh slice culture medium (*see* Subheading 2.2) and place the tube in a 37 °C 5% CO_2 incubator with the cap loose to allow equilibration with CO_2.
3. Fill 90 mm Petri dish and pin dish halfway with ice-cold 1× L-15.
4. Remove electroporated eggs from the incubator.
5. With egg in stand, use large scissors to cut around taped window of egg to reveal embryo.
6. Use small scissors to cut a 10–15 mm square in the extraembryonic tissue around the embryo. Use fine forceps to pick up the embryo by the corner of the square and transfer to the Petri dish of L-15. Soak the tip of the scissors and forceps with L-15 medium before picking up embryos to avoid tissue sticking.
7. Repeat **steps 5** and **6** for each embryo.
8. Viewing through the stereomicroscope, illuminated from below, use both pairs of fine forceps to remove the vitelline membrane from the dorsal side of the embryo. This should peel away easily as a translucent square (Fig. 2a). Remove any yolk by gently washing the embryo in L-15 or stroking with fine forceps, being careful not to rip it. Repeat for each embryo.
9. Using the cut Pasteur pipette, gently transfer the embryos to the large Petri dish of clean L-15 and sink with drops of L-15.
10. Using the cut Pasteur pipette, transfer a clean embryo to the pin dish and sink with drops of L-15.
11. With fine forceps in each hand, use one to hold down a corner of the extraembryonic tissue and the other to pin it down. Repeat until all four corners are pinned and the extraembryonic tissue is stretched into a flat square with the embryo facing dorsal side up (somites visible; Fig. 2b).

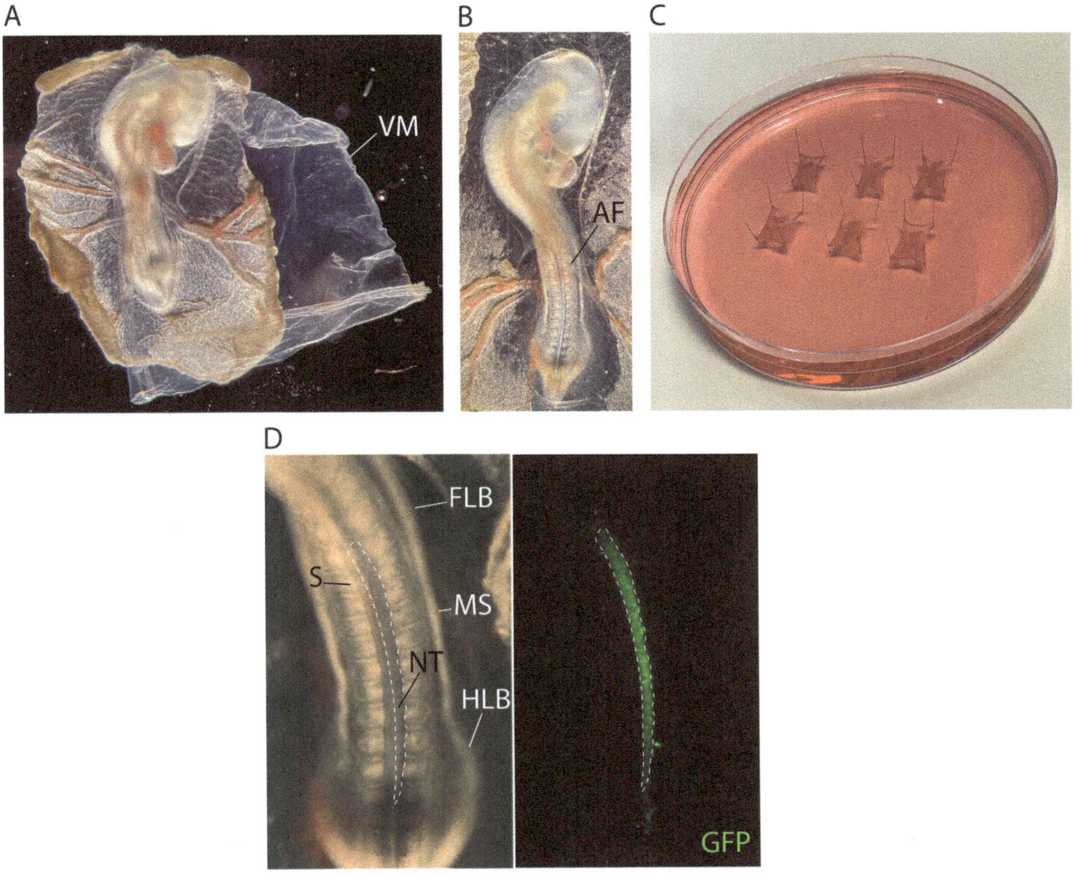

Fig. 2 Preparation for slicing. (**a**) Hamburger–Hamilton stage 17 embryo removed from egg showing partially removed vitelline membrane. (**b**) Pinned embryo, dorsal side up, showing amniotic folds. (**c**) Multiple embryos in pin dish. (**d**) Zoomed region of embryonic trunk showing anatomy (left) and fluorescence indicating transfection (right). Dotted white line indicates border of transfected region. AF amniotic folds, FLB forelimb bud, GFP green fluorescent protein, HLB hindlimb bud, MS midsection, NT neural tube, S somite, VM vitelline membrane

12. Using fine forceps, tear open the amniotic folds (Fig. 2b), if they are covering the spinal cord neural tube.
13. Repeat **steps 10–12** for all embryos (Fig. 2c). Placing them in orderly rows will help you identify them later.
14. Identify successful transfection by viewing embryos under a fluorescence stereomicroscope on high zoom. You should observe a stripe of fluorescence along one side of the developing spinal cord (Fig. 2d). Use transmitted light to identify the transfected regions of the embryo (e.g., forelimb bud, hindlimb bud, or midsection; *see* **Note 18**). Note the region of interest for each embryo for slicing.
15. Proceed to either Hand Slicing (*see* Subheading 3.3) or Vibratome Slicing (*see* Subheading 3.4) depending on your needs (*see* **Note 19**).

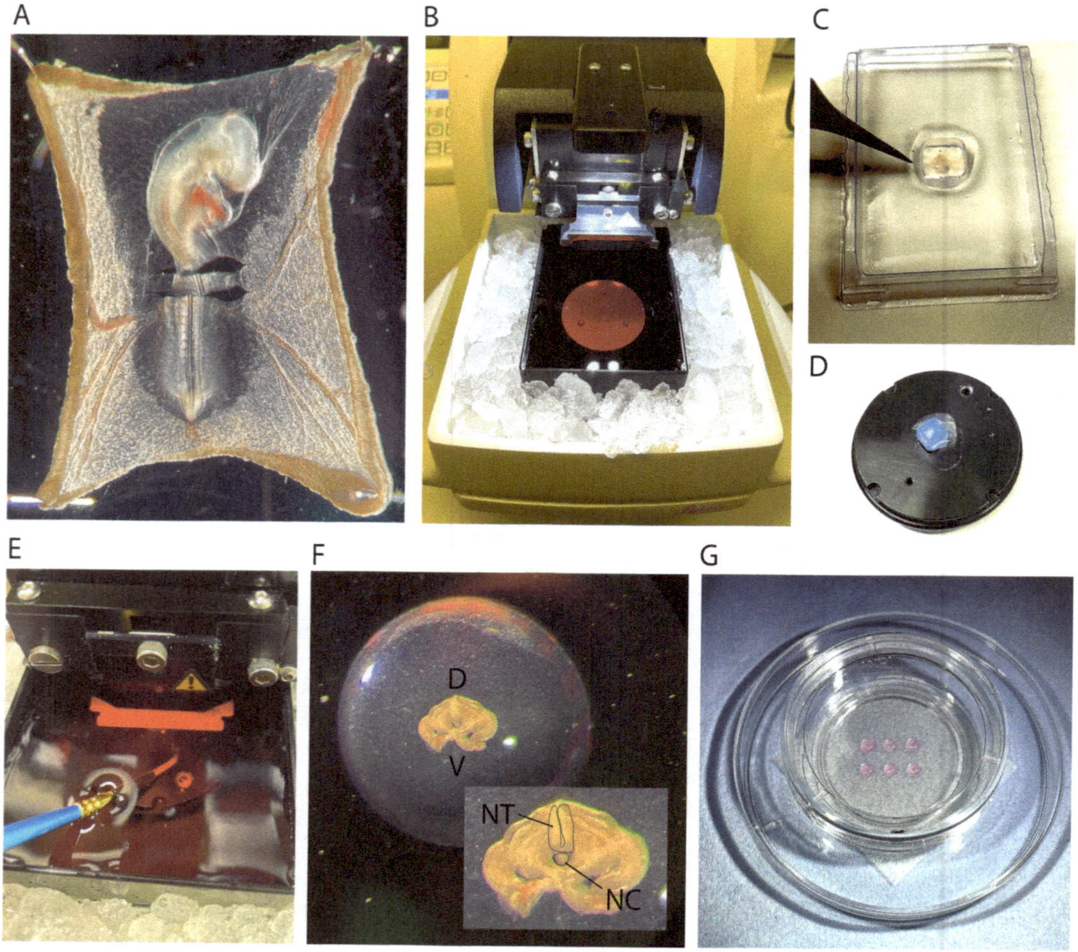

Fig. 3 Embryo slicing. (**a**) Pinned, hand-sliced embryo with slice still attached laterally. (**b**) Vibratome setup. (**c**) Embedding of tissue in agarose within mold. (**d**) Agarose cube glued to specimen holder. (**e**) Removal of sliced tissue from vibratome using paintbrush. (**f**) Slice embedded in collagen droplet with zoomed inset indicating anatomy. (**g**) Final dish setup. D dorsal, V ventral, NC notochord, NT neural tube

3.3 Hand Slicing

1. On the standard stereomicroscope, zoom to bring one embryo into the field of view.
2. Identify the region of interest. Using a microknife, slice the entire width of the embryo as straight as possible from dorsal to ventral in a single, clean, downward motion at intervals approximately two to three somites thick (Fig. 3a). Detach the sides of the transverse slice from the extraembryonic tissue and move to the side using the microknife.
3. Proceed immediately to Mounting (*see* Subheading 3.5).

3.4 Vibratome Slicing

1. Pre-warm 4% low-melting agarose in PBS in a 37 °C water bath.
2. Fill the buffer tray with 1× L-15 and surround with ice (Fig. 3b).

3. Set the following parameters on the vibratome: speed 0.4 mm/s, amplitude 0.45 mm, thickness 200–300 μm.
4. Fit the blade to the vibratome and position close to the buffer tray (*see* **Note 20**). Submerge blade into L-15 by raising the stage.
5. Using a microknife, dissect out the entire region of interest, preserving the whole width of the embryo (*see* **Note 21**). Remove extraembryonic membranes to facilitate mounting.
6. Fill a plastic base mold with warmed agarose solution.
7. Using fine forceps, embed the dissected region of the embryo into the agarose solution (Fig. 3c). Looking through the stereomicroscope, orient the tissue orthogonally to the base of the mold so that the transverse cross section is visible from the top. Repeat for all embryos. Work quickly before the agarose solidifies and avoid over-manipulation, which leads to tissue fragmentation.
8. Place the molds on ice for at least 2 min.
9. Remove solidified cube of agarose from the mold using fine forceps and glue it on the specimen disk using super glue (Fig. 3d).
10. Proceed with vibratome sectioning (according to manufacturer's instructions) to create a slice.
11. Pick up the slice using one or two thin paintbrushes (Fig. 3e) and transfer it into a 50 mm Petri dish containing ice-cold 1× L-15 kept on ice.
12. Continue slicing the embryo.
13. Repeat **steps 9–12** for each embryo using a separate Petri dish for each.
14. Under the stereomicroscope and in the L-15, gently peel away the agarose from the tissue using forceps, taking care not to touch the tissue with the forceps in order to preserve tissue structure.
15. Proceed immediately to Mounting (*see* Subheading 3.5).

3.5 Mounting

1. Prepare micropipette tip for transferring embryo slices by cutting off a small portion (~1 mm) of the tip to widen it. Vortex collagen mixture and draw in 5–10 μL using a P20 micropipette. Leave for approximately 1–2 min to coat the tip to prevent tissue from sticking, discard excess collagen, and wash with L-15. Transfer the tip to a P2 micropipette.
2. Vortex the collagen mixture thoroughly. Using a new tip on a P20 micropipette, dispense a 3–8 μL droplet of collagen mixture into the center of a FluoroDish (*see* **Note 22**).

3. Set the P2 micropipette with the coated tip to 0.6–1 μL. Viewing under the stereomicroscope, pick up a single slice with as little L-15 as possible, pausing the plunger as soon as the slice is in the pipette. Carefully dispense into a droplet of collagen mixture.
4. Move the FluoroDish to the stereomicroscope and, working quickly to avoid the collagen setting, use closed fine forceps to position the slice flat in the middle of the droplet, and flush with the glass dish bottom so that the transverse cross section is visible from the top and the dorsal side of the embryo is oriented to the top of the dish (Fig. 3f).
5. Obtain the next slice as directed by your chosen method.
6. Repeat **steps 3–5** placing up to nine collagen droplets in a grid pattern at the center of FluoroDish (for alternative steps, *see* **Note 23**). Be careful to leave spaces between droplets and avoid knocking the dish so that the droplets do not merge.
7. When all slices have been placed, move the FluoroDish to the stereomicroscope and, working quickly to avoid the collagen setting, use fine forceps to position each slice flat in the middle of the droplet, and in contact with glass coverslip bottom.
8. Carefully place the lid on the FluoroDish. Place the FluoroDish in the lid of a 50 mm Petri dish on top of a small piece of lens cleaning tissue to protect the underside of the coverslip (Fig. 3g).
9. Leave at room temperature in the dark for 45–60 min to allow the collagen to solidify into a gel. Adjust this timing to suit the conditions in your lab (*see* **Note 24**).
10. Once collagen gel has solidified, gently add 2–3 mL of equilibrated warm slice culture medium to the dish by very gently dispensing droplets directly on top of the collagen, taking care not to dislodge it from the dish.
11. Place the 50 mm dish containing the FluoroDish with lid in a humidified plastic box and place into a 37 °C 5% CO_2 incubator.
12. Allow the slices to recover for at least 1 h before imaging.

3.6 Live Imaging

1. Set up a widefield inverted microscope with the chamber set to 37 °C and 5% CO_2 roughly an hour before imaging.
2. Add a generous amount of immersion fluid to the objective (*see* **Note 25**) and position FluoroDish in the stage (*see* **Note 26**). Set up for imaging and locate the first slice using brightfield, focusing on the neural tube. Set LEDs to a low intensity (ideally 20–40%) with no longer than 20 ms exposure time.

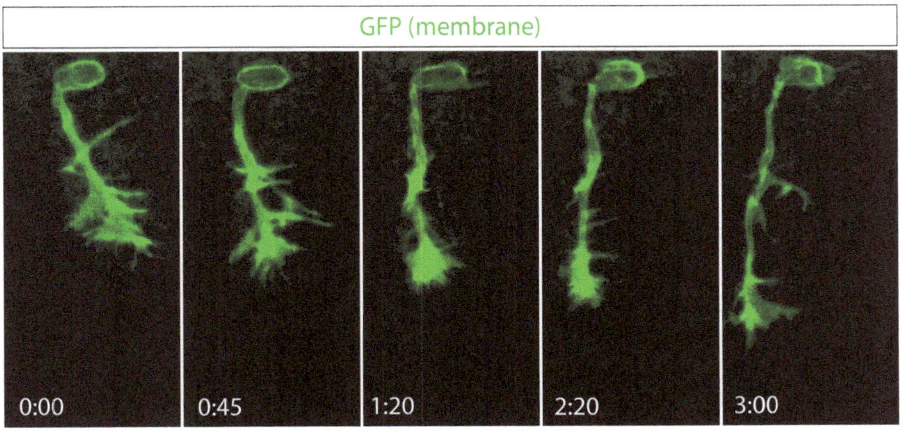

Fig. 4 Stills from an example time-lapse sequence of a differentiating neuron undergoing axon extension obtained using the protocol. Embryos were transfected at Hamburger–Hamilton stage 10–12 with a membrane-targeted GFP construct and hand-sliced at stage 16–17. Time signature shows h:min. GFP green fluorescent protein

3. Locate the transfected side of the tissue by acquiring a single image to visualize fluorescence. Acquire different sections along the Z-plane until you locate an appropriate cell (*see* **Note 27**).
4. Set up a Z-stack with 40–50 optical sections spaced 1.5 μm apart centered on a single cell of interest per slice (*see* **Note 28**) and save the position. Repeat for all tissue slices.
5. Set up a time series, imaging Z-stacks with desired interval and duration, depending on the experiment (*see* **Note 29**). Image whole Z-stacks of individual channels in series to speed up imaging.
6. Use a post-processing deconvolution algorithm to obtain the best possible images of your widefield time-lapse movies (Fig. 4; Movie 1; *see* **Note 30**).

4 Notes

1. Stock of 1% Brilliant Blue kept at room temperature might concentrate over time with evaporation, in which case, re-dilute by eye. The concentration does not need to be precise but should be just dark enough to visualize the DNA mixture once injected. Fast Green FCF can be used as an alternative dye; however, it can fluoresce red and impair imaging.
2. To make a versatile egg stand, slice (a piece of tape can be placed as a guide) and hollow out one side of a tube of polyethylene pipe insulation available from plumbers' merchants. Alternatively, use an upside-down 50 mL conical tube cap; however, this might be too shallow for some eggs.

3. A stereomicroscope with foot pedal-controlled focus is ideal as both hands can be used for dissection while focusing.
4. To reduce plastic waste, Petri dishes and pin dishes can be rinsed with distilled water and reused.
5. Pure collagen stock can be stored in sterile 300 μL aliquots at 4 °C. 5× L-15 is prepared from L-15 medium powder in deionized water and filtered to keep the solution sterile. To avoid contamination, we suggest keeping aliquots of all additive solutions at 4 °C.
6. Optiprep is optional but helps to improve the refractive index of the medium for clearer imaging. Divide neurobasal medium 20% Optiprep into sterile 10 mL aliquots and store at 4 °C for slice culture medium preparation.
7. Sterile 200 μL aliquots of B27 can be stored at −20 °C. Aliquots of gentamicin (10 mg/mL) and GlutaMAX can be stored at 4 °C.
8. We use an Axio Observer Z.1 microscope (Zeiss) equipped with a Colibri 7 LED light source (Zeiss), Flash4 v2 sCMOS camera (Hamamatsu), and an environmentally controlled chamber. To acquire images of the transfected tissue, we typically use a 40X 1.2 NA silicone immersion objective (Zeiss). For more detailed images of individual cells, a higher magnification objective (e.g., 63X) can be used.
9. Use a new microknife for each slicing session to ensure sharpness. To reduce waste, used microknives can be kept for cutting pipette tips.
10. Electroporation of the neural tube can be carried out from Hamburger–Hamilton (HH) stage 10. We usually electroporate at HH stage 10–12 (approximately 33–49 h but test incubation timings for your system). At this point, the neural tube consists mostly of neural progenitor cells.
11. Depending on the promoter used and protein stability, some proteins are expressed more robustly than others. We suggest running preliminary experiments to standardize the concentration of DNA for your transfection. For capturing individual cells, fluorescence observed under the stereomicroscope should be low and sparse rather than solid and bright.
12. If the solution will not draw in, use fine forceps to break off a small amount of the needle tip to widen it. The DNA solution might become stuck in the needle when allowed to dry between injections. Additionally, bubbles can be generated inside the needle during aspiration. To solve these problems, position the needle at a shallow angle on some wet tissue and blow down the aspirator tube assembly until it dispenses.

13. Chicken embryos are relatively resistant to temperature fluctuations. A batch (6–12 eggs) can be kept at room temperature while they are electroporated before being returned to the incubator.
14. The embryo is translucent at early stages and can be difficult to identify against the surrounding tissue, so we suggest to practicing this before an experiment. It can help to initially look at distinctive features such as the brain ventricles or somites alongside the neural tube. Additionally, gently moving the egg while looking for the embryo can help your eyes to pick up on these details.
15. Extending the gap between injection and electroporation reduces the density of cells transfected. This is useful if you want to label isolated cells within the tissue.
16. Some embryos might not be viable following electroporation. Transfect additional embryos to compensate for this.
17. This will depend on what process you want to observe. We typically slice at HH stage 16 (roughly 51 h) or above, at which point neuronal differentiation is underway.
18. Depending on the levels and subcellular localization of the fluorescent protein expressed, the visibility of transfection with a stereomicroscope will vary. For example, a cytoplasmic label will be more visible than that of a small organelle. Take time to study the embryo to check for low levels of fluorescence that are not immediately apparent.
19. For speed and ease, we suggest using hand slicing embryos up to roughly HH stage 17. For later-stage embryos and for timelapse experiments longer than 24 h, we suggest slicing by vibratome to better preserve the tissue structure and extend viability.
20. Setting up the blade before sectioning will save time during the sectioning procedure.
21. Work with a maximum of two embryos to preserve tissue viability, due to the time required for vibratome slicing.
22. Prior to adding collagen, marking the side of the Fluorodish can help in later orienting the dish for imaging. For HH stage 18 and above, use a high volume of collagen mixture (8 μL). Ensure the slice is completely covered and that collagen droplets are confined to the center of the coverslip. Attempting to image slices toward the coverslip edges can cause the objective to push the dish out of place.
23. Alternatively, you can proceed with these steps to speed up the protocol:

(a) In a clean, empty lid of a 90 mm Petri dish, use a 3 mL Pasteur pipette to place large well-separated droplets of clean L-15, one for each embryo you are slicing from. You will place all the slices from one embryo into that droplet to keep them separate from other embryos. Leave large gaps between droplets so that they do not merge. It can be helpful to mark the dish with numbers for each droplet before starting.

(b) Pick up each slice using the collagen-coated micropipette tip on a P2 micropipette and transfer to the appropriate L-15 droplet. Repeat for all slices so that each L-15 droplet contains all the slices from a single embryo.

(c) Vortex the collagen mixture thoroughly. Using a new tip on a P20 micropipette, dispense up to nine 3–8 μL collagen droplets in a grid pattern on the center of the Fluorodish.

(d) Set the P2 micropipette with the coated tip to 0.6–1 μL. Viewing under the stereomicroscope, pick up a single slice and dispense into a droplet of collagen mixture. With the naked eye, you should see a small shadow in the droplet when the slice has been dispensed. Repeat for all slices.

24. Higher temperatures increase the speed of collagen solidification. A reduction in pH can reduce the ability of collagen to set. Ensure the dish is left to set away from any potential sources of CO_2, such as an incubator, and check that the droplets of collagen remain pink during setting.

25. You must add sufficient immersion fluid (e.g., 2–3 drops) to enable the objective to image at various positions repeatedly for an extended time.

26. The collagen will now be colorless and difficult to see, the black mark on the dish can help with orientation. If your dish does not fit tightly in the stage, use a few pieces of Blu tack around the dish (taking care not to seal the gap between the dish and lid, which would prevent gas exchange). This will prevent dish movement during acquisition.

27. In most experiments, many overlapping cells of the neural tube will express the transfected constructs. To observe the behavior of an individual cell, try to choose an isolated cell surrounded by untransfected tissue so that the cell can be clearly visualized. Try to avoid searching with continuous or live mode because constant exposure to fluorescence produces photobleaching and phototoxicity that could damage the tissue and reduce the quality of images produced.

28. Setting up a deep Z-stack is necessary to allow for any cell or tissue movement and the growth of the axon during neuronal differentiation over the course of imaging.
29. For most applications, e.g., following cell behavior like axon extension, intervals of 10 min are sufficient. More dynamic processes might require shorter intervals. For very high temporal resolution (intervals of 1 min or less), we suggest short bursts of imaging (up to 10 min), leaving pauses of approximately an hour in between to enable tissue recovery and avoid photobleaching. The total duration of imaging will depend on the dynamics of the process you are interested in and the viability of the tissue with the slicing method and settings you have used. Cells in hand-sliced sections can be imaged for 12–48 h before significant tissue death is observed. Vibratome slicing improves duration of tissue viability and so longer time series can be acquired. Acquiring a single, centered optical section (not a Z-stack) with the brightfield channel throughout the time series can be used to observe tissue health and determine useable footage.
30. We favor the constrained iterative or fast iterative algorithms offered by ZEN Blue (Zeiss) but recommend testing various algorithms and parameters to suit your images.

Acknowledgments

R.M.D. and G.T.T. are supported by an MRC Transition Support Fellowship (MR/V036386/1) to R.M.D. V.E.H. is supported by a Wellcome Trust PhD studentship (222749/Z/21/Z). H.B.B. is supported by a BBSRC PhD studentship (BB/T008725/1).

References

1. Yogev S, Shen K (2017) Establishing neuronal polarity with environmental and intrinsic mechanisms. Neuron 96:638–650
2. Kaech S, Banker G (2006) Culturing hippocampal neurons. Nat Protoc 1:2406–2415
3. Zolessi FR, Poggi L, Wilkinson CJ, Bin CC, Harris WA (2006) Polarization and orientation of retinal ganglion cells in vivo. Neural Dev 1:2
4. Barnes AP, Polleux F (2009) Establishment of axon–dendrite polarity in developing neurons. Annu Rev Neurosci 32:347–381
5. Lossi L, Merighi A (2018) The use of ex vivo rodent platforms in neuroscience translational research with attention to the 3Rs philosophy. Front Vet Sci 5:1–14
6. Stern CD (2005) The chick: a great model system becomes even greater. Dev Cell 8:9–17
7. Kasioulis I, Storey KG (2018) Cell biological mechanisms regulating chick neurogenesis. Int J Dev Biol 62:163–171
8. Das RM, Storey KG (2012) Mitotic spindle orientation can direct cell fate and bias notch activity in chick neural tube. EMBO Rep 13: 448–454
9. Das RM, Storey KG (2014) Apical abscission alters cell polarity and dismantles the primary cilium during neurogenesis. Science 343:200–204
10. Toro-Tapia G, Das RM (2020) Primary cilium remodeling mediates a cell signaling switch in differentiating neurons. Sci Adv 6:1–11

Chapter 9

In Vitro and In Vivo Analysis of Neuronal Polarity

Mini Jose

Abstract

Neuronal development is characterized by the unidirectional flow of signal from the axon to the dendrites via synapses. Neuronal polarization is a critical step during development that allows the specification of the different neuronal processes as a single axon and multiple dendrites both structurally and functionally, allowing the unidirectional flow of information. Along with extrinsic and intrinsic signaling, a whole network of molecular complexes involved in positive and negative feedback loops play a major role in this critical distinction of neuronal processes. As a result, neuronal morphology is drastically altered during establishment of polarity. In this chapter, we discuss how we can analyze the morphological alterations of neurons in vitro in culture to assess the development and polarity status of the neuron. We also discuss how these studies can be conducted in vivo, where polarity studies pose a greater challenge with promising results for addressing multiple pathological conditions. Our experimental model is limited to rodent hippocampal/cortical neurons in culture and cortical neurons in brain tissues, which are well-characterized model systems for understanding neuronal polarization.

Key words Primary rodent hippocampal/cortical neuronal cultures, Rodent brain slices, Neuronal development and polarity, Axon, Dendrites, Immunocytochemistry, Immunohistochemistry, In utero electroporation, Antibodies, Microscopy, Imaging

1 Introduction

During development, neurons undergo huge alterations in their morphology. Numerous intrinsic and extrinsic factors contribute to this modification, which vary significantly in vitro and in vivo. In the Subheading 1, we discuss briefly about our understanding on neuronal development in vitro, which is a simpler setting where only intrinsic signaling mechanisms are involved, whereas in the Subheading 2, the involvement of extrinsic cues for development in vivo is discussed.

For studying neuronal polarization and development in vitro, the most characterized model systems are the embryonic or postnatal rodent hippocampal and cortical neurons in culture. In culture, neuronal development has been divided into different

Kazuhito Toyooka (ed.), *Neuronal Morphogenesis: Methods and Protocols*, Methods in Molecular Biology, vol. 2831,
https://doi.org/10.1007/978-1-0716-3969-6_9,

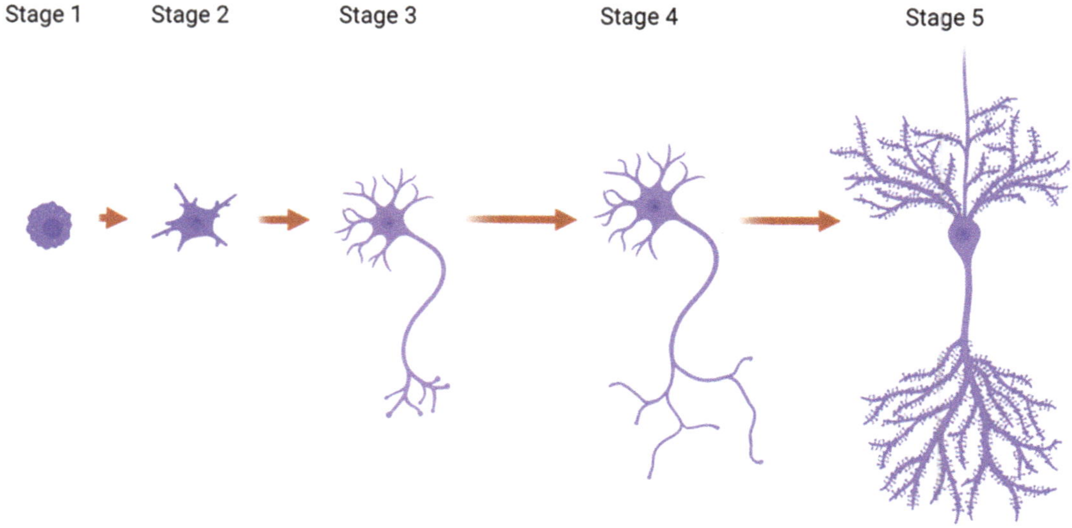

Fig. 1 Developmental profile of rodent hippocampal neurons in vitro. The different developmental stages from 1 to 5 depict how the neuron modifies its structure from a simple round cell to a highly complex and differentiated cell, forming connections. The different stages show (1) lamellipodia formation, (2) neurite outgrowth, (3) axon extension, (4) dendrite extension, and (5) synapse formation. (This image is created with Biorender.com)

subphases, namely phases 1–5 (Fig. 1) [1]. Neurons develop from a round cell to phase 1, where they start to form actin-based projections around the cell body named as lamellipodia. In phase 2, the cell sprouts small projections called neurites, which display a symmetric growth. According to the current model, there is a bulk flow of molecules to different neurites during this phase, enabling symmetric growth [2]. Phase 3 is specified by an asymmetric growth pattern, where one of the neurites breaks symmetry and grows much rapidly and longer compared to the other neurites, developing as the axon, whereas the rest of the neurites develop as the dendrites. Positive feedback from the length of the neurite with restricted localization of polarity molecules is thought to play a critical role in this molecular and functional distinction of a neurite as the axon [3–5]. On the other hand, negative feedback loops are thought to play a prominent role in inhibition or retarded growth of the minor neurites that develop as the dendrites [5, 6]. This distinction or specification of the different neuronal processes as a single axon and multiple dendrites is referred to as neuronal polarity. Selective filtering of molecules into the axon compared to the somato-dendritic compartments, aided via the complementary organization of the molecules at the axon initial segment in the proximal axon, is also thought to be critical for complete neuronal polarity establishment [7–9]. It is challenging to pinpoint the exact period of polarity establishment since generation of neuronal polarity or the specification of the axo-dendritic processes occurs over a certain duration during development, initiating in phase 2/3 and finishing by phase 3/4 (Fig. 1). The minor neurites start to grow rapidly to develop as the dendrites during phase 4. In phase 5, the mature neurons form connections via synapses, allowing flow of information from axons to dendrites.

Since neuronal polarization occurs over a certain duration during development, it makes it challenging to assess the polarity status of the neurons during this period of transition. Conventionally, qualitative methods using antibody labeling of axonal and somato-dendritic compartments and imaging by microscopy is the most common strategy to understand this distinction [10]. Tau, a common axonal marker, and Map 2, a somato-dendritic marker, are the conventional proteins labeled for polarity assessment. Tau protein distribution gets restricted to the axon once polarity is established, whereas Map 2 remains in the soma and dendrites (Fig. 2). In neuronal cultures, Map 2 is also observed in a restricted area of the proximal axon, namely, the axon initial segment (AIS), which also serves to distinguish the axon from the rest of the neurites in culture (*see* Subheading 3.1) (Fig. 2). Additionally, a new method for polarity assessment has been proposed recently, where quantitative measurements can be made based on the relative ratio of the length of the axon to the dendrites, which has been shown to be a reliable quantitative estimate of the polarity

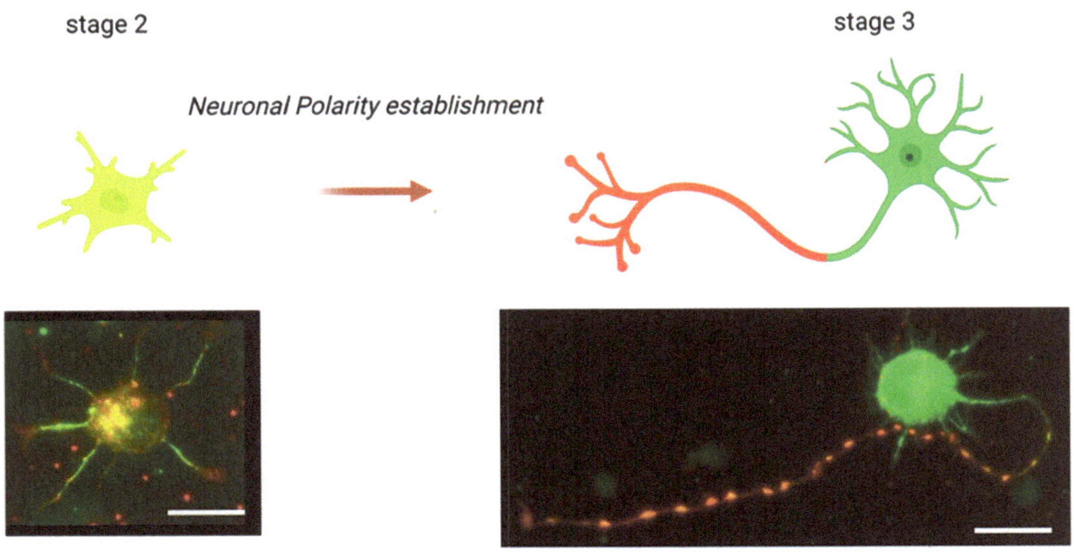

Fig. 2 Neuronal polarity establishment in vitro. The top panel shows schemes of developmental stages (2) and (3) where neurons display symmetric and asymmetric growth, respectively. Red and green colors label axonal marker Tau and somato-dendritic marker Map2 distributions, respectively. In stage 2, since both Tau and Map2 are present in all neurites, the overlay is shown in yellow. However, in stage 3, restricted distribution of Tau and Map2 is observed in the axonal and dendritic compartments, respectively. Note that the proximal axon of the polarized neuron of stage 3 is labeled with Map2. The bottom panel shows neurons of stages (2) and (3) immunolabeled by antibodies targeting Tau and Map2 marked by Alexa fluor 647 (red) and Alexa fluor 488 (green), respectively. The scale bars represent 20 μm

status of the neuron [10]. Therefore, a combination of these multiple techniques would provide a cumulative understanding of the development as well the polarity status of the neurons in culture.

Two strategies, namely positive or negative labeling, can be used for qualitative polarity assessment. For positive labeling, an axonal marker such as an antibody against Tau and a marker for somato-dendritic compartments such as Map 2 are used in combination with spectrally separated secondary antibodies such as Alexa fluor 647 and 488, respectively (Fig. 2). From the fluorescence distribution where Tau is restricted to the axon and Map 2 labels the soma, dendrites, and the proximal axon, polarity status of the neuron can be evaluated. On the contrary, negative labeling refers to labeling all the neurites and soma with a neuron-specific uniform marker such as β3 Tubulin in combination with Map 2, which marks the somato-dendritic compartments. The neurite where Map 2 labeling is absent or is restricted to the proximal part alone would be considered as the axon. Though both of these labeling strategies provide similar results, for the majority of the studies, positive labeling method might be preferred for the ease of understanding and is discussed here.

Neuronal polarization in vivo puts forth a more complex scenario, where polarization involves two different phases of development such as neuronal migration and axo-dendritic specification, which are overlapping and interdependent. The external cues provided by the non-neuronal or glial cells have a prominent role in neuronal migration followed by extension and growth of the neuronal projections into the different brain layers as axons and dendrites. Hippocampal pyramidal cells originate from neural stem cells or radial glial cells in the subventricular zone that do not retain their polarity [11]. Therefore, the inherited polarity is lost during development and is established later. These cells start to develop extensions similar to phase 2 neurons in culture, as discussed previously, which are multipolar in nature (Fig. 1). A majority of these cells develop a tangential trailing process, potentially developing as the axon and then a leading process that develops as dendrites. The multipolar nature of the neurons shifts to bipolar nature at this stage, which is critical during neuronal polarization as well as for migration to the different cortical regions via radial glia-dependent and -independent mechanisms such as somal translocation and locomotion [12, 13]. To understand defects in neuronal migration and polarization, the conventional method is to perform immunohistochemical studies using specific markers of different brain layers along with polarization markers, though this is developmentally regulated. In a rodent brain, where neuronal migration is complete, a genetic diversity is observed where distinct proteins are expressed in cells localized to specific cortical layers, which allow us to use these proteins as reference markers to distinguish any defects in migration (Fig. 3). The standard procedure for assessing migration

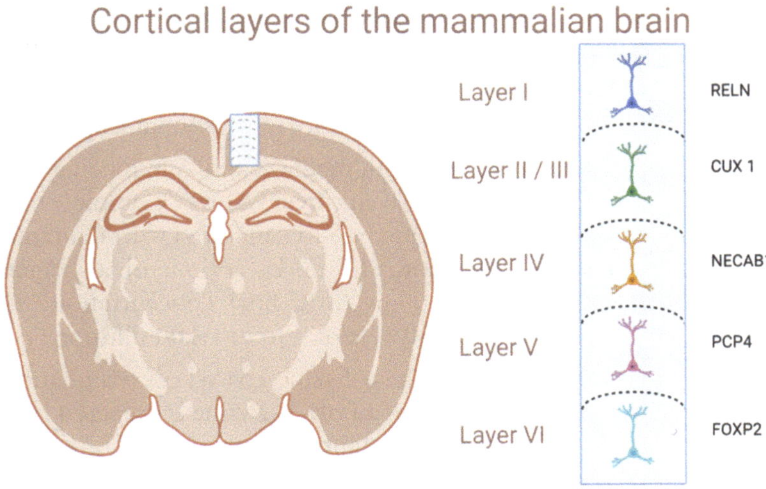

Fig. 3 Cortical layers of the mammalian brain in vivo. The left panel shows the coronal section of a rodent brain. The cortex lies above the hippocampus. The C-shaped hippocampus is displayed in a darker shade on both lobes. The cortex can be divided into six layers labeled by layer specific markers, though their distribution is developmentally regulated. Few of the markers show overlapping distributions between layers. A small section of the cortex (boxed) is zoomed and shown on the right panel along with the specific protein markers for each layer of the cortex once the migration is completed. (This image is created with Biorender.com)

and polarity in vivo is discussed in the second part of the Subheading 3 (*see* Subheading 3.2), at a developmental stage when the migration is complete in the brain. Alternatively, dissociated neuronal cultures prepared from brain tissues under study can also provide similar polarity assessment using antibodies against polarity markers as discussed in Subheading 3.1.

2 Materials

2.1 In Vitro Analysis

2.1.1 Preparation of Rodent Primary Hippocampal/Cortical Cultures

1. Animals (P0/P1 pups; rat/mouse).
2. Dissection scope: For dissection of the brain and hippocampus/cortex.
3. Dissection media for primary hippocampal neuronal cultures: Dissection media consists of Hibernate A supplemented with Glutamax (100×, final concentration: 0.5 mM), antimicrobial reagent Normocin (Invivogen, 50 mg/mL, final concentration: 100 μg/mL), and B-27™ (50×, serum free, 1:100 dilution) (*see* **Notes 1** and **2**).
4. Trypsin 2.5%: For dissociation of brain tissue.

5. Syringes and Needles: Of differing sizes to generate a homogeneous population of neurons. 5/10 mL syringes and 21 G and 25 G needles (*see* **Note 3**).
6. Hemacytometer: For counting viable cells.
7. Plating media/Complete media for primary hippocampal neuronal cultures: Plating/Complete media consists of Neurobasal A media supplemented with Glutamax (100×, final concentration: 0.5 mM), antimicrobial reagent Normocin (Invivogen, 50 mg/mL, final concentration: 100 μg/mL), and B-27™ (50×, serum free, 1:100 dilution) (*see* **Notes 4** and **5**).
8. Poly-L-Lysine-coated coverslips and well plates: Working solution of Poly-L-Lysine is prepared from the stock (1 mg/mL) at 1:50 dilution with ddH_2O. 18 mm coverslips are placed in 12-well cell culture plates and coated with Poly-L-Lysine for 1 h at 37 °C. The coverslips are washed thrice with ddH_2O and dried under the hood (*see* **Notes 6** and **7**).
9. Water/Dry bath maintained at 37 °C.
10. Microcentrifuge.
11. Bright-field microscope.
12. Cell incubator equipped with CO_2 and humidity.

2.1.2 Immunocytochemistry (ICC) and Imaging of Rodent Primary Hippocampal/Cortical Cultures

1. 1 × PBS: NaCl 137 mM, KCl 2.7 mM, Na_2HPO_4 10 mM, KH_2PO_4 1.8 mM. Prepared in autoclaved ddH_2O and calibrated to pH 7.4 (*see* **Note 8**).
2. Fixation buffer: 4% Paraformaldehyde + 4% Sucrose in 1 × PBS.
3. Quenching buffer: 0.1 M Glycine in 1 × PBS.
4. Permeabilization buffer: 0.25% Triton 1 × PBS.
5. Blocking buffer: 10% BSA in 1 × PBS.
6. Antibody incubation buffer: 3% BSA in 1 × PBS.
7. Antibodies: Conventional markers labeling axonal and somatodendritic compartments.
   (i) Primary antibodies: anti-Tau, anti-Map2 (*see* **Note 9**).
   (ii) Secondary antibodies: Alexa fluor 647, Alexa fluor 488 (*see* **Note 10**).
8. ProLong™ Antifade mounting medium.
9. Glass slides.
10. Slide box: Can be used as a dark humidified chamber for labeling and storing immunolabeled slides.
11. Wide field fluorescence/Confocal Microscope: For imaging primary cultures.

2.2 In Vivo Analysis

2.2.1 In Utero Electroporation

1. Dam with embryos of stage E 14.5 (*see* **Note 11**).
2. Sterile Surgical kit.
3. Surgical chamber and mask.
4. Isoflurane (1%).
5. Glass microcapillaries.
6. Mouth aspirator tube.
7. Plasmid preparation of a GFP construct expressed under a CAG or CMV promoter in nuclease-free water (Endotoxin-free).
8. Fast Green (1:20 dilution in ddH_2O): Dye to visualize the site of injection.
9. In utero electroporation unit (Nepagene/ECM830): For electroporation of a GFP construct into embryos for imaging individual neurons in brain tissue sections.
10. Sterile Suture kit.

2.2.2 Preparation of Rodent Brain Sections

1. Animals (P0 rat/mouse pups) (*see* **Note 12**).
2. Dissection scope.
3. Syringes and Needles: 50 mL, 30 G.
4. Perfusion buffer: 4% Paraformaldehyde in 1 × PBS (*see* **Note 13**).
5. 1 × PBS: NaCl 137 mM, KCl 2.7 mM, Na_2HPO_4 10 mM, KH_2PO_4 1.8 mM. Prepared in autoclaved ddH_2O and calibrated to pH 7.4.
6. 0.9% saline.
7. 30% sucrose solution.
8. Isopentane or OCT (optimum cutting temperature) embedding resin for cryofixation.
9. Cryostat/Cryomicrotome.

2.2.3 Immunohistochemistry and Imaging of Rodent Brain Sections

1. Blocking and permeabilization buffer: 3% BSA,0.3% Triton X-100, 1% Normal serum in 1 × PBS (*see* **Note 14**).
2. Primary antibody incubation buffer: 3% BSA in blocking solution.
3. Secondary antibody incubation buffer: 3% BSA, 1% Normal serum in 1 × PBS.
4. Antibodies: Two different sets of antibodies are used for studying neuronal migration and polarity.
   (a) Conventional markers labeling the different brain layers (Fig. 3). For example, primary antibodies against RELN, CUX1, NECAB1, PCP4, and FOXP2 mainly labels layers

1, 2/3, 4, 5, and 6, respectively. These primary antibodies can be used in pairs or triplets for comparison. Corresponding secondary antibodies are selected to avoid cross-reactivity and spectral overlap. For example, Alexa fluor 647/555/488/350 for multispectral imaging.

(b) Polarity markers labeling Tau/Neurofilament and Map2 in tissue sections for labeling of axonal and somato-dendritic compartments (*see* **Note 15**).

   (i) Primary antibodies: anti-Tau/anti-Neurofilament, Map2.

   (ii) Secondary antibodies: Alexa fluor 647, Alexa fluor 555 (*see* **Note 16**).

5. ProLong™ Antifade mounting medium supplemented with DAPI for nuclear labeling.
6. Glass slides and coverslips.
7. Slide box: Can be used as a dark humidified chamber for labeling and for storing immunolabeled slides.
8. Confocal Microscope: For 3D imaging of brain tissue sections.

3 Methods

3.1 In Vitro Analysis

3.1.1 Preparation of Primary Cultures of Rodent Hippocampal/Cortical Neurons

Primary cultures of neurons are prepared from P0/P1 (rat/mouse brain) as per the protocols proposed by Goslin and Banker [14].

1. The pups are decapitated, and the brain dissected out.
2. The hippocampi/cortex (as per requirement) is extracted under the dissection scope.
3. The brain tissue is dissociated in Trypsin (0.25%) for 10 min at 37 °C. The tissue suspension is mixed halfway through for a thorough dissociation. Hippocampi is directly incubated in trypsin solution for dissociation, whereas cortex is sliced into small pieces prior to incubation. 2 mL of trypsin is used for tissue dissociation of hippocampi or cortex in separate tubes from two pups.
4. After trypsin incubation, the tissue is centrifuged for 5 min at $800\times g$.
5. The tissue is then gently washed thrice with ice-cold dissection media along with centrifuging.
6. After the final wash, the tissue is suspended in 5 mL cold dissection media for dissociation into single cells.
7. For dissociation, the tissue suspension is initially passed drop-wise gently and very slowly through a syringe attached with a

needle of large pore size (21 G) 3 times followed by 3 times through a needle of smaller pore size (25 G) (*see* **Note 17**).

8. Once the tissue is completely dissociated, the suspension is centrifuged once again for 5 min at 800× *g*.
9. The supernatant is carefully discarded, and the cells are resuspended in 1 mL Plating/Complete media.
10. 10 μL of the cell suspension is taken in a hemacytometer for counting viable cells (*see* **Note 18**).
11. The cell suspension is diluted in Plating/Complete media and seeded in pre-prepared well plates containing Poly-L-Lysine coated coverslips at a plating density of 40–50 cells/mm^2 (*see* **Note 19**).
12. The cells are grown at 37 °C in 5% CO_2 and 5% humidity.
13. The neuronal cultures are supplemented every week with 200 μL of Plating/Complete media for maintenance of neurons in culture (*see* **Note 20**).

3.1.2 Immunocytochemistry of Rodent Primary Hippocampal/Cortical Cultures

1. Neuronal cultures are grown until DIV 3 (days in vitro) or longer to assess for polarity (*see* **Note 21**).
2. The neurons grown on coverslips are fixed with the fixation buffer for 10 min at 4 °C.
3. The cells are quenched with the quenching buffer for an additional 3–5 min (*see* **Note 22**).
4. The fixed and quenched cells are quickly washed thrice with cold 1 × PBS.
5. Then, the cells are permeabilized with the permeabilization buffer for 5 min (*see* **Note 23**).
6. The permeabilized cells are washed with 1 × PBS, 2 × 1 min.
7. The cells are blocked with the blocking buffer for 30 min prior to antibody incubation (*see* **Note 24**).
8. Working solutions of primary antibodies against Tau and Map2 are prepared fresh according to manufacturer's protocols in cold 3% BSA.
9. The cells are incubated with the primary antibody mix for 1 h at room temperature in a humidified chamber (*see* **Notes 25** and **26**).
10. After the primary antibody incubation, the coverslips are washed 4 × 5 min each in 3% BSA in 1 × PBS.
11. The secondary antibodies targeting the primary antibodies (Alexa fluor 647/488) are freshly prepared in 3% BSA (dilution according to manufacturer's protocols) and the cells are incubated in the secondary antibody mix for 45 min at room temperature in a dark humidified chamber.

12. After the secondary antibody incubation, the coverslips are washed 4 × 5 min each in 1 × PBS.
13. The coverslips are mounted on glass slides with ProLong™ Antifade mounting medium (*see* **Note 27**).

3.1.3 Fluorescence Imaging of Labeled Rodent Primary Hippocampal/ Cortical Cultures

The immunolabeled cells on coverslips mounted on glass slides can be imaged using a wide-field fluorescence microscope or a confocal microscope. Since neuronal cultures at young stages typically grow as a single layer of cells and the processes are thin, a wide field fluorescence inverted microscope is used for imaging, as shown in Fig. 2.

1. The slides are loaded on an inverted microscope equipped with a halogen lamp or laser and a detector (ScMOS/EMCCD) camera for acquisition of images.
2. 3D stacks of images are acquired sequentially in green and red channels with a total volume thickness of 3 μm and a single plane thickness of 200/250 nm.
3. The images are exported as *.tif files.
4. All image processing and analyses are performed using an image processing software such as Metamorph (Molecular Devices) or ImageJ (Fiji).

3.1.4 Image Analysis of Neuronal Development and Polarity In Vitro

Assessing neuronal polarity gets challenging due to the differential labeling of the markers for polarity. Since neuronal polarity is established over a certain developmental duration, cells display differential distribution for the polarity markers such as Tau and Map2. In very young neurons, both markers are distributed in all the neurites and soma, whereas Map2 gets restricted to the somatodendritic compartments and Tau to the axon in later stages (Fig. 2). Therefore, depending on the developmental status of the cells, their fluorescence distribution might vary. A combination of both quantitative and qualitative assessments as described below would be ideal for studying polarity in vitro, utilizing the advantages of both strategies. Image analysis for assessment of neuronal development and polarity in vitro can be performed both qualitatively and quantitatively as follows.

Qualitative Analysis of Polarity

1. The maximum intensity projections of the microscopy images are generated for green and red channels.
2. The overlay of maximum intensity projections of both channels is generated.
3. The number of neurons displaying a single neurite labeled with axon or with multiple axons and total number of neurons assessed are quantified.

4. The ratio of number of neurons displaying single or supernumerary axons to the total number of neurons assessed is evaluated.

Quantitative Analysis of Neuronal Development and Polarity

For analysis of neuronal development profile, multiple parameters need to be considered. The conventional way to study neuronal development is by assessing the following parameters, which give an overview of its developmental profile:

(a) The length of the longest neurite or the length of the potential axon;
(b) The length of dendrites; and
(c) The number of neurites per cell;

To quantitatively assess the polarity status of the neuron, an independent parameter called the *neuronal polarity ratio* has been introduced [10]. Neuronal polarity ratio refers to the ratio of the longest neurite (axon) to the average length of the dendrites. During early stages of development, neurons display a symmetric growth where this ratio tends to be 1. Alternatively, once polarity is established, neurons exhibit an asymmetric growth with the axon much longer compared to the short dendrites, bringing the neuronal polarity ratio to higher values. Thus, the neuronal polarity ratio provides a robust and reliable quantitative assessment of the polarity status of the neuron, independent of the qualitative labeling of the cells with the polarity markers. This has been successfully used for assessing polarity perturbation, where the neurons tend to break the singularity of axon and grow multiple axons, bringing the neuronal polarity ratio close to 1, similar to symmetric growth [10]. The quantitative analysis of neuronal development and polarity in vitro is performed as follows:

1. The maximum intensity projections of the microscopy images are generated for both channels.
2. The overlay of maximum intensity projections of both channels is generated.
3. Using a tracing tool in an image processing software (Metamorph or ImageJ), the primary neurites are traced from the soma to the tip of the neurites (*see* **Note 28**).
4. Secondary and tertiary neurites are traced from the nodes of the primary or secondary neurites, respectively.
5. The length of the neurites is tabulated in an excel sheet continuously for all cells of the sample.
6. The total length of a neurite is calculated as the sum of the primary, secondary, and tertiary neurites of the same (*see* **Note 29**).

7. The neurite with the longest length is considered as the axon and the rest of the neurites as the dendrites.
8. The length of the longest neurite (axon), average length of the dendrites and number of neurites per cell are determined.
9. The neuronal polarity ratio is calculated for all cells as the ratio of the length of the longest neurite to the average length of the dendrites of the respective cell.
10. The data of the different parameters are imported into a data analysis software such as GraphPad Prism and plotted as a histogram for statistical analysis as described below.

3.1.5 Statistical Analysis of Neuronal Development and Polarity In Vitro

Statistics play a very important role in quantitative biology. Multiple technical and biological repeats are crucial to account for the variability between the culturing and experimental conditions and for the reproducibility of results.

1. All the calculated parameters are assessed for their normality of distribution to select the optimal test for assessing the significance of deviation between control and experimental conditions.
2. D'Agostino–Pearson Omnibus test and Shapiro–Wilk tests in statistical analysis software (e.g., GraphPad Prism) are typically used for evaluating the normality of data distribution.
3. The statistical values are represented as mean ± SEM/SD for normally distributed data or as median (IQR 25–75% interval) for non-normally distributed data.
4. For normally distributed data, different tests such as one-way analysis of variance (ANOVA) test followed by multiple comparison tests are used to test significance.
5. For non-normally distributed data, asymmetric tests such as the Kruskal–Wallis test followed by multiple comparison tests or the Mann–Whitney test are opted.
6. Significance is indicated by p values, where $p \leq 0.05$ is considered significant.

3.2 In Vivo Analysis

3.2.1 In Utero Electroporation

For assessing both cell migration and polarity in vivo, expression of volume markers such as GFP is very useful for imaging individual neurons in tissues along with immunolabeling against specific proteins. In this protocol, since cell migration and polarity are assessed for P0 pups, in utero electroporation is performed at embryonic day E 14.5 for generating a transient expression of GFP in newborn brain cells. The embryonic stage is selected to overlap with the stage of neurogenesis and migration from the subventricular zone to the different brain layers. The detailed protocol of embryonic in utero electroporation is described elsewhere [15]. Therefore, a summary of the procedure is provided in this section.

1. The plasmid mix for injection is freshly prepared by mixing the endotoxin-free plasmid prep of the GFP construct (1–2 μg) with Fast Green dye (1:20 dilution).
2. The dam is anesthetized inside a surgical chamber supplied with an airflow of 1 L/min along with 1–5% Isoflurane.
3. The hair around the umbilicus is removed and the area is sterilized.
4. The abdomen of the dam is incised below the umbilicus (*see* **Note 30**).
5. The uterine horn is pulled on one side up to the ovary using forceps.
6. The embryos are monitored for their viability and selected for electroporation (*see* **Note 31**).
7. The selected embryos within the amniotic sac are taken in one hand and DNA injection is done one after the other to save time by fixing each of them between the thumb and index finger.
8. The plasmid mix is injected gently into the lateral ventricles of the embryonic brain using a pulled microcapillary glass tube connected to a mouth aspirator.
9. Once all the embryos are injected with DNA, they are electroporated using the following parameters (*see* **Notes 32** and **33**):

   For ECM830: Voltage 45 V, 4 pulses, Pulse duration 50 ms, Pulse interval 500 ms, Polarity: Unipolar.

   For Nepagene: Voltage 50 V, 5 pulses, Pulse duration 50 ms, Pulse interval 100 ms, Polarity: Unipolar.
10. Electroporation is performed in such a way that the anode is placed on the side of DNA injection and the cathode on the opposite side of the head (*see* **Note 34**).
11. Once all the embryos are electroporated on one side, the uterine horn is carefully placed back in the abdomen.
12. Similar procedure is followed for the embryos on the other side as well (**steps 5–11**).
13. Once the electroporation is complete, the muscle and skin of the dam are stitched using sterile forceps and needle. The dam is allowed to recover after the surgical procedure.
14. After electroporation, the embryos are allowed to develop until birth.
15. The brain sections from P0 pups expressing the GFP construct are prepared as described in the following section and proceeded with immunohistochemistry and confocal imaging.

3.2.2 Preparation of Postnatal Brain Sections from Rodents

1. The P0 rat/mouse pup is kept on ice for 2–3 min to provide a hypothermic shock to anesthetize the animal.
2. Once the animal is numb, the animal is taken under a dissection scope and the thoracic cavity skin is cut open.
3. A needle (30 G) is inserted into the left ventricle of the heart and a small cut is made in the right atrium.
4. Ice-cold 1 × PBS is flushed very slowly through the syringe.
5. Once the body fluid is clear from blood, ice-cold perfusion buffer is passed through the syringe for whole animal fixation (*see* **Note 35**).
6. Once the brain is fixed, the animal is decapitated and the whole brain is extracted.
7. The whole brain is postfixed in 4% PFA overnight.
8. The next day, the brain is washed once in 1 × PBS and transferred to 30% sucrose solution for 1–2 days until the brain sinks to the bottom (*see* **Note 36**).
9. The brain is quickly frozen using isopentane maintained at −80 °C or using an OCT embedding resin and dry ice.
10. The frozen brain is stored at −80 °C until used for sectioning.
11. 40 μm-thick brain sections are made using a cryostat and the sections are processed for immunohistochemistry.

3.2.3 Immunohistochemistry for Floating Sections

1. The brain sections are washed 2 × 15 min in 1 × PBS.
2. Blocking is performed by incubating the tissue sections in blocking buffer for 1–1.5 h on a shaker at room temperature.
3. Then, the slices are incubated with the primary antibody incubation media at 4 °C overnight on a shaker. The sections are labeled with brain layer specific antibodies for migration assay and with antibodies targeting the axonal and dendritic compartments for polarity assays (*see* **Note 37**).
4. The slices are washed 6 × 10 min in 1 × PBS. The slices undergo constant shaking during the washes.
5. Secondary antibody incubation is performed for 1 h at room temperature.
6. The sections are washed 4 × 10 min in 1 × PBS with constant shaking.
7. The floating sections are transferred onto glass slides and mounted with cover glass using ProLong™ Antifade mounting medium supplemented with DAPI.
8. The immunolabeled tissue sections are imaged using a confocal microscope.

3.2.4 3D Confocal Imaging

Due to the considerable thickness of the brain sections, 3D confocal imaging is a preferred choice for studying migration and polarity in vivo.

1. The glass slides with brain sections are loaded onto a confocal microscope equipped with a laser bench and a standard PMT or hybrid detectors (*see* **Note 38**).
2. Multispectral imaging allows a comparative study of up to three to four fluorophores simultaneously. The excitation and detection are adjusted for least bleed-through between the channels.
3. The pinhole is adjusted to 1 Airy unit for the best signal-to-noise ratio and least background.
4. The laser power and detector gain are balanced to have the least photobleaching and the best signal without image saturation.
5. Imaging neuronal migration:
   - (i) GFP expressing neurons are imaged by 488 nm excitation.
   - (ii) Two different cortical layer markers labeled with Alexa fluor 555 and Alexa fluor 647 are imaged by 555 and 647 nm excitation, respectively.
   - (iii) The cells labeled by DAPI are imaged by 350 nm excitation.
6. Imaging neuronal polarity in vivo:
   - (i) GFP expressing neurons are imaged by 488 nm excitation.
   - (ii) Tau/Neurofilament and Map2 labeling are imaged by 647 and 555 nm excitation, respectively.
   - (iii) The cells labeled by DAPI are imaged by 350 nm excitation.
7. 3D sections of approximately 250 nm for a total volume of approximately 40 μm of the section are acquired for the different channels.
8. Maximum intensity projections of the stacks are generated for further analysis.

3.2.5 Analysis

Migration assay:
The brain sections labeled with brain layer specific antibodies are imaged and analyzed for migration assay as described below.

1. The maximum intensity projections of the 3D stacks of different channels (GFP, cortical layers marked by Alexa fluor 555, 647, DAPI) are generated.
2. The localization and distribution of the GFP expressing neurons are compared with reference to the layer specific markers.
3. Neuronal migration is conventionally evaluated from the number of GFP-labeled neurons (electroporated) present in different layers (*see* **Note 39**).

4. Different regions of interest are selected from different layers labeled by the markers.
5. The number of GFP expressing neurons is assessed in different regions of interest of the brain sections with reference to the specific markers to calculate the number of cells per unit area.
6. The neuronal morphological alterations during migration can also be evaluated from the images of the GFP expressing neurons.

Polarity assay:
The brain sections labeled with axonal and dendritic markers are imaged and analyzed for polarity assay as described below.

1. The maximum intensity projections of the 3D stacks of different channels (GFP, axons and dendrites marked by Alexa fluor 647 and 555, respectively, DAPI) are generated.
2. The distribution of the polarity markers Tau and Map2 is analyzed for GFP expressing neurons (*see* **Notes 39** and **40**).
3. From imaging, any alterations in neuronal polarity can be evaluated based on the number of axons generated from individual neurons.
4. Neuronal development can also be evaluated based on the length and branching of the axonal and dendritic compartments as described previously for in vitro analysis.
5. Sholl analysis (also available as an ImageJ/Fiji plugin) is conventionally used for quantitative assessment of dendritic branching where concentric circles of uniformly increasing diameter are drawn around each neuronal soma and the number of dendritic nodes crossing these circles are evaluated [16].

3.2.6 Statistics

Technical and biological repeats conducted in multiple animals is crucial to evaluate the credibility of results. Similar to the in vitro studies, according to the distribution of the evaluated parameters, symmetric or asymmetric tests are opted to evaluate the significance of the results for quantitative assessment to evaluate the migration or polarity defects in the sample with reference to the control.

4 Notes

1. The dissection media is designed to have better media stability with least pH variability during dissection of the brain.
2. Since neuronal cultures are highly sensitive to their microenvironment, Normocin is preferred to conventional Penicillin/Streptomycin combination to avoid fungal, bacterial, and mycoplasma contamination.

3. The pore sizes of needles are selected such that the viability of cells is not compromised if trituration is performed gently but would generate a near pure homogenous population of neurons in culture.
4. Plating/Complete media is highly suitable for neuronal growth and viability. However, this media is highly unstable and sensitive to its microenvironment and shifts the pH balance rapidly, which very strongly affects the state of the neurons. Therefore, it needs to be prepared fresh before use.
5. Similar to the dissection media, Plating/Complete media is supplemented with Normocin, which is preferred to Penicillin/Streptomycin to avoid fungal, bacterial, and mycoplasma contamination.
6. Usage of thickness corrected coverslips improves quality of fluorescence images.
7. If the Poly-L-Lysine-coated coverslips are not used immediately, the well plates can be stored in the incubator and used the next day. Though the parafilmed well plates can be stored at 4 degrees, it is not preferred to use coated coverslips later than 24 h for growing primary neuronal cultures, since the cells may not adhere properly and die.
8. The base of all buffers used for ICC is 1 × PBS calibrated to pH 7.4 and is used for washing steps as well unless otherwise mentioned.
9. The species of the primary antibodies are chosen to avoid cross-reactivity.
10. The species of the secondary antibodies are chosen to match with that of the primary antibodies, avoiding cross-reactivity and spectral bleed-through.
11. The embryonic age is selected to overlap with the stage of neurogenesis in rodents.
12. The age of the animal is chosen such that neuronal migration is largely complete.
13. The whole animal fixation by perfusion is preferred to overnight fixation of the whole brain by submerging the brain in the fixative to significantly reduce background noise during imaging.
14. The host species of the serum is chosen to match the host species of the secondary antibody for better specificity of labeling.
15. The same polarity markers can also be used to label the axonal and somato-dendritic compartments of dissociated neurons in cultures prepared from tissue sections (Subheading 3.1).

16. The secondary antibodies are chosen to accommodate fluorescence imaging of GFP-labeled neurons during multispectral imaging in brain tissues.
17. Extra care needs to be given when using the syringe since the suspension should pass dropwise in order to remove all the neurites from cells to generate round cells and thereby a homogenous population of neurons in culture. Any extra pressure or fast syringing would immediately kill the cells when passing through the small pore.
18. An equal amount of Trypan blue can be added to easily distinguish the viable cells. Trypan blue is taken in only by dead cells whose plasma membrane is ruptured. The live cells would be visualized as transparent/white cells under a bright-field microscope.
19. A plating density of 40–50 cells/mm^2 is ideal for studying polarity establishment and development of well-separated neurons during the early stages of development. However, if the neuronal cultures are used for other studies during the mature stages, it is optimal to keep the densities lower, though very low densities can compromise viability and the cells tend to form ectopic synapses.
20. It is best to prepare the Plating/Complete media fresh for supplementing the neuronal cultures every week for better viability and maintenance of neurons due to the unstable nature of the media.
21. Neurons in vitro establish polarity approximately around DIV 3. Therefore, majority of the neurons display a single axon and multiple short dendrites at this stage [10]. It is ideal to study polarity in vitro around DIV 3–4 since at later stages, the neuronal processes entangle with each other, making it difficult to distinguish the neuronal processes belonging to individual cells.
22. Quenching before antibody labeling significantly reduces the autofluorescence contribution within cells, maximizing the signal-to-noise ratio of fluorescence during imaging.
23. Permeabilization is limited to 5 min to keep the cytoskeleton intact. Cells undergoing longer permeabilization display disrupted cytoskeletal labeling.
24. Blocking step significantly reduces unspecific labeling of antibodies.
25. Approximately 100 μL of antibody mix is calculated to cover an 18 mm coverslip when incubated inverted on a parafilm. The antibody is mixed well in 3% BSA and spotted on the parafilm as a blob on which the coverslip is incubated to avoid drying within a dark humidified chamber. Antibody consumption is significantly reduced in this type of labeling.

26. Primary antibody incubation can also be performed overnight, which requires a slightly larger amount of antibody mix per coverslip, say 150–200 μL, to avoid drying of coverslips.
27. The mounted coverslips are dried at room temperature and stored at 4 °C prior to imaging.
28. Though automated tracing tools such as single neurite tracer might be time saving, although laborsome, it is best to do neurite tracing in vitro manually due to the thin structure of the neurites which provides less fluorescence contrast while imaging. Therefore, significant loss of data while performing automated neurite tracing might occur.
29. A threshold for size filtering can be added to avoid inclusion of filopodia in neurite length calculation.
30. The uterus is kept hydrated with warm 1 × PBS or 0.9% saline during the whole procedure.
31. The healthy embryos are pink with clear amniotic fluid.
32. Injecting all the embryos on one side followed by their electroporation significantly reduces the total duration required for the whole procedure instead of sequentially performing for each embryo, and this also increases the viability of embryos.
33. The electroporation parameters need to be optimized according to the precise experimental conditions by the user. An estimate of the parameters is provided.
34. During electroporation, the electrodes can also be swiped slowly from one lobe to the other to ensure a wider area of DNA spread, increasing the number of cells electroporated.
35. The perfusion volume is calculated as roughly 2–3 times the weight of the animal. Flushing out the blood completely is very important to avoid autofluorescence and background from endothelial cells during imaging. This can be understood from the color of the body fluid flushed out.
36. Sucrose solution being a hypertonic solution acts as a cryoprotectant.
37. Map2 especially labels the thick apical dendrites of neurons, which can be prominently observed during imaging of brain sections, while axonal labeling and imaging is more challenging due to their thin structure.
38. Hybrid detectors provide a better signal-to-noise ratio compared to standard PMT.
39. Assessing the overlay of nuclei labeling via DAPI with neuronal GFP fluorescence provides a relative estimation of individual neuronal localization with neighboring cells, though for statistical analysis, quantification of GFP expressed neurons alone would suffice.

40. Expression of GFP in selected neurons allows for better imaging of individual neurons along with endogenous labeling of axonal and dendritic compartments.

Acknowledgments

The author acknowledges Dr. Soumen Das, Dr. Surya Chandra Rao, and Dr. Narendrakumar Ramanan, Centre for Neuroscience, Indian Institute of Science, Bangalore, for their valuable input in preparing the chapter. This work was supported by DBT Ramanlingaswami Fellowship, DBT Har Gobind Khorana Innovative Young Biotechnologist Award, and DST-SERB Early Career Research Award to M.J. The author acknowledges DBT–IISc Partnership Program for funding the microscopy imaging system used in this work.

References

1. Dotti CG, Banker GA (1987) Experimentally induced alteration in the polarity of developing neurons. Nature 330:254–256
2. Schelski M, Bradke F (2017) Neuronal polarization: from spatiotemporal signaling to cytoskeletal dynamics. Mol Cell Neurosci 84:11–28
3. Toriyama M, Shimada T, Kim KB, Mitsuba M, Nomura E, Katsuta K, Sakumura Y, Roepstorff P, Inagaki N (2006) Shootin1: a protein involved in the organization of an asymmetric signal for neuronal polarization. J Cell Biol 175:147–157
4. Yoshimura T, Arimura N, Kawano Y, Kawabata S, Wang S, Kaibuchi K (2006) Ras regulates neuronal polarity via the PI3-kinase/Akt/GSK-3beta/CRMP-2 pathway. Biochem Biophys Res Commun 340:62–68
5. Cheng PL, Song AH, Wong YH, Wang S, Zhang X, Poo MM (2011) Self-amplifying autocrine actions of BDNF in axon development. Proc Natl Acad Sci USA 108:18430–18435
6. Shelly M, Lim BK, Cancedda L, Heilshorn SC, Gao H, Poo MM (2010) Local and long-range reciprocal regulation of cAMP and cGMP in axon/dendrite formation. Science 327:547–552
7. Leterrier C, Clerc N, Rueda-Boroni F, Montersino A, Dargent B, Castets F (2017) Ankyrin G membrane partners drive the establishment and maintenance of the axon initial segment. Front Cell Neurosci 11:6
8. Leterrier C, Dubey P, Roy S (2017) The nano-architecture of the axonal cytoskeleton. Nat Rev Neurosci 18:713–726
9. Leterrier C, Potier J, Caillol G, Debarnot C, Rueda Boroni F, Dargent B (2015) Nanoscale architecture of the axon initial segment reveals an organized and robust scaffold. Cell Rep 13:2781–2793
10. Jose M, Sivanand A, Channakeshava C (2021) Membrane cholesterol is a critical determinant for hippocampal neuronal polarity. Front Mol Neurosci 14:746211
11. Miyata T, Kawaguchi A, Okano H, Ogawa M (2001) Asymmetric inheritance of radial glial fibers by cortical neurons. Neuron 31:727–741
12. Nadarajah B (2003) Radial glia and somal translocation of radial neurons in the developing cerebral cortex. Glia 43:33–36
13. Namba T, Funahashi Y, Nakamuta S, Xu C, Takano T, Kaibuchi K (2015) Extracellular and intracellular signaling for neuronal polarity. Physiol Rev 95:995–1024
14. Banker G, Goslin K (1988) Developments in neuronal cell culture. Nature 336:185–186
15. Meyer-Dilhet G, Courchet J (2020) In utero cortical electroporation of plasmids in the mouse embryo. STAR Protoc 1:100027
16. Binley KE, Ng WS, Tribble JR, Song B, Morgan JE (2014) Sholl analysis: a quantitative comparison of semi-automated methods. J Neurosci Methods 225:65–70

Chapter 10

Proteomic Identification and Label-Free Quantification of Proteins Implicated in Neurite and Spine Formation

Kanika Sharma, Prashant Kaushal, and Vikas Kumar

Abstract

The molecular mechanisms underlying neurite formation include multiple crosstalk between pathways such as membrane trafficking, intracellular signaling, and actin cytoskeletal rearrangement. To study the proteins involved in such complex pathways, we present a detailed workflow of the sample preparation for mass spectrometry-based proteomics and data analysis. We have also included steps to perform label-free quantification of proteins that will help researchers quantify changes in the expression levels of key regulators of neuronal morphogenesis on a global scale.

Key words Neurite, Dendritic spine, Proteomics, Liquid chromatography, Mass spectrometry, Label-free quantitative analysis, Differentially expressed proteins

1 Introduction

Neuritogenesis is a complex biological phenomenon that involves the extension of neurites from the soma of immature neurons into axons or dendrites to form an intricate circuit of neurons in the central nervous system. It is a crucial aspect of embryonic development and adult neurogenesis and plays a central role in memory formation through synaptic plasticity [1, 2]. Significant progress has been made in identifying genes and proteins involved in the precise spatiotemporal mechanisms of neurite outgrowth and dendritic spine formation [3–5]. Researchers have demonstrated the involvement of multiple intracellular signaling pathways, such as cytoskeleton-associated protein rearrangement and extracellular matrix remodeling, in this complex mechanism [6–11]. To profile the proteins involved in these intricate mechanisms and to check their differential expression, mass spectrometry (MS)-based proteomics could prove as a beneficial tool [12]. In addition, global profiling of posttranslational modifications (PTMs), such as phosphorylation, can help identify novel signaling molecules during

Kazuhito Toyooka (ed.), *Neuronal Morphogenesis: Methods and Protocols*, Methods in Molecular Biology, vol. 2831,
https://doi.org/10.1007/978-1-0716-3969-6_10,

neuronal development and network formation [13–15]. Indeed, the depth of proteome coverage has significantly increased in the last decade owing to technological advancements in high-resolution Orbitrap and high-speed time-of-flight (TOF) mass spectrometers [16]. With these technical improvements, the MS range has now been extended to the single-cell level, making this technology best suited for quantifying proteins at the subcellular level [17], which could help advance our understanding of the underlying molecular mechanisms in great detail.

The most common approach for global protein profiling is bottom-up (or shotgun) proteomics, which involves proteolytic digestion of a protein into multiple small peptides using a proteolytic enzyme [18]. Trypsin is one such protease that cleaves at the C-terminus of lysine and arginine residues. After proteolytic cleavage, peptides are further fractionated using reverse-phase liquid chromatography (LC) according to their hydrophobicity and injected into a mass spectrometer in the form of charged molecular ions using either of two soft ionization modes: electrospray ionization (ESI) or matrix-assisted laser desorption/ionization (MALDI) [19]. These ions are subsequently fragmented and detected using an Orbitrap or TOF mass detector (Fig. 1). To identify the proteins present in the sample, the fragmentation (MS/MS) spectra of peptides were matched with similar ions generated upon in silico digestion of proteins available in a database using peptide search algorithms, such as Sequest (Thermo Scientific), Mascot (Matrix Science), and Andromeda [20]. For protein quantification, both label-free and label-based approaches can be used. Isobaric tags, such as tandem mass tags (TMT; Thermo Scientific) and isobaric tags for relative and absolute quantitation (iTRAQ; Applied Biosystems), have been used extensively to label free amino groups present at the peptide N-terminus and lysine side chain [21–25]. Such isobaric tags generate a unique reporter ion tag upon fragmentation, and their abundance measurement leads to relative peptide and protein quantification [26]. In contrast, label-free quantification (LFQ) relies on either peptide precursor ion peak intensity or spectral count measurement [27, 28], and relative change in expression is calculated based on the protein quantification values. Finally, statistical and other bioinformatic analyses, such as gene ontology and pathway analyses of differentially expressed proteins, are performed to determine their biological significance. This chapter provides a detailed step-by-step procedure for sample preparation for mass spectrometry-based proteomics and data analysis using vendor-specific and open-source proteomic software packages. Detailed steps to conduct label-free quantitative proteomic analysis are also included to identify and functionally annotate the differentially expressed proteins.

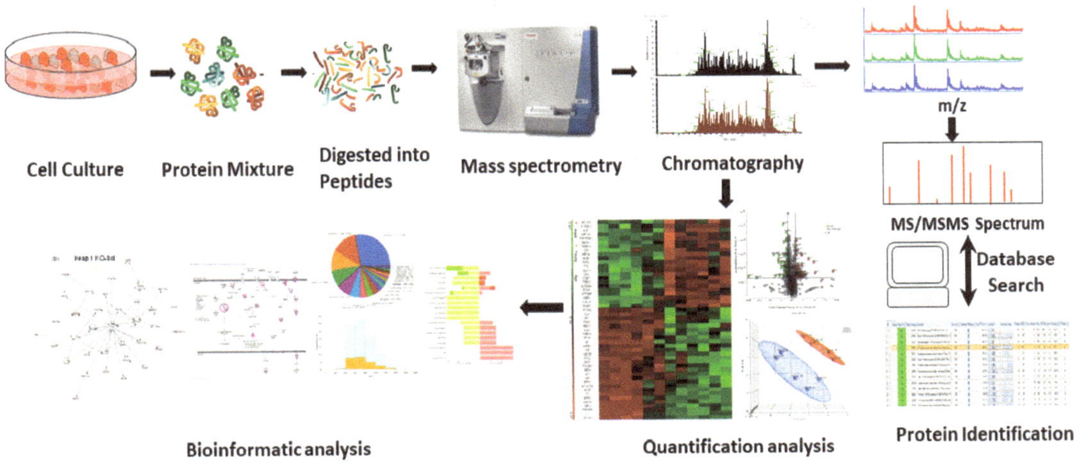

Fig. 1 A mass spectrometry-based proteomic workflow for the sample preparation and data analysis

2 Materials

Unless otherwise mentioned, all chemicals used in this protocol are LC–MS grade or the highest available quality.

2.1 Cell Collection and Protein Extraction

1. RIPA lysis buffer, pH 7.4, 50 mM Tris–HCl, 150 mM NaCl, 1.0% (v/v) NP-40, 0.5% (w/v) Sodium Deoxycholate, 1.0 mM EDTA, 0.1% (w/v) SDS, and 0.01% (w/v) sodium azide at a pH of 7.4.
2. Cell scraper.
3. Phosphate buffer saline (PBS, 1×), pH 7.2.
4. Trypsin–EDTA solution (0.25%).
5. Protease inhibitor cocktail.
6. Phosphatase inhibitor cocktail (optional).
7. Tissue homogenizer with 1/8 inch probe for low sample volumes.
8. Protein concentration estimation kit (Bradford or BCA assay kit).
9. Microcentrifuge with rotor speed up to 20,000× *g*.
10. Dulbecco's modified eagle medium (DMEM) supplemented with FBS (10%).

2.2 Reduction and Alkylation

1. DTT (Dithiothreitol) or Tris(2-carboxyethyl)phosphine hydrochloride (TCEP-HCl).
2. Iodoacetamide.

2.3 Protein Precipitation

1. Acetone.
2. Methanol.
3. Chloroform.
4. Water.
5. Vacuum concentrator.

2.4 Trypsin Digestion of Proteins

1. Ammonium bicarbonate or similar buffer such as TEAB and HEPES (100 mM, pH 8).
2. Trypsin protease.

2.5 Peptide Desalting

1. Peptide desalting spin columns.
2. Trifluoroacetic acid (TFA).
3. Water with 0.1% (v/v) Trifluoroacetic acid.
4. Acetonitrile (ACN).
5. Low protein binding microcentrifuge tubes.

2.6 Liquid Chromatography and Tandem Mass Spectrometry

1. Peptide retention time calibration mixture.
2. HeLa protein digest standard.
3. High-performance liquid chromatography system.
4. High-resolution mass spectrometer.
5. LC Buffer A: 2% ACN, 0.1% formic acid.

2.7 Data Analysis

1. Database search software (Proteome Discoverer or MaxQuant).
2. Perseus software package for graphical visualization and statistical data analysis.
3. Cytoscape for visualizing protein–protein interactions.

3 Methods

3.1 Cell Collection and Protein Extraction

1. Grow HeLa cells in DMEM media up to 70–80% confluency for adhered cell cultures in a 15 mm-diameter Petri dish. Wash the plate surface twice gently with 10 mL ice-cold PBS (1×) to remove excess media.
2. Add 1 mL of Trypsin–EDTA (0.25%) and spread it along the whole plate by repeated aspiration of the cell suspension. Manual scrapping can be done alternatively with the help of a cell scraper.
3. Immediately transfer the cell suspension to a 1.5 mL microcentrifuge tube and centrifuge at 400× *g* for 5 min at 4 °C.

4. To eliminate excess Trypsin–EDTA, add 500 μL of ice-cold PBS (1×) and resuspend cells by slow aspiration and release cycles using a pipette. Centrifuge at 400× *g* for 5 min and discard the supernatant. Repeat this step.
5. Add 200 μL of RIPA lysis buffer containing protease inhibitor cocktail (1×) and mix well with repeated pipetting or vertexing (*see* **Notes 1** and **4**).
6. Place the sample tube in an ice bucket and probe sonicate the cell suspension. Set cycle time to 3 s ON and 2 s OFF at 40% amplitude for five cycles (*see* **Note 5**). Repeat once again if the sample is not thoroughly homogenized.
7. Boil at 95 °C for 5 min and allow it to cool at RT (*see* **Note 6**).
8. Spin at 20,000× *g* for 10 min at 4 °C and transfer the supernatant to a fresh 1.5 mL microcentrifuge tube.
9. Per manufacturer's instructions, estimate protein concentration with BCA or Bradford protein assay kit.
10. Adjust protein concentration to 1 mg/mL with lysis buffer and store 100 μg aliquots in the −80 °C freezer until further use (*see* **Note 7**).

3.2 Reduction and Alkylation

1. Prepare 0.2 M stock of DTT by dissolving 15.4 mg in 500 μL of 0.1 M ammonium bicarbonate buffer, pH 8.
2. Add 5.25 μL of 0.2 M DTT stock (final concentration 10 mM) to the 100 μL sample and incubate at 55 °C for 60 min.
3. Let the sample tube cool down at RT and spin briefly to settle the water vapors on the lid.
4. Freshly prepare 375 mM stock of iodoacetamide by dissolving 9 mg in 132 μL of 0.1 M ammonium bicarbonate buffer, pH 8. Avoid prolonged exposure of iodoacetamide stock solution to light.
5. Add 6 μL of iodoacetamide stock (final concentration 20 mM) and incubate in the dark for 30 min (*see* **Notes 8** and **9**).

3.3 Protein Precipitation

1. Add 6× volume (660 μL) of prechilled acetone to the sample and keep overnight at −20 °C (*see* **Notes 10** and **11**).
2. Centrifuge at 10,000× *g* for 5 min and discard the supernatant containing salts and detergents.
3. Wash the protein pellet with 500 μL of prechilled acetone, vortex, and spin tube briefly. Discard the supernatant. Repeat this step one more time.
4. Air or vacuum dry the protein pellet briefly (*see* **Note 12**).

3.4 Trypsin Digestion of Proteins

1. Redissolve protein pellet in 100 μL of 0.1 M ammonium bicarbonate buffer, pH 8. The use of probe sonication may improve pellet re-solubility.
2. Prepare aliquots of trypsin protease in 0.1% (v/v) formic acid in water and store them at −80 °C. Avoid multiple freeze–thaw cycles.
3. Add 4 μg trypsin (1:25 protease-to-proteins ratio) to the sample and incubate for 16 h at 37 °C on a vortexer (*see* **Note 13**).

3.5 Peptide Desalting

1. Vacuum dry the tryptic digest and resuspend in 0.1% (v/v) TFA in water. Vortex well to ensure complete solubilization of peptides (*see* **Note 14**).
2. Place a new peptide desalting spin desalting column into a 2 mL microcentrifuge tube and centrifuge per manufacturer's instructions to remove the storage buffer and set the C18 resin (*see* **Note 15**). Discard the flowthrough.
3. Activate the resin twice by adding 50% ACN and discard the flow through.
4. Equilibrate the resin twice by adding 0.1% (v/v) TFA in water and discard the flow through.
5. Transfer the equilibrated column to a fresh 2 mL microcentrifuge tube and add the sample (*see* **Note 16**).
6. Wash resin twice with 0.1% (v/v) TFA in water and discard the flow through.
7. Place the column in a fresh 2 mL protein LoBind tube and elute peptides with 50% ACN in water, 0.1% formic acid. Repeat this step once for maximum peptide recovery.
8. Dry the eluted peptides in a vacuum concentrator and store at −80 °C until further use.

3.6 Liquid Chromatography and Tandem Mass Spectrometry

1. Redissolve the peptides in 20 μL of 0.1% (v/v) formic acid in water and estimate the peptide concentration per manufacturer's instructions.
2. Transfer the sample to an autosampler vial and inject 1 μg peptide for high-resolution LC-MS/MS analysis.
3. Load peptides in LC Buffer A (2% ACN, 0.1% formic acid) onto a pre-column (or trap column) and wash for 10 min at 4 μL/min flow rate.
4. Switch the position of the port-valve and elute the peptides using a linear gradient of 9–30% of Buffer B (80% ACN, 0.1% formic acid) over the next 80 min at a flow rate of 300 nL/min. Ramp up the Buffer B to 99% in the next 5 min, hold it for another 5 min, and then bring the gradient to the original composition for column reequilibration. The total method time is 120 min.

5. Set the capillary temperature at 275 °C and spray voltage at 1900 V.
6. Acquire raw data in data-dependent mode on a high-resolution MS system and collect the precursor ion (MS1) spectra in the 375–1500 m/z range with a resolution of 60,000 at m/z 200.
7. Select the top 10 most intense precursor ions for further fragmentation (MS2) with higher-energy C-trap dissociation (HCD) or collision-induced dissociation (CID) fragmentation using a dynamic exclusion setting of 30 s.
8. For quality check, inject 500 fmol of Pierce™ peptide retention time calibration mixture to check the elution profile of peptides.
9. Also run Pierce™ HeLa protein digest standard to monitor the MS performance.

3.7 Data Analysis

1. Submit the MS generated raw data files to database search with software packages such as Proteome Discoverer (Thermo Scientific) or MaxQuant.
2. In the Proteome Discoverer, create a new study and select in-built processing and consensus workflows for raw data processing.
3. Select the appropriate protein database and common contaminants (like the common Repository of Adventitious Proteins, cRAP) and keep precursor and fragment ion mass tolerance at 10 ppm and 0.02 Da (0.5 Da for the CID data), respectively.
4. Set Carbamidomethylation of Cysteines (C) (+57.021464 Da) as static modification, and oxidation of Methionine (+15.994915 Da), deamidation of Asparagine (N) and Glutamine (Q) (+0.984016 Da), and acetylation of protein N-terminus (+42.010565 Da) as variable modifications during database search.
5. Select trypsin (full) as enzyme specificity, with a maximum of two missed cleavages allowed.
6. Apply a 1% false discovery rate at both peptide spectrum match (PSM) and peptide levels to obtain high-confidence protein identifications (*see* **Note 17**).
7. Use consensus workflow with default parameters.
8. For label-free quantitation, choose the Proteome Discoverer in-built Minora feature detector-based precursor quantification and LFQ workflow. To enable this feature in MaxQuant [29], select the LFQ option under the "Group-specific parameters" tab while conducting a database search to obtain raw and normalized precursor ion intensities.
9. For MaxQuant, use the "proteins.txt" file and import the LFQ results in the Perseus software [30].

10. Filter protein groups matched as "Only identification by site," "Reverse," or "Potential contaminants."
11. Log transform data and plot a histogram to see if the data follows a normal distribution (*see* **Note 17**).
12. Perform categorical annotation of rows and filter rows based on valid values.
13. If required, perform the imputation of missing values with random numbers drawn from the normal distribution of data, representing low abundance measurements using Perseus default parameters. Again, check the distribution of imputed values with histogram.
14. Check correlation between the biological groups and within their replicates by plotting a multi-scatter plot displaying the Pearson correlation coefficient value.
15. Perform principal component analysis (PCA) on the normalized intensities to check if different biological groups are well separated on first two or three principal components of the data.
16. To check the similarity between samples and see if there is any batch-effect in data, convert the normalized intensity values of proteins on rows into a Z-score, and perform hierarchical clustering on columns.
17. Perform student's t-test or ANOVA to check the significance of expression fold change between biological groups. Estimate false discovery rate (FDR) based on either permutation-based or Benjamini–Hochberg correction for multiple hypothesis testing method and set FDR <0.05 as cutoff for significance.
18. Plot a volcano plot with default parameters to see the two-sample t-test results. If required, adjust the S0 and FDR parameters to see statistically significant differentially expressed proteins (DEP).
19. Perform hierarchical clustering of DEPs (like **step 16**, on both rows and columns) to visualize clusters of proteins and their expression patterns between biological groups.
20. Annotate proteins with Gene Ontology (GO) terms and KEGG database-based protein pathway annotations using species-specific annotations built in Perseus.
21. Perform Fisher's exact test to check significantly enriched GO terms and pathways (*q*-value <0.05).
22. The protein–protein interaction networks can be visualized with the STRING database (https://string-db.org/). Interactions with a confidence score greater than 0.7 can be exported and analyzed further in Cytoscape [31] software.

4 Notes

1. 1 μL of 1X Nuclease can also be added to the lysis buffer to reduce the sample viscosity.
2. Centrifugation at high speed may result in cells bursting and loss of proteins.
3. Add phosphatase inhibitor cocktail (1×) for phosphoproteomic analysis.
4. Ensure the tip is always dipped in the sample to avoid sample foaming. This may lead to protein degradation.
5. Do not boil or increase the temperature above 37 °C for proteins present in urea containing buffer. This may lead to a significant increase in lysine and protein N-terminus carbamylation.
6. Do not freeze–thaw proteins multiple times, as this leads to protein degradation.
7. Quenching extra iodoacetamide, after the alkylation step, with 10 mM DTT is preferred if the alkylation step is not followed by protein precipitation.
8. Incubating samples in higher concentrations of iodoacetamide may lead to overalkylation at the N-terminus of lysine and histidine residues [32].
9. Alternatively, Chloroform–Methanol extraction of proteins can be performed by adding MeOH, Chloroform, and water to the sample in a 4:3:1 ratio. Vortex after each step of addition, and centrifuge at 12,000 × *g* for 5 min at 4 °C. Carefully discard the top and bottom layers without disturbing the sandwich protein layer. Wash with prechilled methanol, vortex, and centrifuge at 12,000 × *g* for 5 min. Repeat this wash step.
10. Trichloroacetic acid (TCA) precipitation can also be used for higher-volume samples. Add TCA to a final concentration of 20% and incubate the sample at −20 °C for 2 h. Centrifuge at 12,000 × *g* for 5 min at 4 °C and discard the supernatant. Wash protein pellet with prechilled acetone, vortex, and centrifuge at 12,000 × *g* for 5 min. Repeat this wash step.
11. Excessive drying under vacuum may result in incomplete resuspension of the pellet in the buffer. However, overnight trypsin digestion on a vortexer will result in the complete resuspension of proteins (peptides).
12. Adding $CaCl_2$ in 1–10 mM final concentration may improve trypsin digestion efficiency [33].
13. The pH should be less than 3 at this stage. Add more TFA stock to reduce the sample pH.

14. For low peptide amounts sample, use C18 desalting tips of appropriate binding capacity to perform desalting.
15. Repeating this step once more with the collected flow through may increase the peptide recovery.
16. Add an additional IMP-ptmRS node to localize the probability of PTM sites, such as phosphorylation on serine, threonine, tyrosine, and histidine residues.
17. Other normalization methods, such as mean, median, or quantile normalization, available in Perseus software can also be explored for the best-fit data normalization method.

References

1. Naeve GS, Ramakrishnan M, Kramer R et al (1997) Neuritin: a gene induced by neural activity and neurotrophins that promotes neuritogenesis. Proc Natl Acad Sci 94: 2648–2653. https://doi.org/10.1073/pnas.94.6.2648
2. Jacovina AT, Zhong F, Khazanova E et al (2001) Neuritogenesis and the nerve growth factor-induced differentiation of PC-12 cells requires annexin II-mediated plasmin generation*. J Biol Chem 276:49350–49358. https://doi.org/10.1074/jbc.M106289200
3. Park JH, Lee SB et al (2012) Nuclear Akt promotes neurite outgrowth in the early stage of neuritogenesis. BMB Rep 45:521–525
4. Ujike H, Takaki M, Kodama M, Kuroda S (2002) Gene expression related to synaptogenesis, neuritogenesis, and MAP kinase in behavioral sensitization to psychostimulants. Ann N Y Acad Sci 965:55–67. https://doi.org/10.1111/j.1749-6632.2002.tb04151.x
5. Arimura N, Kaibuchi K (2005) Key regulators in neuronal polarity. Neuron 48:881–884. https://doi.org/10.1016/j.neuron.2005.11.007
6. Powers RM, Daza R, Koehler AE et al (2022) Growth cone macropinocytosis of neurotrophin receptor and neuritogenesis are regulated by neuron navigator 1. Mol Biol Cell 33:ar64. https://doi.org/10.1091/mbc.E21-12-0623
7. Da Silva JS, Dotti CG (2002) Breaking the neuronal sphere: regulation of the actin cytoskeleton in neuritogenesis. Nat Rev Neurosci 3:694–704. https://doi.org/10.1038/nrn918
8. Rout UK (2013) Roles of integrins and intracellular molecules in the migration and neuritogenesis of fetal cortical neurons: MEK regulates only the neuritogenesis. Neurosci J 2013:e859257. https://doi.org/10.1155/2013/859257
9. Tewari BP, Chaunsali L, Prim CE, Sontheimer H (2022) A glial perspective on the extracellular matrix and perineuronal net remodeling in the central nervous system. Front Cell Neurosci 16:1022754
10. Wu G, Lu ZH, Ledeen RW (1995) Induced and spontaneous neuritogenesis are associated with enhanced expression of ganglioside GM1 in the nuclear membrane. J Neurosci Off J Soc Neurosci 15:3739–3746. https://doi.org/10.1523/jneurosci.15-05-03739.1995
11. Dong X, Liu OW, Howell AS, Shen K (2013) An extracellular adhesion molecule complex patterns dendritic branching and morphogenesis. Cell 155:296–307. https://doi.org/10.1016/j.cell.2013.08.059
12. Frese CK, Mikhaylova M, Stucchi R et al (2017) Quantitative map of proteome dynamics during neuronal differentiation. Cell Rep 18:1527–1542. https://doi.org/10.1016/j.celrep.2017.01.025
13. Igarashi M, Kawasaki A, Ishikawa Y et al (2020) Phosphoproteomic and bioinformatic methods for analyzing signaling in vertebrate axon growth and regeneration. J Neurosci Methods 339:108723. https://doi.org/10.1016/j.jneumeth.2020.108723
14. Triplett JC, Zhang Z, Sultana R et al (2015) Quantitative expression proteomics and phosphoproteomics profile of brain from PINK1 knockout mice: insights into mechanisms of familial Parkinson's disease. J Neurochem 133:750–765. https://doi.org/10.1111/jnc.13039
15. Yu P, Pisitkun T, Wang G et al (2013) Global analysis of neuronal phosphoproteome regulation by chondroitin sulfate proteoglycans. PLoS One 8:e59285. https://doi.org/10.1371/journal.pone.0059285
16. Hosp F, Mann M (2017) A primer on concepts and applications of proteomics in neuroscience.

Neuron 96:558–571. https://doi.org/10.1016/j.neuron.2017.09.025
17. Nelson RS, Dammer EB, Santiago JV et al (2022) Brain cell type-specific nuclear proteomics is imperative to resolve neurodegenerative disease mechanisms. Front Neurosci 16:902146
18. Dupree EJ, Jayathirtha M, Yorkey H et al (2020) A critical review of bottom-up proteomics: the good, the bad, and the future of this field. Proteomes 8:14. https://doi.org/10.3390/proteomes8030014
19. Nadler WM, Waidelich D, Kerner A et al (2017) MALDI versus ESI: the impact of the ion source on peptide identification. J Proteome Res 16:1207–1215. https://doi.org/10.1021/acs.jproteome.6b00805
20. Cox J, Neuhauser N, Michalski A et al (2011) Andromeda: a peptide search engine integrated into the MaxQuant environment. J Proteome Res 10:1794–1805. https://doi.org/10.1021/pr101065j
21. Lindhout FW, Kooistra R, Portegies S et al (2020) Quantitative mapping of transcriptome and proteome dynamics during polarization of human iPSC-derived neurons. elife 9:e58124. https://doi.org/10.7554/eLife.58124
22. Dumrongprechachan V, Salisbury RB, Soto G et al (2021) Cell-type and subcellular compartment-specific APEX2 proximity labeling reveals activity-dependent nuclear proteome dynamics in the striatum. Nat Commun 12:4855. https://doi.org/10.1038/s41467-021-25144-y
23. Shuster SA, Li J, Ur C et al (2022) In situ cell-type-specific cell-surface proteomic profiling in mice. Neuron 110:3882–3896.e9. https://doi.org/10.1016/j.neuron.2022.09.025
24. Varderidou-Minasian S, Verheijen BM, Schätzle P et al (2020) Deciphering the proteome dynamics during development of neurons derived from induced pluripotent stem cells. J Proteome Res 19:2391–2403. https://doi.org/10.1021/acs.jproteome.0c00070
25. Gonzalez-Lozano MA, Klemmer P, Gebuis T et al (2016) Dynamics of the mouse brain cortical synaptic proteome during postnatal brain development. Sci Rep 6:35456. https://doi.org/10.1038/srep35456
26. Zecha J, Satpathy S, Kanashova T et al (2019) TMT labeling for the masses: a robust and cost-efficient, in-solution labeling approach * [S]. Mol Cell Proteomics 18:1468–1478. https://doi.org/10.1074/mcp.TIR119.001385
27. Ramus C, Hovasse A, Marcellin M et al (2016) Benchmarking quantitative label-free LC–MS data processing workflows using a complex spiked proteomic standard dataset. J Proteome 132:51–62. https://doi.org/10.1016/j.jprot.2015.11.011
28. Dowell JA, Wright LJ, Armstrong EA, Denu JM (2021) Benchmarking quantitative performance in label-free proteomics. ACS Omega 6:2494–2504. https://doi.org/10.1021/acsomega.0c04030
29. Tyanova S, Temu T, Cox J (2016) The MaxQuant computational platform for mass spectrometry-based shotgun proteomics. Nat Protoc 11:2301–2319. https://doi.org/10.1038/nprot.2016.136
30. Tyanova S, Temu T, Sinitcyn P et al (2016) The Perseus computational platform for comprehensive analysis of (prote)omics data. Nat Methods 13:731–740. https://doi.org/10.1038/nmeth.3901
31. Shannon P, Markiel A, Ozier O et al (2003) Cytoscape: a software environment for integrated models of biomolecular interaction networks. Genome Res 13:2498–2504. https://doi.org/10.1101/gr.1239303
32. Boja ES, Fales HM (2001) Overalkylation of a protein digest with iodoacetamide. Anal Chem 73:3576–3582. https://doi.org/10.1021/ac0103423
33. Nickerson JL, Doucette AA (2022) Maximizing cumulative trypsin activity with calcium at elevated temperature for enhanced bottom-up proteome analysis. Biology (Basel) 11:1444. https://doi.org/10.3390/biology11101444

Chapter 11

3D Reconstruction of the Mitochondrial Network within the Neuronal Soma from SBF-SEM Volume Data

Jane Tweedy, Ross Laws, George Merces, Tracey Davey, Amy K. Reeve, and Amy E. Vincent

Abstract

Neurons contain three compartments, the soma, long axon, and dendrites, which have distinct energetic and biochemical requirements. Mitochondria feature in all compartments and regulate neuronal activity and survival, including energy generation and calcium buffering alongside other roles including proapoptotic signaling and steroid synthesis. Their dynamicity allows them to undergo constant fusion and fission events in response to the changing energy and biochemical requirements. These events, termed mitochondrial dynamics, impact their morphology and a variety of three-dimensional (3D) morphologies exist within the neuronal mitochondrial network. Distortions in the morphological profile alongside mitochondrial dysfunction may begin in the neuronal soma in ageing and common neurodegenerative disorders. However, 3D morphology cannot be comprehensively examined in flat, two-dimensional (2D) images. This highlights a need to segment mitochondria within volume data to provide a representative snapshot of the processes underpinning mitochondrial dynamics and mitophagy within healthy and diseased neurons. The advent of automated high-resolution volumetric imaging methods such as Serial Block Face Scanning Electron Microscopy (SBF-SEM) as well as the range of image software packages allow this to be performed.

We describe and evaluate a method for randomly sampling mitochondria and manually segmenting their whole morphologies within randomly generated regions of interest of the neuronal soma from SBF-SEM image stacks. These 3D reconstructions can then be used to generate quantitative data about mitochondrial and cellular morphologies. We further describe the use of a macro that automatically dissects the soma and localizes 3D mitochondria into the subregions created.

Key words Mitochondria, Morphology, 3D reconstruction, SBF-SEM, Neurodegeneration, mtDNA, Ageing, Three-dimensional, Two-dimensional

1 Introduction

Neurons are highly polarized cells where their processes, known as the axon and dendrites (i.e., neurites), extend from the soma and transmit information rapidly throughout the central nervous system. Mitochondria play an important role within neurons including

Kazuhito Toyooka (ed.), *Neuronal Morphogenesis: Methods and Protocols*, Methods in Molecular Biology, vol. 2831, https://doi.org/10.1007/978-1-0716-3969-6_11,

their major bioenergetic role of synthesizing adenosine triphosphate (ATP). Since mitochondria are predominantly synthesized and destroyed within the soma, but also feature at the Nodes of Ranvier and synapses, between 10% and 20% mitochondria require bidirectional transportation to maintain mitochondrial homeostasis and must separate from the connected network within the soma [1, 2]. Continuous shape (i.e., morphology) and size transformations occur not only to facilitate their distribution but also to perform their different roles within the neuron. Therefore, neuronal mitochondria are extremely varied in shape, size, and connectivity, existing within highly branched networks in the soma to isolated and less complex shapes, e.g., elongated rods in the neurites [3, 4].

Mitochondrial morphology is governed by continuous cycles of fission and fusion events within the soma. Fusion is the process where mitochondria merge to exchange their internal and membranal components, forming elongated morphologies associated with a higher bioenergetic capacity [5, 6]. In contrast, fission is where mitochondria split into smaller bodies, facilitating the clearance of dysfunctional mitochondria by mitophagy [7, 8]. Distortions in the morphological profile caused by an imbalance in fission and fusion may occur in ageing, but may also contribute to the common neurodegenerative disorders such as Alzheimer's disease and Parkinson's disease [9, 10]. Fragmented networks consisting of either punctate or swollen phenotypes have been shown in models that use neurotoxins or contain mutations linked to Parkinson's disease [11–13]. Literature suggests that mitochondria situated close to the nucleus, i.e., perinuclear mitochondria, are actively replicating and are therefore more likely to acquire damaged mitochondrial DNA (mtDNA) [14–16]. Indeed, abnormal mitochondrial morphologies were associated with the mtDNA extracted from both Alzheimer's disease and Parkinson's disease donors in the perinuclear region of cybrid cell lines [17]. However, no statistical quantitative assessment was made on mitochondrial morphology between the perinuclear and distal regions of the neuronal soma.

Primary literature has extensively quantified neuronal mitochondrial morphology in models of the common neurodegenerative disorders and their controls using either low- or high-resolution 2D images, particularly within whole neurons or their neurites [18–22]. However, mitochondrial morphology has rarely been examined in high-resolution 3D images of the neuronal soma [3]. The use of 2D images and shape descriptors, e.g., aspect ratio, will not provide an accurate description of the complex mitochondrial morphologies that exist in 3D. Their measurement depends on the relative orientation of the mitochondrion to the 2D slice, and nanoscale structures interconnecting at various points in 3D remain unidentifiable in a 2D image. Indeed, a novel phenotype, termed "mitochondria-on-a-string," or "MOAS" was only revealed in 3D images of brain tissue from Alzheimer's disease donors [10]. Therefore, segmentation of the mitochondrial network within 3D volumetric datasets, ideally at nanoscale resolution, is required to make accurate conclusions.

There are several volumetric electron microscopy methods for producing high-resolution 3D datasets comprised of voxels (3D pixels). These include serial sectioning Transmission Electron Microscopy (ssTEM), Focused Ion Bean Scanning Electron Microscopy (FIB-SEM), and Serial Block Face Scanning Electron Microscopy (SBF-SEM) [23–26]. In ssTEM, hundreds of ultrathin tissue sections (50–100 nm) are manually cut, individually imaged via TEM with a spatial resolution <1 μm and then aligned. Manual processing is time-consuming and requires a high level of expertise in both ultra-microtomy and microscopy. Further, staining artifacts may be introduced, and sections may become damaged or lost. These problems reduce the accuracy of the 3D measurement for each mitochondrion spanning those sections.

The advent of SEM-based volumetric methods, FIB-SEM and SBF-SEM, avoid these drawbacks as they automatically generate digitalized 3D datasets with a near-TEM resolution in a reliable and repeatable way. In both techniques, the tissue block surface is illuminated with an electron beam and an image is generated through the detection of back-scattered electrons. Each resin embedded tissue block is progressively destroyed as an ultrathin surface layer is removed in each cycle to reveal a new face for imaging. Therefore, one disadvantage when compared to ssTEM is that the same tissue will no longer be available for further study. For FIB-SEM, the focused gallium ion beam consistently erodes fine surface layers, yielding high-resolution images that contain isotropic voxel dimensions (~3–20 nm^3). In contrast, for SBF-SEM the voxel dimension is anisotropic since the image resolution in the direction perpendicular to the block-face (*Z*-axis) is not equidimensional to the spatial resolution in the *X*–*Y* image face (~ 3–30 nm^2). The minimum slice thickness cut by the in-chamber ultramicrotome in SBF-SEM is ~25 nm, termed the Z-step. However, a Z-step selected in the range between 50 and 100 nm is preferred to minimize data storage and reduce the associated technical difficulties with finer slices (25–50 nm). Beam damage can soften the resin, resulting in irregular cuts of the block-face with the generation of images that lack uniformity in thickness or contain wave-like cutting artifacts [27]. While the relatively large Z-dimension means the resolution is inferior for the discrimination of features in the third dimension when compared to FIB-SEM, the field of view and volume imaged in SBF-SEM are both greater than in FIB-SEM. Furthermore, SBF-SEM images have the resolution to identify mitochondrial structures in the nanometer scale within the soma where mitochondrial complexity and density are both high [3]. This makes it feasible to quantify mitochondrial morphology in whole compartments of large neurons, detectable in larger volumes of brain samples up to 1 mm^3. Due to the volume required, impregnation with heavy metals en bloc is required prior to resin embedding to provide adequate signal generation and contrast of the lipid-rich membranes throughout. The resin embedded samples are glued onto a metal pin, trimmed to approximately 0.5 mm^2 and coated with gold. The block is placed on the ultramicrotome, whose height is automatically raised in the Z-step following each image cycle. Since the cutting speed in SBF-SEM is faster than ion beam milling, hundreds of images can be generated in a few hours versus days for FIB-SEM. The discussed advantages of SBF-SEM combined with the fact that it is the least labor-intensive technique supported its use in our project.

Several imaging software packages allow the identification of the mitochondrial objects and their segmentation within the 3D stacks. These include commercialized software such as Amira and Imaris, as well as free and open-source software such as FIJI, ImageJ, IMOD, Ilastik, and Microscopy Image Browser (MIB) [28–33]. While MIB is not the most user-friendly program when compared to FIJI, ImageJ, and Ilastik, it combines both manual and automated workflows, e.g., preprocessing of large datasets and segmentation using a range of tools and methods that can be applied to anisotropic datasets. Since MIB has limited visualization options, the range of different file saving formats allows easy exportations of data to other software, IMOD or Amira, for visualization and quantification. Amira was chosen over IMOD as the 3D visualizations are more refined and there are a wider range of tools in Amira to optimize 3D surfaces, as well as additional scripting options and extensibility for more tailored and complex operations. Since FIJI offers a large variety of analysis options, it was used to run a macro to partition the soma into different regions for the automated localization of mitochondria.

This chapter provides an instruction for performing image analysis in SBF-SEM-generated images of the neuronal soma and their mitochondria. This includes image preprocessing in FIJI and MIB (*see* Subheadings 3.2, 3.3, 3.4, and 3.5) to segmentation of the neuronal soma, and random sampling of the mitochondria followed by their segmentation in MIB (*see* Subheading 3.6). Reconstruction of the 3D surfaces and data extraction are performed in Amira (*see* Subheading 3.7). Finally, the neuronal soma is divided into subregions followed by mitochondrial classification into each region using a customizable macro in FIJI (*see* Subheading 3.8). The image analysis methods presented will be adaptable to any neuron imaged by the other volume EM techniques discussed.

2 Materials

Analysis can be conducted on a Windows 64-bit operating system with Intel® Core™ i7-10750H CPU, 2.60 GHz, 16 GB RAM or a device of similar specification. Segmentation requires the use of a mouse.

1. FIJI (Install from: https://fiji.sc/) and ensure the MorpholibJ package is installed: Within FIJI (Fig. 1), go to [**"Help"** > **"Update"** > **"Manage update sites"** and select **"IJPB-Plugins"**] and close the dialog. Select **"Apply changes"** and close FIJI. MorpholibJ is a mathematical package that performs erosion and dilation tasks required in Subheading 3.8.
2. Microscopy Image Browser (MIB) (Install from: http://mib.helsinki.fi/). MIB can be installed as the stand-alone version if you do not require MATLAB for the generation of coordinates to randomly sample mitochondria (*see* Subheading 3.6.3).
3. MATLAB R2020b (Install from: https://www.mathworks.com/help/install/) is required to import models from MIB to randomly generate coordinates. These coordinates are required in Subheading 3.8. Note that a license is required for MATLAB.
4. Amira-Avizo3D Pro 2021.2 (Thermo Fisher) and Amira XL Volume Extension 2021.2, both of which contain a batch processing package. Although a license is required for each software, a 10-day free trial is available for Amira-Avizo3D Pro 2021.2 before purchasing: (https://www.fei-software-center.com/avizo/amira-avizo-packagedownload).
5. Microsoft Excel 365.

Codes can be downloaded from NCL-Image-Analysis repository on GitHub: https://github.com/NCL-ImageAnalysis/3D-reconstruction-of-the-mitochondrial-network-within-the-neuronal-soma-from-SBF-SEM-volume-data.

Fig. 1 FIJI Interface, also referred to in Subheading 3.8

3 Methods

3.1 Dataset Parameters

Since SBF-SEM images of large neurons are time-consuming to collect, and it is difficult to locate whole neurons throughout the tissue block, a minimum of three partial neuronal soma volumes were generated per sample. Segmentation and reconstruction of the neuronal soma and their mitochondria were performed within 3D image stacks of postmortem human brain tissue containing between 100 and 500 slices, sectioned at a thickness of 70 nm (*Z*-axis). An image resolution of either 3000 × 3000 pixels or 4500 × 4500 pixels were chosen due to the differences in the size of individual neurons, with one neuronal soma imaged per stack. A pixel dimension of 12 nm × 12 nm (*X*–*Y* axes) was selected so that the mitochondrial network could be reconstructed. A pixel dwell time of 20 μs was used.

3.2 Image Conversion in FIJI

The image stacks require conversion from a (.dm^3) to an 8-bit Tiff format (*see* **Note 1**).

1. Open FIJI (Fig. 1): Go to [**"Process"** > **"Batch"** > **"Convert"**].
2. A dialog box will appear with several components that require actioning:
   (a) **Input:** Choose the folder containing the dm^3 stack to be converted.
   (b) **Output:** This requires the selection of the folder to save the converted Tiff stack (*see* **Note 2**).
   (c) **Output format >** select **"8-bit TIFF"**.
   (d) Select **"Convert"**.

3.3 Opening MIB and a Tiff Stack

1. Open up MIB within MATLAB: Type mib, in lowercase, into the **"Command Window"** (*see* **Note 3**). A separate MIB interface will show (Fig. 2).
2. Locate the path containing all data files by navigating to the icon showing three dots [. . .] within the **"Path"** panel.
3. Locate the folder containing the 3D dataset, saved as a Tiff image series, by using the icon [..] in **"Directory contents"**.
4. To view the imaged slices in the 3D stack, you need to highlight all of the images; left-click on the first image within the panel, then press the keyboard.
5. Shortcut **"Shift"** and click on the last image within the panel. Using the right mouse button, right-click on the highlighted files and select **"Combine selected dataset"**. The images will show in the **"Image View"** panel and a scroll bar will appear to the left of the image. You can now navigate through the *Z*-axis of the 3D stack using this scroll bar or the mouse wheel.

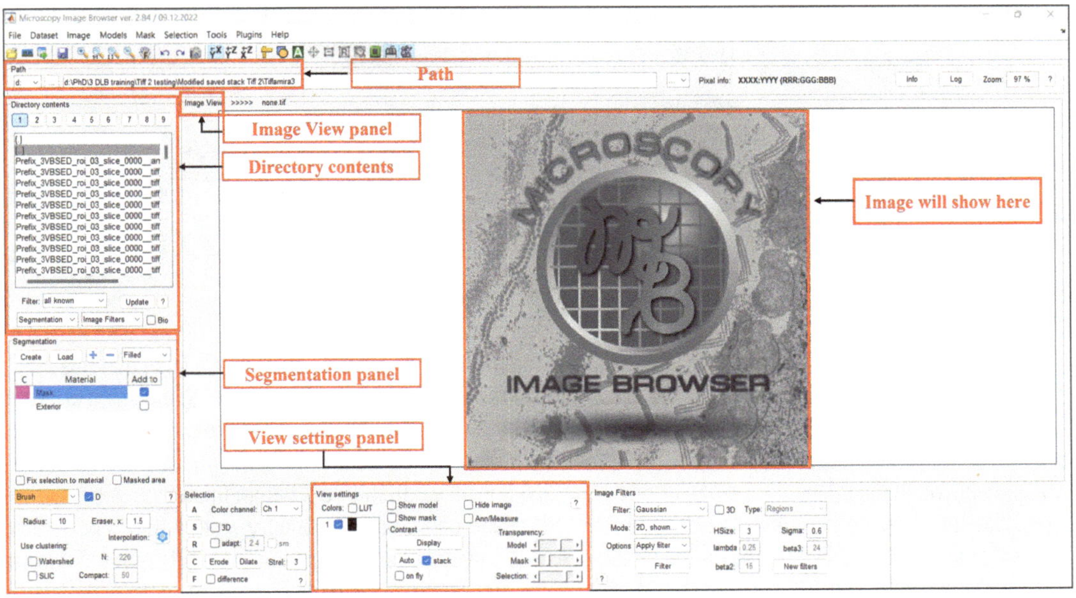

Fig. 2 Microscopy Image Browser (MIB) user interface. Each panel tab will be referred to throughout this guide. The series of Tiff images to be opened are shown to the left under **"Directory contents"** and the image will show under the **"Image View"** panel. The main panels used in this guide are labelled in red.

3.4 Performing Quality Checks in MIB

Use the mouse scroll, or left scroll bar, to navigate through the dataset so that each image can be visually examined for quality. Two quality control checks should be performed:

1. The first quality control checks for the presence of artifacts through the stack with examples shown in Fig. 3 (*see* **Note 4**).
2. The second quality control checks for the presence of a soma (*see* **Note 5**).

3.5 Image Preprocessing in MIB

Image preprocessing is required to ensure adequate contrast of the neuronal and mitochondrial membranes as well as aligning images to correct for drift within the 3D stack.

1. Normalize the contrast through the Z stack: Go to the Menu and select [**"Image"** > **"Contrast"** > **"Normalize layers"** > **"Z-stack"**]. A dialog will show **"Z-stack normalization settings"** and will prompt you to select the settings. Leave the default settings and select **"OK"**.
2. Adjust the contrast for optimal display of the membranes: Select **"Display"** in the **"View Settings"** panel. Use the **"Min"** and **"Max"** sliders to choose the intensity range (*see* **Note 6**). Apply this range to all imaged slices by selecting **"All slices"**.

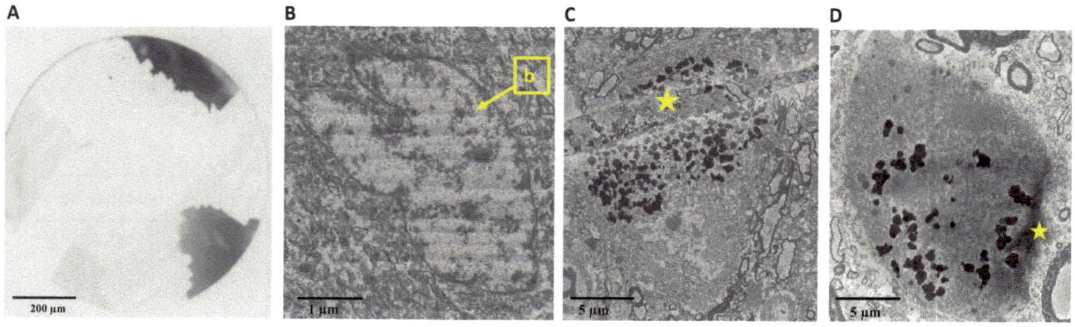

Fig. 3 Potential quality issues within the image stacks generated by SBF-SEM. (**A**) Shows a slice that has fallen onto the camera during imaging. (**B**) Shows a 2D representation of the 3D image containing a cutting artefact, observed by the horizontal lines (one line is labelled **b**). The irregular cutting by the microtome knife is caused by beam damage which softens the resin. (**C**) Shows the region of interest (soma) in the image partly obscured by a section which has fallen on the block-face (star). (**D**) Shows a dark artefact within the image (star) caused by a build-up of electrons within the sample, termed 'tissue charging'. Secondary electrons are generated alongside back-scattered electrons during imaging. If these electrons accumulate, local electrostatic fields are created at the block-face. This can cause electrons to be emitted from the block-face, and combined with the back-scattered electrons a noisy signal is detected, reducing image quality.

3. Align the dataset to correct any drifts that may have occurred during SBF-SEM imaging: go to the Menu and select [**"Dataset" > "Alignment tool"**]. A dialog box will appear and within the **"Align"** panel, select **"Current dataset"** and then **"Drift correction"**. This is the default algorithm and is what we recommended for all image processing. Press **"Continue"**. A graph will show the drifts, then select **'yes'** in the dialog box.
4. Save the dataset as an uncompressed stack of 2D Tiff files: go to the Menu [**"File"> "Save Image As" > "Tiff Format LZW compression"**]. Also save the image stack as an **"Amira Mesh binary file** (.am)". During saving, a dialog box will prompt you to set the image voxel dimensions based on the image resolution for the *X*, *Y*, and *Z* axes. Ensure the corresponding unit is set by selecting the appropriate unit in the drop-down box (*see* **Note 7**).

3.6 Segmentation in MIB

Once the image has been preprocessed and the contrast adjusted to clearly view the membranes, we can proceed to segmentation (*see* **Note 8**). A useful summary of the keyboard shortcuts used to facilitate segmentation in MIB is shown in Table 1.

3.6.1 Segmenting the Neuronal Soma in MIB

1. Create a new model following preprocessing of the image: go to the Menu and select [**"Model" > "New model"**]. A dialog box will appear labeled **"Model Type"**. Select **"255 materials"** and then **"OK"** (*see* **Note 9**).

Table 1
A summary of the keyboard shortcuts

Keyboard shortcut	Function[a]
"W"	To zoom in
"Q"	To zoom out
"**Ctrl + Scroll**"	Changes brush size
"**F**" (in 2D) OR "**Shift + F**" (in 3D)	Fills the holes in the traced selection
"**I**"	Interpolates between two traced selections. Materials are unaffected
"**A**" (in 2D) or "**Shift + A**" (in 3D)	Adds the selected pixels (2D) or voxels (3D) to a material
"**Ctrl + A**" (in 2D) or "**Alt + A**" (in 3D)	Converts the added material back to the green tracing. This is useful if the pixels (2D) or voxels (3D) have been added to the wrong object
"**R**" (in 2D) or "**Shift + R**" (in 3D)	Replaces a selected material with the current traced selection
"**S**" (in 2D) or "**Shift + S**" (in 3D)	Subtracts the pixels (2D) or voxels (3D) from the added material
"**C**" (in 2D) or "**Shift + C**" (in 3D)	Clears the traced selection. Materials are unaffected
"**Ctrl + Z**"	To undo the last action

[a] If there is no response to the keyboard function during segmentation tasks, right-click back on the image within the "**Image View**" panel and then implement the function again

2. Add a new material to the model: go to the "**Segmentation**" panel and select the [+] icon. A dialog box will show, prompting you to name the object, i.e., neuronal soma.
3. Within the "**Segmentation**" panel, click on the drop-down list and select "**Brush**". The size of the brush tool is changed by altering the number in the box located next to "**Radius**".
4. Use the mouse scroll wheel to navigate through the *Z*-axis to locate the first appearance of the neuronal soma.
5. To obtain the membrane of the neuronal soma in an optimal view for segmentation, use the hand tool and row of magnification icons located below the Menu bar.
6. Using the left mouse button, trace the membrane of the soma. The traced outline of the membrane will show up green. This is performed in each consecutive image (*see* **Note 10**).
7. Press the keyboard shortcut "**F**" to fill the traced outline in 2D (*see* **Note 11**). The whole soma will appear green in the *X–Y* view within the image.

8. Add the green traced selection to the material using the keyboard shortcut **"A"** for each image (*see* **Note 12**). The soma will change color to match the color located next to the material, under **"C"** within the **"Segmentation"** panel.
9. Repeat **steps 2–8** to add the nucleus to the model (*see* **Note 13**).
10. Save the model: Click [**"Models"** > **"Save model as"**] to save the neuronal soma model in the **"Matlab format** (.model)". This allows reopening of the model within MIB if the segmentation is still in progress. When the segmentation is complete save as an **"Amira mesh binary file** (.am)" ready to export to Amira™ 3D Pro for 3D surface rendering and quantitative analysis.

3.6.2 Reopening a Model in MIB

1. To reopen the model, the MIB preprocessed Tiff image stack will need to be opened first as explained in Subheading 3.3, **steps 2–5**.
2. Go to the Menu and select [**"Models"** > **"Load Model"**].-Choose the folder containing the desired model.

3.6.3 Generating ROI Coordinates Using MATLAB and MIB

Next generate 20 coordinates, termed regions of interest (ROIs), within MATLAB to randomly sample mitochondria within the segmented neuronal soma (*see* **Note 14**).

1. Generate the sampling coordinates for the segmented model of the neuronal soma (created in Subheading 3.6.1): Within MIB, go to [**"Models"** > **"Export Model to Matlab"**]. A dialog box will appear labeled **"input a destination variable"**. The variable for the structure will show up as **"O"**. Click **"OK"**.
2. Open the code to generate coordinates by double clicking the file. The code will automatically appear in the **"Editor"** section of MATLAB.
3. Within MATLAB, click on the **"Editor"** section and select **"Run"**.
4. The randomly generated numerical coordinate values for both X (labeled x1) and Y (labeled y1) axes will show in the **"Command Window"** of MATLAB, located below the **"Editor"** section (*see* **Note 15**).
5. Switch back to MIB and under the **"Directory contents"** panel, change the **"Segmentation"** option to **"ROI"** in the drop-down box. The **"ROI"** panel will show.
6. Within the **"ROI"** panel manually enter the coordinates: first, select the **"Manually"** box and enter the coordinates within the corresponding X1 and Y1 boxes. A yellow outline of a box will appear (size: 50 pixels × 50 pixels) within the image and

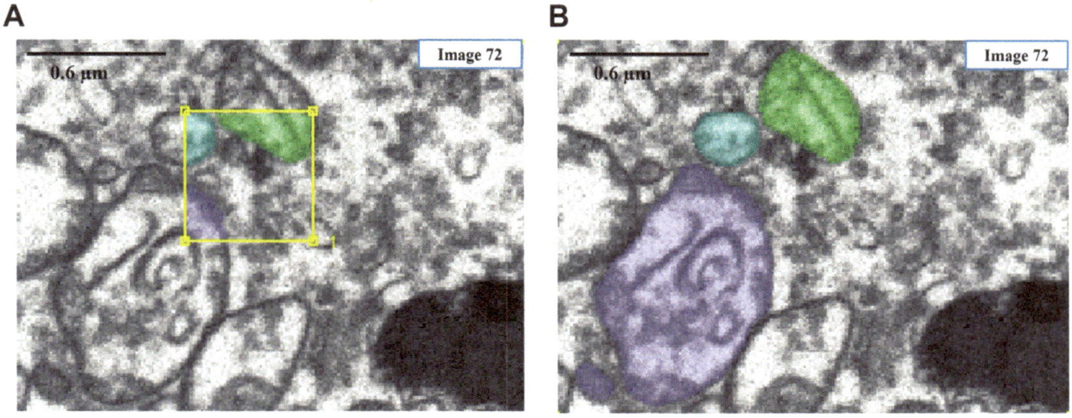

C

$$\textit{Volume density}(\%) = \frac{\textit{Total Mitochondrial Volume per ROI}}{\textit{Cytoplasmic Volume Occupied}}\ x\ 100$$

$$\textit{Sphericity} = \frac{\left(\pi^{1/3}(6V)^{2/3}\right)}{SA}$$

$$MCI = \left(\frac{SA^{3/2}}{4\pi V}\right)$$

Fig. 4 Example segmentations of mitochondrial morphology in SBF-SEM images. (**a**) A two-dimensional representation of three mitochondrial objects segmented within the yellow boundary created by a randomly generated region of interest (n*on-extended version*). (**b**) Segmented mitochondrial extensions (*extended version*). Notice that there are two purple objects segmented which eventually connect in subsequent images. These objects could easily be mistaken for two different mitochondria if they were analysed in a 2D image. (**c**) Definitions of volume density and 3D shape descriptors. Notice volume density concerns the segmentation in (**a**).

throughout the entire stack. The coordinate system will be listed in numerical order in the **"ROI List"**, starting from number 1.

7. Repeat **steps 3–6** to generate 19 more coordinates within the image.
8. Save the yellow boxes corresponding to the numerical coordinates in the (.roi) format for future use by selecting **"Save"** located within the **"ROI"** panel (*see* **Note 16**).

3.6.4 Segmenting Mitochondria in MIB

1. Identify the mitochondria that appear within the first ROI, i.e., ROI 1, in the neuronal model (*see* **Note 17** and Fig. 4a).
2. **Segment the first model containing all mitochondria** found **within the ROI** (*non-extended version*) and stopping at the RPI boundry: add each material by selecting the [+] icon in the **"Segmentation"** panel and label the materials in

numerical order, i.e., Mitochondrion 1, 2, etc. Using the **"Brush"** tool, segment the mitochondria one by one (*see* **Note 18**). Refer to Subheading 3.6, **#1** for segmentation details.

3. Save the model as both (.am) and (.model) files, ensuring you rename the original model file (*see* **Note 19**).
4. **Segment the second model containing mitochondrial extensions** (*extended version*): remove the ROI coordinate system [**"ROI list"** > **"All"** > **"Remove"** > **"Delete"**] and extend each mitochondrial segmentation to ensure all mitochondria are whole and include regions of each mitochondrion not contained within the ROI, starting at mitochondrion 1 (Fig. 4b and *see* **Note 20**). Once all mitochondria are segmented, save the new extended model under a different name.
5. Repeat **steps 1–4** for each ROI and two models containing segmented mitochondria will result.

3.6.5 Segmenting the Volume of Cytoplasm within the ROI in MIB

Next, segment the volume of cytoplasm within each ROI generated in Subheading 3.6.3 (*see* **Note 21**).

1. Load the original model containing the neuronal soma created in Subheading 3.6.1.
2. Upload the 20 yellow ROIs by switching **"Segmentation"** to **"ROI"** and select **"Load"**. Segment each ROI individually as follows.
3. Switch back to the **"Segmentation"** panel and add each material by selecting the [+] icon and name the material, e.g., ROI 1, ROI 2, etc. Change **"Brush"** in the **"Segmentation"** panel to **"Lasso"** using the drop-down box.
4. Navigate, using the mouse wheel or left scroll bar, to the minimum image number where the first segmented mitochondrion appeared within the ROI (*values attained in* **step 2**, *see* Subheading 3.6.4). Within this image, drag the **"Lasso"** tool to accurately match the yellow box boundaries. Using the left mouse button, double-click within the yellow ROI box. The interior of the box will appear green. Go to the Menu bar [**"Selection"** > **"Selection to buffer"** > **"Copy"**] to copy the green box.
5. Navigate, using the mouse wheel or left scroll bar, to the maximum image number where the last segmented mitochondrion appeared within the ROI. Go to the Menu bar and select [**"Selection"** > **"Selection to buffer"** > **"Paste"**]. A green box will show within the ROI.
6. Use the keyboard shortcut **"I"** to interpolate the two green colored regions. All boxes between the min-max images within the ROI will appear green (*see* **Note 22**).

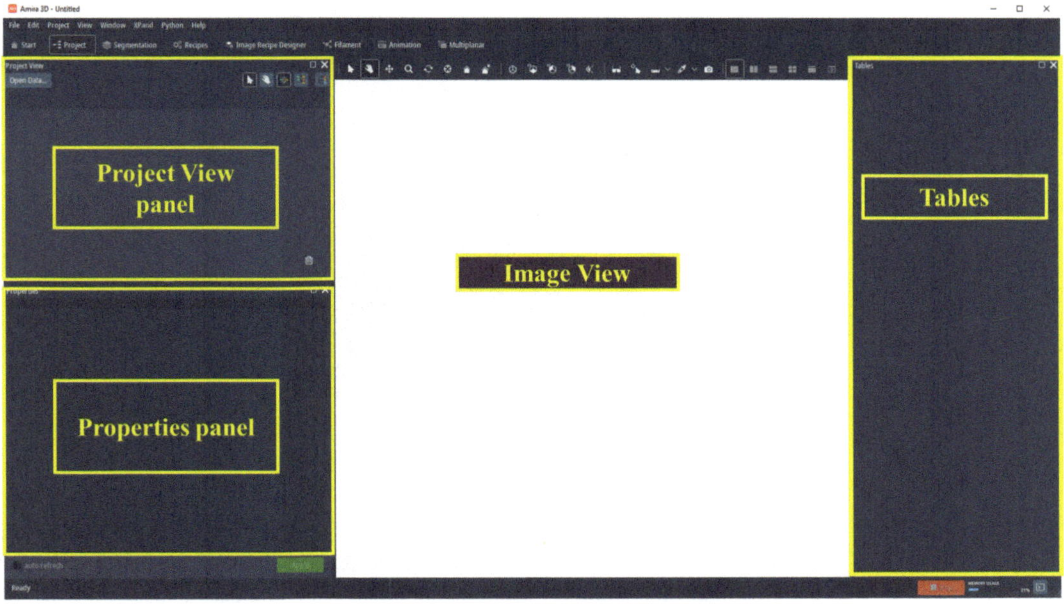

Fig. 5 Amira™ 3D Pro user interface. Each panel tab highlighted in yellow will be referred to throughout this guide. The Amira image will show in the **"Project View"** panel and the part associated with each module, linked to the image, will show in the **"Properties"** panel. The 3D surface-volume rendering will be viewed in the blank white space labelled **"Image View"**. The quantitative results, following application of the **"Surface Area Volume"** module will show in **"Tables"** (use **"Window"** on the tool bar to select **"Tables"** first).

7. Add the traced selection to the correct material selected in the **"Materials"** list by pressing **"Shift + A"**. The ROI selected will change color in 3D.
8. Repeat **steps 3–7** to segment the next 19 ROIs, one by one. Save as both (.model) and (.am) files.

3.7 3D Polygonal Surface Reconstruction in Amira™ 3D Pro

The segmented image, in the .am file format, is ready to be transformed into a closed polygonal surface (*see* **Note 23**) within Amira™ 3D Pro (Fig. 5).

1. Open Amira™ 3D Pro and ensure the correct units are set: [**"Edit"** > **"Preferences"** > **"Units"**]. Change the units to **μm** and do this before loading the (.am) file.
2. Open the (.am) file by clicking on **"Open Data"** (*see* **Note 24**). The label image will be displayed as a green rectangle in the **"Project View"** panel tab (*see* **Note 25**).
3. Within **"Project View"**, right click on the (.am) file, go to [**"Compute"** > **"Generate Surface"** > **"Create"**]. A red rectangle labeled **'Generate Surface'** will appear connected to the (.am) file.
4. Click on the red **"Generate Surface"** rectangle and within the **"Properties"** panel tab, a port will show containing all of the options associated with the module.

Fig. 6 Example 3D surface-rendered-volume of all the smoothed mitochondria (different colours) from 20 ROIs (*Extended version*) in the surrounding portion of neuronal soma (semi-transparent blue), featuring the nucleus (solid purple). The reconstruction overlays a series of tilted ortho slices using the original SBF-SEM image.

5. Select the type of smoothing required in the port: next to the **"Smoothing"** option, use the drop-down box to change **"Unconstrained smoothing"** to **"None"** and click **"Apply"**.
6. Apply Taubin smoothing to the extended mitochondrial models only (*extended model*) (*see* **Note 26**): right-click on the resulting green (.surf) file in the **"Project View"** panel. Locate **"Surface Transforms"** > **"Smooth surface"** and press **"Create"**. In the smooth surface port, enter **"5"** iterations and leave the lambda value at **"0.6"**. Press **"Apply"** and a new green (.smooth) file will be generated.
7. To view the 3D rendering, right-click on the desired green (.surf) or (.smooth) file and select **"Surface View"** (*see* **Note 27**). The 3D structure will appear over a black ortho slice unless the image stack (.am) is uploaded and **"ortho slice"** is selected (Fig. 6). Hide the **"ortho slice"** by clicking on the toggle button (blue square) within the orange rectangle labeled **"ortho slice"**.
8. Save the whole project: Click an empty space within **"Project View"** panel so that no files are selected. Go to **"File"** and select [**"Save as"** > **"Minimize project computation"** > **"Amira project and data pack and go** (.hx)"].

3.7.1 Data Extraction in Amira™ 3D Pro

Surface area and volume measurements are generated from the polygonal surfaces in Amira™ 3D Pro.

1. Right click on the green (.surf) file or (.smooth) file (*see* **Note 28**).
2. Select [**"Measure and Analyze"** > **"Surface area volume"** > **"Create"**] and use the default settings.
3. Click **"Apply"** at the bottom right of the **"Properties"** screen.
4. Click on the (.statistics) file then select **"Show"** in the **"Properties"** screen. The results will show in the **"Tables"** panel (*see* **Note 29**).
5. Export the (.statistics) file as a (.csv) file: [**"File"** > **"Export data as, (.csv)"**] (*see* **Note 30**).

3.7.2 Automated Batch Processing in Amira™ 3D Pro Using the Xtra Library

All steps from opening of the (.am) file, generation of a surface, to quantification are automated using a recipe created for batch processing with the data output contained in a spreadsheet.

1. Right click the (.statistics) file generated and select **"Create Recipe"**.
2. Check the recipe steps under the **"Recipe"** section. A list of steps will show. Save the recipe by clicking the save icon.
3. Run the recipe: right-click in an empty region of **"Project View"**. Type in [**"Xtra"** > **"BatchApply_RecipePlayer"** > **"Create"**]. A blue **"BatchApply_RecipePlayer"** rectangle will show.
4. Go to the **"Properties"** pane linked to **"BatchApply_RecipePlayer"** and select the desired recipe file for the first entry in **"Recipe files"**, followed by the files to be processed for **"File (s) to Process"**, and then the output folder for **"Output Directory"** where the statistics (.csv) will be saved. Leave **"Grayscale"** checked and select **"Apply"**.

3.7.3 Identifying and Measuring Potential Nanotunnels in MIB and Amira™ 3D Pro

1. Identify potential nanotunnels in MIB for the segmented mitochondria within each ROI as seen in Fig. 7 and detailed in **Note 31**. Ensure you export them as an (.am) file as previously explained in Subheading 3.6.1, **step 10**.
2. Open Amira™ 3D Pro and go to [**"Edit"** > **"Preferences"** > **"Units"** > **"Spatial information only"**] to change the unit to **nm** under **"Display units"** and **"Working units"**. Select **"Apply"** and then **"OK"**. Ensure you check the voxel size of the image is in **nm** (*refer to instructions in* **Note 25**).
3. Open the (.am) file in Amira™ 3D Pro to generate a Taubin smoothed surface and view it using **"Surface View"** (explained in Subheading 3.7).
4. Select the ruler icon above the image viewing panel and then the **"3D Length"** tool in the drop-down box (*see* **Note 32**). A yellow box labeled **"Measurement"** will show in **"Project View"**.

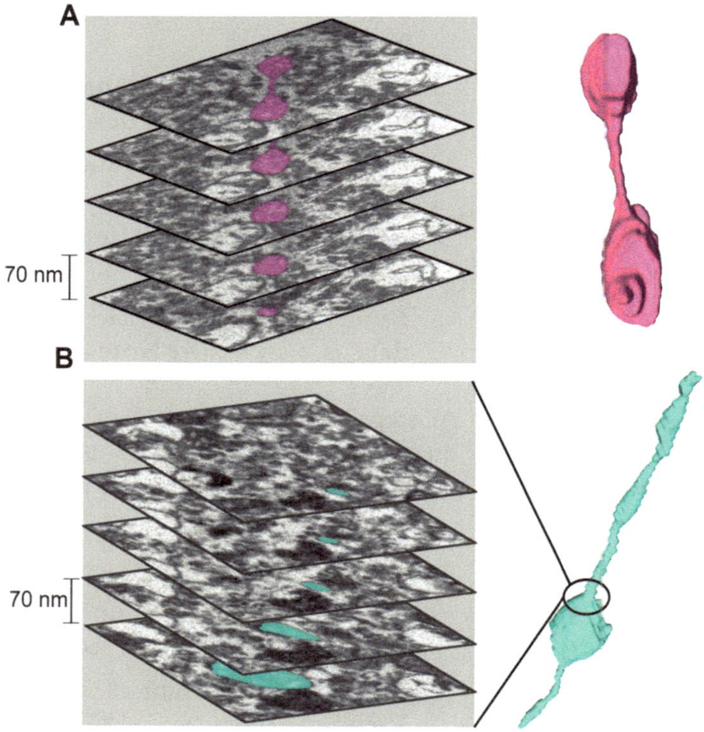

Fig. 7 The identification of mitochondrial nanotunnels within the neuronal soma using SBF-SEM volume data. (**a**) Pink coloured mitochondria connected via a nanotunnel viewed in the longitudinal plane. (**b**) Turquoise mitochondrion with two free ended nanotunnels viewed in the transverse plane. For this mitochondrion, five of the sixty-nine segmented SBF-SEM images show the emergence of a nanotunnel from the nanotunnel hillock. This portion is included in the circled outline of the complete 3D surface-volume rendering.

5. Measure the length and diameter of the nanotunnel (*see* **Note 33**): Click on **"Measurement"** and go to the **"Properties"** panel. Select both the **"3D"** box and **"Line"** tool. Choose the hand icon and using the left mouse button rotate the 3D surface in the best orientation to view the nanotunnel.
6. Use the **"Line"** tool to select the two points to be measured and the corresponding distance will show in **"Measurement"**.
7. Export as a (.csv) file: Select **"Export"** located below the list of measurements. A corresponding green file labeled **"Measurement.Spreadsheet"** will show in **"Project View"**. Click on this file and export by going to [**"File"** > **"Export Data as, (.csv)"**].

3.8 Subdividing the Soma into Regions Using FIJI

The coordinates for each ROI (previously generated in Subheading 3.6.3) are automatically categorized into perinuclear and distal regions of the partially segmented neuronal soma using a customizable macro that runs in FIJI. The macro creates these subregions

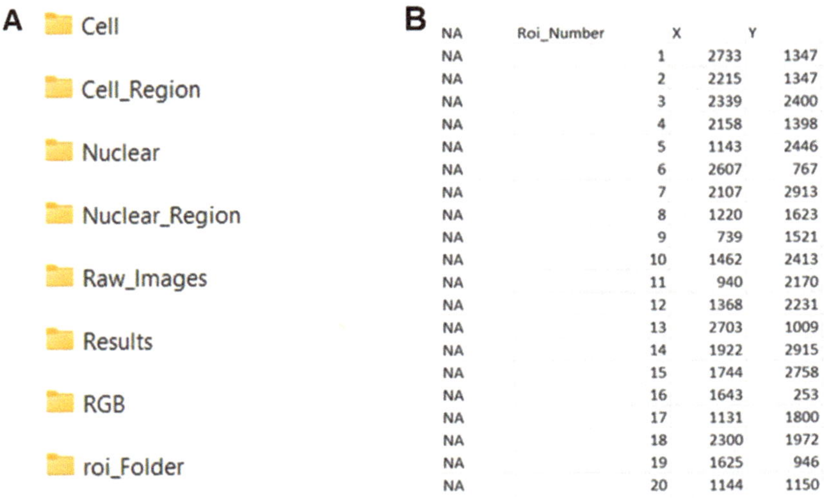

NA	Roi_Number	X	Y
NA	1	2733	1347
NA	2	2215	1347
NA	3	2339	2400
NA	4	2158	1398
NA	5	1143	2446
NA	6	2607	767
NA	7	2107	2913
NA	8	1220	1623
NA	9	739	1521
NA	10	1462	2413
NA	11	940	2170
NA	12	1368	2231
NA	13	2703	1009
NA	14	1922	2915
NA	15	1744	2758
NA	16	1643	253
NA	17	1131	1800
NA	18	2300	1972
NA	19	1625	946
NA	20	1144	1150

Fig. 8 (**a**) Folder system for Macro functioning. (**b**) CSV file layout for the 20 ROI coordinates generated to sample one neuronal soma.

using predefined 3D radial factor (*RF*) values for the perinuclear and perimembrane regions (*calculated in* Subheading 3.8.2 *and used within the macro in* Subheading 3.8.3) (*see* **Note 34**). Mitochondrial volume density and morphology can then be compared between the subregions.

3.8.1 Creating a Folder System

1. First, create a folder system exactly as shown in Fig. 8a so that the results from the macro are filtered correctly.
2. Open Excel and enter the 20 ROI coordinates exactly as shown in Fig. 8b. These coordinates were generated within the relevant neuronal segmentation in Subheading 3.6.3. Save as a (.csv) file within the ***roi_folder*** using exactly the same naming system as the neuronal model (.tiff), e.g., "*Case_004_Neuronal_Soma_1.*"

3.8.2 Calculating the Radial Factor Using MIB and Amira™ 3D Pro

The *RF* for both the perinuclear and peri-membranal regions need to be defined in 3D. To derive the *RF*, you need to set a distance for both regions. Distance is calculated in an unbiased manner since it takes into account the radius of both the neuronal soma and the nucleus. To calculate the radii, height and volume measurements are obtained in MIB and Amira™ 3D Pro for a minimum of 20 models containing the segmented neuronal soma (*see* **Note 35**).

1. Setup an Excel spreadsheet with the column headings shown in Fig. 9F.
2. *Calculate the height*: Inspect the segmented neuronal soma within MIB to ensure the peak of the nucleus shows (*demonstrated in* Fig. 9A and B). Count the number of images of

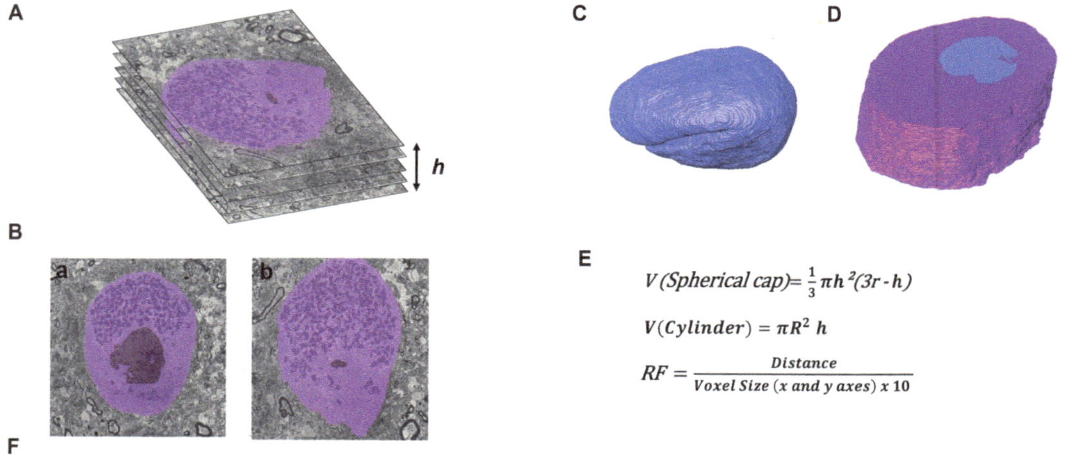

	A	B	C	D	E	F	G	H	I
1	Case/Neuronal	Cytoplasm Volume (µm³)	Nuclear volume (µm³)	Total volume (µm³)	h (µm)	R (µm)	r (µm)	Region size (µm)	RF X/Y
2	004/01	7058.81	718.72	7777.53	10.36	15.45845313	5.584857921	3.291198402	27.42665335

Fig. 9 Visualisation of the components required for the calculation of the radii and *RF* using height and volume. (**A**) Example SBF-SEM stack showing the soma and peak of the nucleus to demonstrate height. (**B**) The corresponding first image (**a**) and last image (**b**) of the stack containing the nucleus. (**C**) Example surface-volume-rendering of the portion of nucleus analogous to a spherical cap and (**D**) the portion of neuronal soma analogous to a cylindrical shape. (**E**) Equations used to calculate the radius of a spherical cap and cylinder (using volume and height measurements) which are later used to calculate the radial factor (*RF*) in the X/Y axes. (**F**) Example Excel sheet setup to calculate the *RF* in the X and Y axes for one neuronal soma.

neuronal soma that contain the portion of nucleus. Then multiply the total number of images by the image resolution in the *Z*-axis, e.g., *169 images* × *0.07 μm*. The resulting value represents the height (***h***). Enter ***h*** into column E of the Excel spreadsheet.

3. Save the segmentation of the neuronal soma as a Tiff within the **Raw_Images** folder (Fig. 8a): [**"Model"** > **"Save model as"** > **"TIF format** (*.tif)"] then select **"3D TIF"** in the dialog that appears. Also save as an (.am) file in a separate folder (*see* **Note 36**).

4. *Extract volume measurements*: Open the neuronal model (.am) into Amira™ 3D Pro and determine the volume of the nucleus and the volume of the cytoplasm within the soma (*extraction of statistics previously explained in* Subheading 3.7). Enter these values into columns B and C of the Excel spreadsheet. A surface-volume-rendering of the neuronal soma components required are shown in Fig. 9C and D.

5. To calculate *total volume* of the neuronal soma, enter the formula = *SUM(B2 + C2)* into column D of the Excel spreadsheet.

6. To calculate the *external radius* of the soma (***R***), enter the rearranged formula for a whole cylinder = *SQRT(D2/(E2*PI ()))* into column F of the Excel spreadsheet. The original equation is provided in Fig. 9E.
7. To calculate the *radius* of a spherical cap (***r***), enter the formula = *(C2/(PI()*E2^2) + (1/3)*(E2))* into column G of the Excel spreadsheet.
8. To calculate *region size*, enter the formula = *(F2-G2)/3* into column H of the Excel spreadsheet (*see* **Note 37**). The resulting value is the distance used to find the *RF* in the *X*/*Y* axes.
9. To calculate the *radial factor (RF)* for the *X*–*Y* axes only, populate column I in the Excel spreadsheet using the formula in Fig. 9E (*see* **Note 38**).
10. Repeat **steps 2–9** for each neuronal soma to gain a minimum of 20 *RF* values for the *X*/*Y* axes.
11. Calculate the *RF* value corresponding to the 90th percentile in the dataset (*see* **Note 39**).
12. Calculate one corresponding *RF* value in the *Z*-axis by dividing the *RF* in the *X*/*Y* axes by the anisotropic factor (*see* **Note 40**).

3.8.3 Running the Macro in FIJI

Once the *RF* values are defined in Subheading 3.8.2 for the *X*–*Y*–*Z* axes, the macro is ready to run.

1. Open FIJI (refer to Fig. 1).
2. Within FIJI, open the label image of the neuronal soma (*previously saved in* Subheading 3.8.2): go to [**"File"** > **"Open"**] and select the 3D Tiff from the **Raw_Images** folder (Fig. 8). The label image will appear (*see* **Note 41**). Set the voxel size of the image in FIJI: [**"Image"** > **"Properties"**]. A dialog box will show the Pixel width (*X* size), Pixel height (*Y*), and Voxel depth (*Z* size). Enter the relevant image pixel size/s here and save over the original Tiff image [**"File"** > **"Save As"** > **"Tiff"**] (*see* **Note 42**).
3. Open the macro: drag and drop the file into FIJI. The code will automatically appear.
4. Enter the three calculated *RF* into the macro (Subheading 3.8.2). Add the *RF* values in **Line 57** to dilate the nuclei segmentation in the *X*–*Y*–*Z* axes. Also enter the same *RF* values into **Line 80** to erode the cytoplasm inwards from the neuronal membrane in the *X*–*Y*–*Z* axes.
5. Click **"Run"** and a dialog box automatically appears three times, prompting you to select the correct folders. The folders must be selected in the stated order:

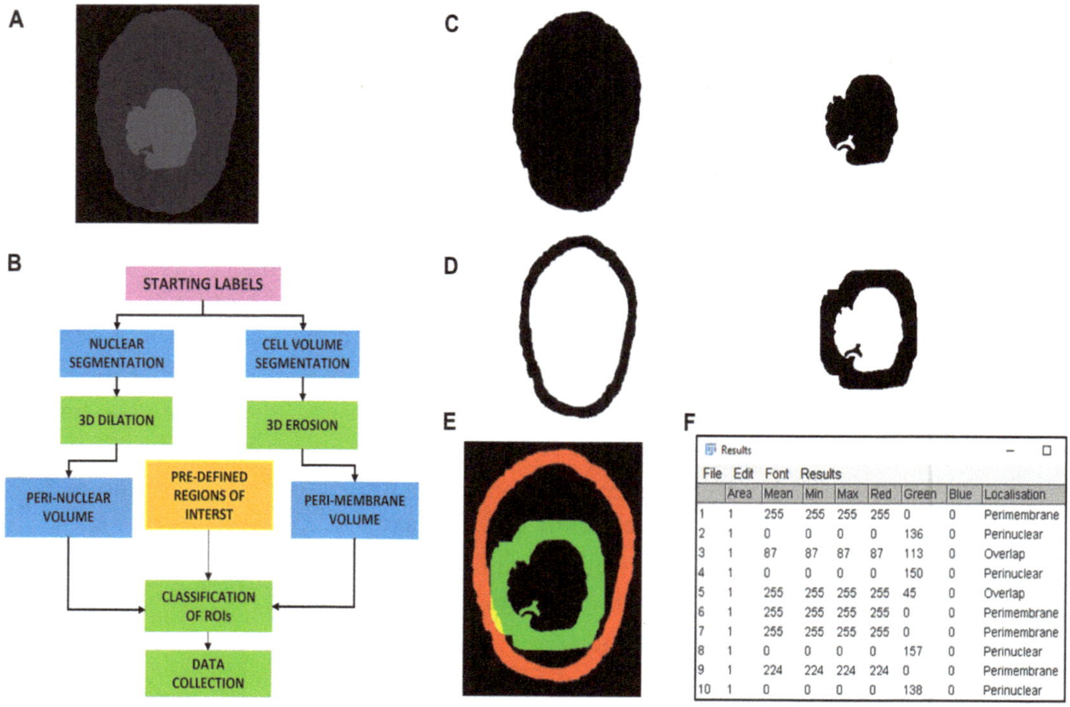

	Area	Mean	Min	Max	Red	Green	Blue	Localisation
1	1	255	255	255	255	0	0	Perimembrane
2	1	0	0	0	0	136	0	Perinuclear
3	1	87	87	87	87	113	0	Overlap
4	1	0	0	0	0	150	0	Perinuclear
5	1	255	255	255	255	45	0	Overlap
6	1	255	255	255	255	0	0	Perimembrane
7	1	255	255	255	255	0	0	Perimembrane
8	1	0	0	0	0	157	0	Perinuclear
9	1	224	224	224	224	0	0	Perimembrane
10	1	0	0	0	0	138	0	Perinuclear

Fig. 10 Image processing pipeline of the neuronal soma and output in FIJI. (**a**) A 2D representation of the segmented neuronal soma saved as a label image. (**b**) A flowchart of the macro pipeline which uses the label image to isolate the components of the soma. (**c**) A 2D representation of both the somal volume segmentation and nuclear volume segmentation. (**d**) A 2D representation of the eroded neuronal membrane area (peri-membrane region) and the dilated nuclear area (peri-nuclear region) defined by the radial factor (*RF*); These areas are compared to the original label image and the original segmentations are subtracted. (**e**) The resulting Red-Blue-Green (RBG) image shows the peri-membrane (red) and peri-nuclear regions (green). The region between these is classed as the intermediate region (remains black), and since the nucleus can be asymmetrically positioned an overlapping region is also generated when the peri-nuclear and peri-membrane regions intersect (yellow). (**f**) The macro automatically categorises the coordinates for each ROI into one of the regions as shown by the example window output for the first 10 of the 20 ROIs. This allows mitochondrial morphology and mitochondrial volume density within each ROI to be compared between the sub-regions in an unbiased manner.

(a) Select the ***Home Directory*** first (*see* **Note 43**).

(b) Select the ***Raw image folder*** containing the label image of the neuronal soma (.tiff).

(c) Select the ***roi_folder*** containing the Excel (.csv) file that contains the relevant ROI coordinates.

6. The macro will automatically process the label image and the results will show in several pop-up windows (Fig. 10) but will be automatically saved in the correct folders shown in Fig. 8. The assignment of the ROI coordinates to the different regions will be automatically saved as a (.csv) file in the **Results** folder.
7. Repeat **steps 1–5** for each label image of the neuronal soma (.tiff) using the same *RF* values.

4 Notes

This section contains information on hints and tips, explanations as well as troubleshooting.

1. The raw data from SBF-SEM imaging using the incorporated Gatan 3View system (Gatan inc., Abingdon, UK) generates 3D image files in the .dm^3 format. The image is converted to a Tiff to be usable in MIB. The dataset is also reduced to an 8-bit image for memory-intensive activities such as segmentation.
2. It is recommended to have a folder system for each case where the raw image files, and their different versions, as well as segmented model files are saved separately. A simple example of the folder subdivisions per case is shown:
   (a) **Raw Data**: contains (.dm^3) files for each neuronal soma (saved in different folders).
   (b) **Converted data**: contains the Tiff files following FIJI preprocessing.
   (c) **Modified Tiffs**: contains the Tiff files following MIB preprocessing.
   (d) **Model files**: subdivided into folders:
      (i) *Sub-Folder 1*: contains neuronal soma models saved as both (.model) and (.am) files (the .model file will be used each time a new ROI is created).
      (ii) *Sub-Folder 2*: contains *non-extended* mitochondria models saved as both (.model) and (.am) files (for volume density measurements).
      (iii) *Sub-Folder 3*: contains extended mitochondrial models saved as both (.model) and (.am) files (for 3D shape analysis).
      (iv) *Sub-Folder 4*: contains ROI cytoplasmic segmentations saved as both (.model) and (.am) files (also for volume density measurements).
3. If using the stand-alone version of MIB, this can be opened by clicking the MIB icon.
4. Any slices that fall on to the SBF-SEM camera appear bright white within the image stack (*see* Fig. 3A). It is recommended to discard the image stack generated and restart the run for that neuron since too much detail will be lost. It is also recommended to discard the data if images show the presence of cutting artifacts throughout the stack (*see* Fig. 3B). In contrast, slices that fall on the block face (*see* Fig. 3C) can be removed if they present at the edge of the stack, or replaced if they are in the center. Replacing images will remove the artifact while

retaining the original stack volume when segmenting structures. To remove images: navigate to [**"Dataset"** > **"Slice"** > **"Delete Slices"**]. Enter the image number. If successive images are to be deleted, enter a semicolon between the minimum image number and maximum image number. Conversely, to replace images: go to [**"Dataset"** > **"Slice"** > **"Copy slice"**]. A dialog box will appear. Select **"Replace"** from the first drop-down box, then enter the number of the adjacent image in **"Index of the source slice"**. Lastly enter the number corresponding to the image containing the artifact in **"Index of the destination slice"**. No more than 5% images containing the neuron within the Z-stack should be replaced as this may affect 3D quantitative assessments of mitochondrial morphology and no more than two successive images. If charging is observed in the image (*see* Fig. 3D), there is no preprocessing solution [27]. However, it is recommended to keep the image/s affected so long as the morphology of each structure is still visible for segmentation.

5. The soma is identified by the presence of a nucleus. Check that the soma appears whole throughout the stack. Where the dataset contains the appearance of a partial soma in the *X*/*Y* view, they are included in analysis so long as ≥25% of the Z-stack contains a whole soma. If artifacts surround the neuronal soma, you can crop the image stack in 3D: Go to [**"Dataset"** > **"Crop dataset"**]. A dialog box will appear that prompts you to select your choice of cropping within the **"Crop mode"** panel. It is recommended to use the **"Interactive"** option since automated cropping may lead to loss of important detail. Once selected, use the left mouse button to draw a rectangle around the area to be maintained. Check the region of interest does not fall outside of the rectangle in 3D by scrolling through the image stack. Double-click within the area to crop the image. Note that if you decide to crop the image after the corresponding model has been created in Subheading 3.6.1, both the image stack as well as the model will need to be resaved. The image and model dimensions must be kept the same otherwise the model will not reopen.
6. Using the min and max, sliders will stretch the histogram. Generally, the whole intensity range is used unless significant black or white artifacts are present. Scroll through all images before applying the intensity range to ensure the structures have adequate contrast.
7. The units selected are in μm. The file is saved as a Tiff Format LZW compression file so that the modified images can be reopened within MIB. The image stack is also saved as an Amira binary file so that it can be imported into Amira™ 3D Pro and used as an **"ortho slice"** for visualization purposes.

8. Image segmentation is the process of identifying an object, e.g., the nuclear membrane or mitochondrion, and assigning a label to the image voxels (3D pixels) that belong to an object. This process separates the different objects within a 3D image. Several methods for segmentation exist and it can be performed manually, semiautomatically, or automatically. While automated segmentation speeds up the process, it is still being optimized for anisotropic 3D EM datasets as the current methods cannot handle the significant variations in mitochondrial shape, texture, or touching objects [34, 35]. While semiautomatic is a viable option for segmentation of structures that do not vary considerably throughout the stack (e.g., neuronal soma and nuclear membranes), it cannot be used to segment mitochondria due to their highly varied morphology. These issues necessitate the manual segmentation of the desired mitochondrial objects within the volume of the neuronal soma (the latter segmented by manual or semiautomated methods).
9. Selecting the **"255 materials"** option means there are $\leq$255 materials to segment, which corresponds to an 8-bit image. If there are >255 materials to segment, the number of materials can be increased. This may be important if you prefer to segment all the mitochondria within the soma without the MATLAB random sampling code (Subheading 3.6.3). Go to [**"Models"** > **"Convert Type"** > **"65535"**] and save as both (.model) and (.am) files. Note this action increases the model size from an 8-bit to a 16-bit image, requiring ~1.25 times more memory. There is also the option to select **"4294967295"** materials (32-bit image) but this requires twice more memory than the **"65535"** option. The **"Segmentation"** panel will change for models having >255 materials with only two materials displayed. This is because there would be too many materials to scroll through in the **"Materials"** list. The pixels belonging to one object are traced and added to one of these materials as previously described in Subheading 3.6.1. Following this, select the  icon labeled **"press to find and select the next empty material"**. The current material number will change to a new number, indicating the next material is available to add a selection: the pixels belonging to the next object in question are traced and added to the next empty material. To view any segmented material, use the mouse to hover the cursor over the desired object within the image and press the keyboard shortcut **"Ctrl + F"**. The corresponding material will now show under **"Material"** in the **"Segmentation"** panel. In contrast to models that have $\leq$255 materials, each material is renamed using numerals alone.

10. An alternative method to speed up the segmentation of larger structures such as the membrane of the neuronal soma employs semiautomated segmentation, namely interpolation within MIB. This technique complements manual segmentation but depends on the variability of the membranal shape throughout the 3D image. If the membrane does not considerably change in shape over a few images throughout the Z-stack, it is not necessary to manually segment each image. Instead, manually segment the object in the first and last image (*n*th image) in this section. Use the keyboard shortcut **"I"** or select [**"Selection"** > **"Interpolate as shape"**] to interpolate the images between. The interpolation algorithm calculates the traced shape between the first and *n*th images and fills the gaps between these separated images. Since errors can occur, it is recommended to visually check the accuracy of the semiautomated segmentations between the images and remove any erroneous tracing. If the green colored tracing goes outside the membrane, use the left mouse button and press the keyboard shortcut **"Ctrl"** to erase it. The intensity of green can be modified using the scroll bar next to **"Selection"** under the **"View Settings"** panel (*see* Fig. 2).
11. As an alternative, select **"Shift + F"** to fill the traced outline in 3D.
12. Rather than adding each tracing individually to a material, it is possible to add several tracings of the object to a material at the same time by pressing **"Shift + A"**. If an erroneous tracing of the membrane was unintentionally added to the material, this can be removed by retracing over the added mistake and pressing the keyboard shortcut, **"S"** (2D image) or **"Shift + S"** (in 3D).
13. During segmentation in MIB, a pixel can only be assigned to one material at a time and therefore segmentation needs to be performed in a particular order. It is important to segment the somal membrane first then the nuclear membrane so that they appear as separate entities. The mitochondrial membranes can then be segmented if the generation of coordinates is not required.
14. The large population of mitochondria within the soma of a neuron, coupled with the time-constraints of manual segmentation, necessitates the generation of multiple regions of interest (ROIs). The neuronal soma model can be randomly sampled using the MATLAB code downloaded from GitHub. The selection criteria used to accept each ROI is: (1) The coordinate system (yellow box) must enter the neuronal soma. (2) The ROIs should not physically overlap and since the neuronal size varies considerably, it is recommended to

generate 20 ROIs per neuron. (3) The neuronal membrane that appears within the ROI should be visibly intact for accurate volume density measurements.

15. Make sure you make a copy of the numerical coordinate values as they will not be saved in the last step. They are required for later use in Subheading 3.8 to classify mitochondria into subregions of the neuronal soma. Only the yellow outlines of the ROI boxes that correspond to these coordinates will be saved.
16. The yellow boxes can be reloaded within a model by selecting **"Load"** located in the **"ROI"** panel.
17. Mitochondria are identified for segmentation by the appearance of a thick dark colored membrane and their cristae. Two models containing mitochondria are created for each ROI: the first model (*non-extended version*) contains partial or whole mitochondria that exist fully within the defined boundary created by the ROI. This segmentation is required for the determination of mitochondrial volume density, defined as the percentage of neuronal soma volume (in each ROI) occupied by mitochondria [36]. The mitochondrial segmentations within this model are then extended so that the second model (*extended version*) contains whole mitochondria. All mitochondria that enter the defined boundary created by the ROI, whether in part or fully, are selected for segmentation so long as their total volume can be determined within the stack and their membrane is intact. This allows each mitochondrial object to be analyzed using 3D shape descriptors. Since the model containing the neuronal soma is saved as an 8-bit image that only stores a maximum of 255 materials, it is recommended that you segment the mitochondria within each version (*non-extended* and *extended versions*) as a new model per ROI. This creates a total of 40 models for the 20 ROIs generated. Each dataset will range between 7000 and 12,000 KB and one whole mitochondrion takes between 1 and 45 min to segment depending on its complexity.
18. To make segmentation of the mitochondrial membranes easier within the colored background of the segmented neuronal soma, each mitochondrial object can be isolated by right-clicking on the material under the **"Segmentation"** panel and selecting **"Show selected material only"**. Once all materials have been segmented, make a note of the minimum and maximum image numbers that correspond to first and last appearances of segmented mitochondria within the ROI. This is required for **steps 4** and **5** in Subheading 3.6.5.
19. A recommended naming system for the mitochondria segmented within each ROI is "*Case_Number_Neuron_Number_ROI Number_Nonextended.*" Save these files within a separate model case folder to the neuronal model.

20. Any mitochondria whose volume is incomplete, i.e., presents at the edges of the stack volume, should be excluded for whole mitochondrial analyses even if the mitochondrion was present in the *non-extended model.* This can happen where the mitochondrion "moves" out of the ROI as its morphology changes in 3D. It is recommended to save the model containing the mitochondrial extensions with the following naming system: "*Case_Number_Neuron_Number_ROI Number_Extended.*"
21. A third model containing cytoplasmic volume only is segmented for the calculation of mitochondrial volume density. This is necessary because it is unlikely this volume can be directly calculated for each ROI within the cytoplasm, i.e., by multiplying the dimension of the *Z*-axis within which mitochondria were segmented, by the lateral size of the ROI generated (*lateral size* (μm) = *X/Y pixel size* × *50 pixels*). The randomly generated ROIs may contain nuclear portions or extracellular space as the neuronal soma changes shape in 3D. Since only 20 ROIs will be segmented, this can be completed in one model (saved as a model containing ≤63 or ≤255 materials).
22. If any green portions within the ROI cover other parts of the neuron, including the nucleus or extraneuronal space, use the right button of the mouse and press **"Ctrl"** to retrace and remove the highlighted structure/s that are not required for volume density measurements. The nuclear segmentation is excluded from the total soma volume because mitochondria are only present in the cytoplasm.
23. A triangular approximation of the material is formed. A triangular approximation is considered more suitable to model 3D shapes since analysis from the voxel (cuboid) assembly overestimates the surface area statistic, causing the MCI values to fall below a value of 1. This step is required for all models segmented in MIB.
24. If a pop-up message appears **"out of core data"** when opening the large (.am) dataset, select the third option **"Read complete volume into memory"** and select **"OK"**. Each dataset is expected to range between a size of 792,017 KB and 1,900,325 KB for the (.am) files. If a 16-bit image (containing ≤65,535 materials) or 32-bit image (containing ≤4,294,967,295 materials) is imported into Amira™3D Pro, a warning will automatically show that states: [**"the image requires a lossless conversion to an 8-bit image. The segmentation editor only supports 8-bit label fields. Would you like to continue"** > Select **"yes"**]. All segmented materials will show in the **"Segmentation Editor"**. This contrasts to MIB as only two materials will show in the **"Materials"** list for 16-bit and 32-bit images.

25. Multiple ROIs (.am) can be opened together in one project if required for visualization purposes. At this stage, it is recommended to check the image voxel dimensions and coordinates are correct to ensure the correct behavior when generating surfaces and statistics: select the green (.am) file in the **"Project View"** panel. Go to the **"Properties"** panel and select **"Crop Editor"**, represented by the icon. A dialog will appear and the voxel size in the *X*–*Y*–*Z* axes are located next to **"Voxel size"** within the **"Resolution"** panel. The values located next to **"Min coord"** should be set to [**"0,0,0"**] if they are not already. Select **"OK"** and resave the label image.

26. **Step 6** is only required for the extended mitochondrial models. Smoothing of mitochondria is required since biological structures are naturally smooth. The anisotropic nature of SBF-SEM-generated datasets can lead to the formation of artifacts in the form of plateaus or stair-steps following surface generation. While there are several smoothing algorithms available in Amira™ 3D Pro that can reduce these artifacts (e.g., "unconstrained" smoothing of the 3D surface), a trade-off should be considered between visual quality and preserving the accuracy of the quantitative information. For instance, volume shrinkage and detachment of mitochondria connected by nanoscale structures can occur with **"Unconstrained"** smoothing. The Taubin algorithm was chosen as it preserves topological features, and is associated with the least shrinkage maintaining an MCI value ≥ 1 [37]. However, note that the algorithm does not always completely smooth the larger stair-steps. Smoothing was not applied to the polygonal surface of the neuronal soma or volume density models since volume is the only measurement required that is not significantly affected by smoothing.

27. The neuronal soma segmentation will obstruct the ability to view the mitochondria when **"Surface View"** is applied. If you wish to view the mitochondria alone, go to **"Segmentation Editor"** located next to **"Project"**. Select the unwanted material within the list (e.g., neuronal soma) and then click **"Delete"**. In addition, the transparency of the object can be altered by selecting **"Surface View"**. Go to the drop-down box next to **"Draw Style"** and change **"shaded"** to **"transparent"**. To zoom in or out on the surface created, use the mouse wheel.

28. Select the (.surf) file generated after applying the **"Generate surface"** module for the neuronal and *non-extended* mitochondrial models since only volume measurements are required. Select the (.smooth) file following Taubin smoothing for the whole mitochondria in the *extended model*.

29. The volume and surface area measurements are displayed in one row for each mitochondrial object. The exterior region is usually negative and is ignored.
30. The statistics are copied into Excel. One Excel sheet is used to calculate the 3D shape descriptors, mitochondrial complexity index (MCI), and sphericity for each mitochondrion within the *extended* models (*equations supplied in* Fig. 4c). MCI is derived from the 2D shape descriptor, Form Factor [36]. This 3D descriptor starts from a value of 1 and describes the range of morphologies present within a network that result from the continuous fission and fusion events. The formula used to calculate MCI in Excel is: =$D4^3/((16*PI()^2)*($E4^2)), where D4 is surface area and E4 is the volume of a mitochondrion. A higher MCI value represents more complex and branched, or elongated mitochondria. Sphericity is another unitless measure of 3D shape that ranges between 0 and 1, where a value of 1 describes a perfect sphere. The formula used to calculate sphericity in Excel is: =((PI()^(1/3))*(6*$E4)^(2/3))/$D4, where D4 is surface area and E4 is volume of a mitochondrion. A separate Excel sheet is used to calculate mitochondrial volume density. Volume is the only measurement extracted from the (.statistics) file generated in Amira™ 3D Pro for the *non-extended* models and cytoplasmic volume models. Total mitochondrial volume is calculated per ROI using the Excel formula = SUM(A2:A21), where A2:A21 represents the cells containing mitochondrial volumes. The volume density (%) is then calculated in Excel using the formula = (B2/C2)*100, where B2 is the total mitochondrial volume per ROI and C2 is the corresponding cytoplasmic volume per ROI.
31. A nanotunnel is identified as a long tubular structure connecting two mitochondrial bodies, or a long tubular structure protruding from one body that is freely ended (which can be termed "projection") [38]. There are several criteria to consider when defining a nanotunnel: there will be a continuity in the darkly stained double mitochondrial membrane observed within the image stack in MIB, either in the longitudinal view or transverse view. Within Amira™ 3D Pro, nanotunnels are confirmed and measure a diameter between 40 and 200 nm and a total length of up to 30 μm [38]. These structures have been theorized to form where the stress level, e.g., high intracellular calcium, or mtDNA mutations exceed a biochemical threshold where function declines, but where the mitochondria are unable to partake in fusion, particularly where they are stationary in muscle [36]. A similar structure was detected in the hippocampus from rodents and Alzheimer's donors using ssTEM, introduced earlier as MOAS [10, 39]. MOAS were

described as teardrop-shaped mitochondria (~0.5 μm in diameter) connected by a double membrane, up to 5 μm long in rodents, and were theorized to occur because of arrested fission. Their formation may counteract stress by allowing them to share their contents, while inducing a size-related "steric-hindrance" effect to prevent clearance by mitophagy.

32. If the 3D length icon does not show, type in the following command into the Tcl console: measure useNewMeasureTools1. Restart Amira™ 3D Pro. If the Tcl console does not show, use **"Window"** on the tool bar to select **"Consoles"**.

33. It is recommended that you take three measurements for each value. For the measurement of each nanotunnel diameter, this should include minimum and maximum values by visual inspection of the diameter and taking the lowest (minimum) or highest (maximum) value of three.

34. The *RF* is a unitless 3D value calculated by dividing a predefined distance by each dimension of the voxel. For SBF-SEM-generated data, the image voxel dimension is anisotropic (unequal). This means that the *RF* values for the *X* and *Y* axes are equal but greater than the *RF* for the *Z*-axis because of the higher resolution in the lateral dimensions. There are two lines of code that can be modified within the macro; the *RF* for the perinuclear region and the *RF* for the perimembrane region in all axes (*X*–*Y*–*Z*). The *RF* defines the extent of morphological dilation from the nuclear membrane in 3D to create the perinuclear region. The same *RF* values also define the amount of morphological erosion from the membrane of the neuronal soma in 3D to create the perimembrane region. The space between these regions is classified as the intermediate region. Any overlap between the perinuclear and peri-membranal regions creates a fourth "overlapping region."

35. Twenty segmentations are used to consider the natural variation in size of the neuronal soma. These models are selected from the neuronal soma segmentations created in Subheading 3.6.1. Each soma must be visibly whole in the *X*/*Y* axis and must contain a portion of nucleus that is complete at one end, i.e., shows a peak. Since a portion of the neuronal soma is segmented, the nucleus is treated as a spherical cap since evidence has shown neuronal nuclei are highly spherical in 3D [40]. The portion of soma is treated as cylindrical shape. A limitation is that the geometries chosen may not fit all partially segmented somal morphologies. The nucleus may take on an ellipsoidal morphology where the radius is not symmetrical in all axes, to a convoluted morphology under processes such as apoptosis [9]. In addition, the segmentation

of the somal membrane may include protrusions, or dendrites, causing a deviation from the cylindrical shape. These issues are partly negated since the macro maintains the original shape of the neuronal soma membrane and nuclear membrane to generate the subregions.

36. The label image (.tiff), also termed (.tif), is used in FIJI. The (.am) image file will be required for the extraction of volume measurements for the nuclei and cytoplasm. Do not save the (.am) file in the same folder as the Tiff since it is not required by the macro.
37. Subtracting the radius of the nucleus (***r***) from the total radius of the soma (***R***) gives one distance for the *X* and *Y* dimensions, radiating outwards from the nuclear membrane up to the membrane of the soma. A further division by *3* ensures there are *three* regions of equal distance: the perinuclear, intermediate region, and perimembrane region.
38. The *RF* for both *X* and *Y* axes are the same value and since models can be several kilobytes large, the macro scales the image down by a factor of 10 to decrease processing time. This is taken into consideration in the formula for the calculation of the *RF* value. The *RF* value in the *X*/*Y* axes will differ to the *RF* in the *Z*-axis due to the anisotropic voxel dimension. The *RF* for the *Z*-axis is calculated separately in the final step (**step 12** in Subheading 3.8.2). This would be the case for ssTEM data. In /contrast, for FIB-SEM-generated images the *RF* values will be the same in all axes (*X*/*Y* and *Z*) due to the isotropic voxel dimensions. Therefore, **step 12** in Subheading 3.8.2 will not be required.
39. The *RF* value corresponding to the 90th percentile can be calculated in Excel. Type the formula = *(PERCENTILE.INC (array, k)*, where array will be the range of *RF values* (*X*/*Y*) and *k* is the percentage required, written in decimal form. An example is = *(PERCENTILE.INC(J2:J21,0.9))*, where *J2:J21* represents the cells in Fig. 9F. The *RF* for the *X*/*Y* axes is now defined, and can be used for every dataset containing a neuronal soma so that the subregions can be created for the subsequent localization of mitochondria in Subheading 3.8.3.
40. The anisotropic factor is a unitless value calculated by dividing the difference between the image voxel dimension in the *X*/*Y* axis and in the *Z*-axis, e.g., *0.07 (Z-dimension)/ 0.012 (X/Y dimension)* = *5.8*. The *RF* for the *Z*-axis could also be calculated in the same way as the *RF* for the *X*/*Y* axes. However, it is unlikely the value corresponding to the 90th percentile will differ to the *RF* in *X*/*Y* axes by the correct anisotropic factor.
41. The opened label image will appear black. To view the structures in the image (*example shown in* Fig. 10a), the pixel intensity will need to be altered: go to

[**"Image"** > **"Brightness/Contrast"**] and use the minimum and maximum scrolls to change the contrast until the structure can be viewed. Do not apply this setting to the image.

42. The Tiff is resaved over the original Tiff since the voxel size will need resetting to ensure the correct behavior within FIJI. Having two versions of the same image in the **Raw_Images** folder may cause the macro to malfunction.

43. The ***Home Directory*** contains folders for the input files and folders for the output files.

Acknowledgments

This work is currently supported by the Medical Research Council (MRC) and Discovery Medicine North (DiMeN) DTP. The work is also supported by Electron Microscopy Research Services (Newcastle University), Biotechnology and Biological Sciences Research Council (BBSRC) grant reference BB/M012093/1, Wellcome Centre for Mitochondrial Research (Newcastle University), and the Newcastle Brain Tissue Resource [41]. A.E.V. is supported by a Henry Wellcome Postdoctoral Fellowship (215888/Z/19/Z).

References

1. Misgeld T, Schwarz TL (2017) Mitostasis in neurons: maintaining mitochondria in an extended cellular architecture. Neuron 96(3): 651–666
2. Cheng X-T, Zhou B, Lin M-Y, Cai Q, Sheng Z-H (2015) Axonal autophagosomes use the ride-on service for retrograde transport toward the soma. Autophagy 11(8):1434–1436
3. Faitg J, Lacefield C, Davey T, White K, Laws R, Kosmidis S et al (2021) 3D neuronal mitochondrial morphology in axons, dendrites, and somata of the aging mouse hippocampus. Cell Rep 36(6):109509
4. Mukherjee K, Clark HR, Chavan V, Benson EK, Kidd GJ, Srivastava S (2016) Analysis of brain mitochondria using serial block-face scanning electron microscopy. J Vis Exp 113: e54214
5. Chaudhari SN, Kipreos ET (2017) Increased mitochondrial fusion allows the survival of older animals in diverse C. elegans longevity pathways. Nat Commun 8(1):182
6. Detmer SA, Chan DC (2007) Functions and dysfunctions of mitochondrial dynamics. Nat Rev Mol Cell Biol 8(11):870–879
7. Kobayashi T, Kuroda S, Tada M, Houkin K, Iwasaki Y, Abe H (2003) Calcium-induced mitochondrial swelling and cytochrome c release in the brain: its biochemical characteristics and implication in ischemic neuronal injury. Brain Res 960(1):62–70
8. Verburg J, Hollenbeck PJ (2008) Mitochondrial membrane potential in axons increases with local nerve growth factor or semaphorin signaling. J Neurosci 28(33):8306–8315
9. Anglade P, Vyas S, Hirsch EC, Agid Y (1997) Apoptosis in dopaminergic neurons of the human substantia nigra during normal aging. Histol Histopathol 12(3):603–610
10. Zhang L, Trushin S, Christensen TA, Bachmeier BV, Gateno B, Schroeder A et al (2016) Altered brain energetics induces mitochondrial fission arrest in Alzheimer's disease. Sci Rep 6(1):18725
11. Sterky FH, Lee S, Wibom R, Olson L, Larsson N-G (2011) Impaired mitochondrial transport and Parkin-independent degeneration of respiratory chain-deficient dopamine neurons in vivo. Proc Natl Acad Sci 108(31): 12937–12942
12. Park J, Seo J, Won J, Yeo H-G, Ahn Y-J, Kim K et al (2019) Abnormal mitochondria in a non-human primate model of MPTP-induced Parkinson's disease: Drp1 and CDK5/p25 signaling. Exp Neurobiol 28(3):414–424

13. Okarmus J, Bogetofte H, Schmidt SI, Ryding M, García-López S, Ryan BJ et al (2020) Lysosomal perturbations in human dopaminergic neurons derived from induced pluripotent stem cells with PARK2 mutation. Sci Rep 10(1):10278
14. Calkins MJ, Manczak M, Mao P, Shirendeb U, Reddy PH (2011) Impaired mitochondrial biogenesis, defective axonal transport of mitochondria, abnormal mitochondrial dynamics and synaptic degeneration in a mouse model of Alzheimer's disease. Hum Mol Genet 20(23):4515–4529
15. Davis AF, Clayton DA (1996) In situ localization of mitochondrial DNA replication in intact mammalian cells. J Cell Biol 135(4): 883–893
16. Lawless C, Greaves L, Reeve AK, Turnbull DM, Vincent AE (2020) The rise and rise of mitochondrial DNA mutations. Open Biol 10(5):200061
17. Trimmer PA, Swerdlow RH, Parks JK, Keeney P, Bennett JP Jr, Miller SW et al (2000) Abnormal mitochondrial morphology in sporadic Parkinson's and Alzheimer's disease cybrid cell lines. Exp Neurol 162(1):37–50
18. Little D, Luft C, Mosaku O, Lorvellec M, Yao Z, Paillusson S et al (2018) A single cell high content assay detects mitochondrial dysfunction in iPSC-derived neurons with mutations in SNCA. Sci Rep 8(1):1–16
19. Bolognin S, Fossépré M, Qing X, Jarazo J, Ščančar J, Moreno EL et al (2018) 3D cultures of Parkinson's disease-specific dopaminergic neurons for high content phenotyping and drug testing. Adv Sci (Weinh) 6(1):1800927
20. Wiemerslage L, Lee D (2016) Quantification of mitochondrial morphology in neurites of dopaminergic neurons using multiple parameters. J Neurosci Methods 262:56–65
21. Wang W, Yin J, Ma X, Zhao F, Siedlak SL, Wang Z et al (2017) Inhibition of mitochondrial fragmentation protects against Alzheimer's disease in rodent model. Hum Mol Genet 26(21):4118–4131
22. Song DD, Shults CW, Sisk A, Rockenstein E, Masliah E (2004) Enhanced substantia nigra mitochondrial pathology in human α-synuclein transgenic mice after treatment with MPTP111-Methyl-4-phenyl-1,2,3,6-tetrahydropyridine. Exp Neurol 186(2):158–172
23. Denk W, Horstmann H (2004) Serial block-face scanning electron microscopy to reconstruct three-dimensional tissue nanostructure. PLoS Biol 2(11):e329
24. Harris KM, Perry E, Bourne J, Feinberg M, Ostroff L, Hurlburt J (2006) Uniform serial sectioning for transmission electron microscopy. J Neurosci 26(47):12101–12103
25. Bushby AJ, P'ng KMY, Young RD, Pinali C, Knupp C, Quantock AJ (2011) Imaging three-dimensional tissue architectures by focused ion beam scanning electron microscopy. Nat Protoc 6(6):845–858
26. Peddie CJ, Collinson LM (2014) Exploring the third dimension: volume electron microscopy comes of age. Micron 61:9–19
27. Courson JA, Landry PT, Do T, Spehlmann E, Lafontant PJ, Patel N et al (2021) Serial block-face scanning electron microscopy (SBF-SEM) of biological tissue samples. J Vis Exp 169: e62045
28. Belevich I, Joensuu M, Kumar D, Vihinen H, Jokitalo E (2016) Microscopy image browser: a platform for segmentation and analysis of multidimensional datasets. PLoS Biol 14(1): e1002340
29. Schneider CA, Rasband WS, Eliceiri KW (2012) NIH image to ImageJ: 25 years of image analysis. Nat Methods 9(7):671–675
30. Kremer JR, Mastronarde DN, McIntosh JR (1996) Computer visualization of three-dimensional image data using IMOD. J Struct Biol 116(1):71–76
31. Lanika Solutions (2021) Amira software for life & biomedical sciences. ThermoScientific, Bangalore
32. Sommer C, Straehle C, Koethe U, Hamprecht FA (eds) (2011) Ilastik: Interactive learning and segmentation toolkit. 2011 IEEE international symposium on biomedical imaging: from nano to macro. IEEE
33. Schindelin J, Arganda-Carreras I, Frise E, Kaynig V, Longair M, Pietzsch T et al (2012) Fiji: an open-source platform for biological-image analysis. Nat Methods 9(7): 676–682
34. Kaynig V, Fuchs T, Buhmann JM (eds) (2010) Neuron geometry extraction by perceptual grouping in ssTEM images. 2010 IEEE computer society conference on computer vision and pattern recognition; 13–18 June 2010
35. Liu T, Jurrus E, Seyedhosseini M, Ellisman M, Tasdizen T (eds) (2012) Watershed merge tree classification for electron microscopy image segmentation. Proceedings of the 21st International Conference on Pattern Recognition (ICPR2012); 11–15 Nov 2012
36. Vincent AE, White K, Davey T, Philips J, Ogden RT, Lawless C et al (2019) Quantitative 3D mapping of the human skeletal muscle mitochondrial network. Cell Rep 26(4): 996–1009.e4

37. Taubin G (ed) (1995) Curve and surface smoothing without shrinkage. Proceedings of IEEE international conference on computer vision. IEEE
38. Vincent AE, Turnbull DM, Eisner V, Hajnóczky G, Picard M (2017) Mitochondrial nanotunnels. Trends Cell Biol 27(11): 787–799
39. Trushina E (2016) A shape shifting organelle: unusual mitochondrial phenotype determined with three-dimensional electron microscopy reconstruction. Neural Regen Res 11(6):900
40. Calì C, Agus M, Kare K, Boges DJ, Lehväslaiho H, Hadwiger M et al (2019) 3D cellular reconstruction of cortical glia and parenchymal morphometric analysis from serial block-face electron microscopy of juvenile rat. Prog Neurobiol 183:101696
41. Newcastle Brain Tissue Resource. Institute of Neuroscience, Newcastle University, Campus for Ageing and Vitality, Newcastle upon Tyne, NE4 5PL

Chapter 12

DNeuroMAT: A Deep-Learning-Based Neuron Morphology Analysis Toolbox

Min Liu, Zhuangdian Lin, Weixun Chen, Erik Meijering, and Yaonan Wang

Abstract

Digital reconstruction of neuronal structures from 3D neuron microscopy images is critical for the quantitative investigation of brain circuits and functions. Currently, neuron reconstructions are mainly obtained by manual or semiautomatic methods. However, these ways are labor-intensive, especially when handling the huge volume of whole brain microscopy imaging data. Here, we present a **d**eep-learning-based **neuro**n **m**orphology **a**nalysis **t**oolbox (DNeuroMAT) for automated analysis of neuron microscopy images, which consists of three modules: neuron segmentation, neuron reconstruction, and neuron critical points detection.

Key words 3D neuron reconstruction, Deep learning, Image analysis, Critical points detection, Image segmentation

1 Introduction

Neuronal morphology plays a crucial role in understanding the workings of brain circuits and the mechanisms of the nervous system [1, 2]. The study of three-dimensional (3D) neuronal morphology contributes to understanding the neuronal growth, death, regeneration, and connection. It is of great significance to the research and treatment of central nervous system diseases [3]. In order to carry out quantitative measurement and statistical analysis of neuron morphological properties, many software tools have been developed for manually and semiautomatically reconstructing neuronal structures more efficiently and reliably such as Virtual Finger [4] and TeraVR [5]. However, it is still a tedious and labor-intensive task. To speed up the neuron reconstruction process, efforts have been taken to develop more automatic neuron reconstruction algorithms and tools.

Over the past decades, researchers focused on traditional computing techniques and proposed many methods [6–19]. Most of

Kazuhito Toyooka (ed.), *Neuronal Morphogenesis: Methods and Protocols*, Methods in Molecular Biology, vol. 2831,
https://doi.org/10.1007/978-1-0716-3969-6_12,

these methods are based on artificially designed features, and their performances depend on parameter tuning and image preprocessing to a certain extent. Recently, the DIADEM challenge [20] and the BigNeuron project [21] have revealed that existing neuron reconstruction methods still fail to process low-quality images accurately and efficiently. To obtain better neuron morphology reconstructions, many segmentation algorithms for neuron structures have been proposed [22–25]. These methods demonstrate considerable performances when processing images with high contrast and clear background, but fail to maintain their performances when processing images containing tiny neuron structures or strong background noises [26].

In recent years, we have witnessed that deep-learning-based methods outperform traditional methods in a wide variety and still growing number of pattern recognition and computer vision tasks. And many deep-learning-based methods have been proposed for neuron microscopy image analysis and neuron reconstruction, most of which are based on Convolutional Neural Network (CNN). Li and Shen [27] proposed a 3D U-Net-based network that has the ability to segment the target neuron and cope with the interference of peripheral nerve fibers. Combining residual learning and inception learning, Li et al. [28] proposed a novel CNN architecture to segment the neuronal voxels from 3D microscopy images. Yang et al. [29] proposed a voxel-wise multilevel Fully Convolutional Network (FCN) and a Hessian-repair model for 3D neuron image segmentation and broken structures repair, respectively. In ref. [30], a method based on deep reinforcement learning has been proposed for subpixel neural tracking in 2D microscopy images. To reduce the dependence on manual annotations, based on progressive learning, Huang et al. [31] and Zhao et al. [32] iteratively trained the neuron segmentation networks and updated the pseudo-labels produced by automatic neuron reconstruction methods.

Considering many neuron reconstruction methods, such as [15–17], require automatically detected or manually identified seed points, neuronal critical points detection methods have been proposed, such as terminations [3, 33, 34], and branch points [35, 36]. The detected critical points can be utilized as seeds in reconstruction methods to enhance their performance. Besides serving as starting points, critical points are also helpful for repairing and revising neuron reconstructions obtained from automatic reconstruction methods. Yu et al. [37] made use of terminations extracted from initial reconstruction and branch points detected from the microscopy image to repair prematurely terminated neurite traces and untraced branches, respectively. Guo et al. [38] utilized the detected crossover points to eliminate false connections and crossover structures.

Here, we describe DNeuroMAT (**d**eep-learning-based **neuro**n **m**orphology **a**nalysis **t**oolbox), which consists of three modules: neuron segmentation, neuron reconstruction, and critical points detection. The toolbox works with 3D light microscopy images of single neurons. We provide comprehensive step-by-step descriptions of how to use the toolbox along with brief explanations of the theoretical principles of the underlying methods. DNeuroMAT is freely available,[1] and we give an example of its usage.

2 Materials

The presented toolbox is developed as a plugin of Vaa3D [17], an open-source software platform for the visualization and analysis of large-scale multidimensional images. The deep learning frameworks Pytorch and TensorFlow are used to build and train the CNNs for the implementation of neuron analysis modules in DNeuroMAT.

2.1 Vaa3D Platform

As an open-source, cross-platform visualization and analysis system, Vaa3D, provides various released binary versions and plugins for users and source codes for developers. Here, we use the relatively stable version for developing our toolbox (*see* **Note 1**). There are two approaches to get the executable Vaa3D, and we recommend the second one.

1. For users, go to the Vaa3D site on GitHub[2] and download the released binary version from the "release" page.
2. For developers, go to the same site, then download the source code of Vaa3D from the "v3d_external" page and, if needed, additional plugins from the "vaa3d_tools" page, and build the software following the online tutorial related to your operating system.

2.2 DNeuroMAT Plugin

DNeuroMAT contains three neuron microscopy image analysis modules: neuron segmentation, neuron reconstruction, and critical points detection. Besides, we designed a method-editable (add, delete, and move) and parameter-editable (add, delete, and modify) graphic user interface (GUI) for further development while allowing users to add their methods into the toolbox for better performance. Users can easily add DNeuroMAT to Vaa3D following the steps below.

[1] DNeuroMAT can be obtained from https://github.com/lzd0123/DNeuroMAT.git

[2] Here is the Vaa3D site on GitHub: https://github.com/Vaa3D

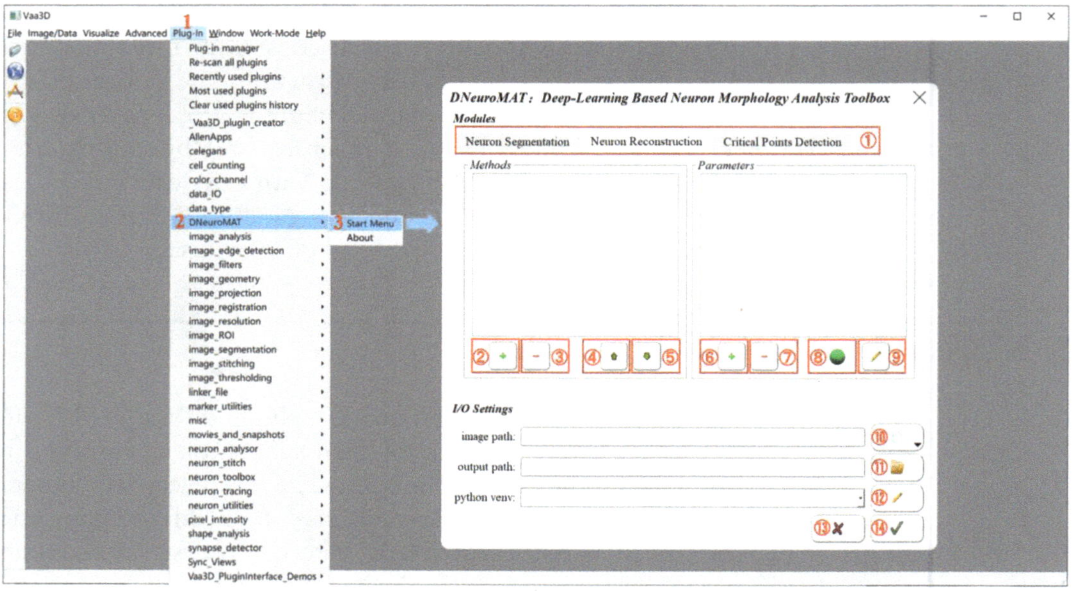

Fig. 1 Activating DNeuroMAT from the main window of Vaa3D

1. Download the source codes of DNeuroMAT and put them into the folder "**{Vaa3DPath}**/vaa3d_tools/released_plugins/v3d_plugins/" where **{Vaa3DPath}** is the installation path of Vaa3D on your computer.
2. Copy the compiled DNeuroMAT plugin file (DNeuroMAT.dll on Windows, libDNeuroMAT.so on Linux) to "**{Vaa3D-Path}**/v3d_external/bin/plugins/DNeuroMAT/" or recompile the DNeuroMAT plugin with CMake (*see* **Note 2**).
3. Start Vaa3D and the plugin will be loaded automatically. Users can find the toolbox in **Plug-in > DNeuroMAT** (shown in Fig. 1, more details in Table 1).

2.3 File Formats

Our toolbox takes 3D microscopy image stacks in TIFF format as input by default. For data in other formats, users can convert them into TIFF format using Vaa3D or other tools. For neuron segmentation, the outputs are stored in the same format as input. The outputs of neuron reconstruction and critical points detection are stored in SWC format [39] and Marker format [17], respectively. SWC format is widely used to digitally represent neuron structures using tree structure, and it consists of number, type, position, radius, and parent of each node. Marker format consists of position, radius, shape, name, comment, and color; here, we use comment and color to distinguish the different kinds of critical points (branch points in red, terminations in green, crossover points in blue).

Table 1
Details of DNeuroMAT GUI

Name	Description
① Module Tabs	Select the module of interest.
② Add Method Button	Add new method to the corresponding module.
③ Delete Method Button	Delete existing method of the corresponding module.
④ Move Up Button	Move the selected method up.
⑤ Move Down Button	Move the selected method down.
⑥ Add Parameter Button	Add parameter for the corresponding method.
⑦ Delete Parameter Button	Delete existing parameter of the corresponding method.
⑧ Segmentation Checker	Indicate whether to run with neuron segmentation (used for neuron reconstruction and critical points detection).
⑨ Edit Parameter Button	Edit the name of the selected parameter.
⑩ Select Image Path Button	Select the path of TIFF image (file or folder).
⑪ Select Output Path Button	Select the path to save outputs (folder).
⑫ Edit Python Environment Button	Edit existing Python virtual environment.
⑬ Cancel Button	Cancel and close.
⑭ Ok Button	Confirm and run.

3 Methods

Here, we detail how to run the three modules in DNeuroMAT. To start, launch Vaa3D and activate DNeuroMAT using the menu item **Plug-in > DNeuroMAT > Start Menu**. Next, choose one method for the module of interest from the ***Methods*** box and specify the corresponding parameters in the ***Parameters*** box. Then, specify a TIFF image or a folder that contains TIFF images through the **Select Image Path Button** in DNeuroMAT, or load a TIFF image through the menu item **File > Open image**... in Vaa3D before activating DNeuroMAT (*see* **Note 3**). After that, assign an existing or new folder where the outputs will be stored in using the **Select Output Path Button** (*see* **Note 4**). Finally, select an existing Python virtual environment satisfying running requirements from the drop-down box or create a new environment and add it through the **Edit Python Environment Button** (*see* **Note 5**). This concludes all the settings and DNeuroMAT is now ready to run.

Table 2
Summary of neuron segmentation parameters

Method	Parameter	Description
SGSNet	Threshold	A threshold for binarizing the probability maps (set to 0.5 by default).
DC-BLSTM	Lambda	A weight factor for calculating local/global intensity features (set to 5 by default).
	Threshold	A threshold for classifying each node of the rays into foreground node or background node (set to 0.5 by default).

3.1 Neuron Segmentation

As the basic module for both neuron reconstruction and critical points detection in DNeuroMAT, we provide two deep-learning-based methods in neuron segmentation, SGSNet [40], and DC-BLSTM [41]. The following is the parameter setting process of those methods, and all parameters have been summarized in Table 2.

3.1.1 SGSNet

Focusing both on the voxel-wise errors and the neuronal structure errors, we proposed a two-branch 3D neuron segmentation network called Structure-Guided Segmentation Network (SGSNet) containing one shared base encoder but two different decoding paths.

1. Select **Neuron Segmentation > SGSNet** (shown in Fig. 2).
2. Specify the expected threshold for binarizing the probability maps derived from SGSNet in the parameter **Threshold** (*see* **Note 6**).
3. Click the **Ok Button**, all expected parameters are saved in a file named **segmentation.yaml** in the output folder, and then the neuron segmentation algorithm will start automatically. All results are stored in TIFF format in a subfolder (named **segmentations**) of the output folder after running. If only one TIFF image is taken as input, the result will automatically display.

3.1.2 DC-BLSTM

Based on the ray-shooting model, we proposed a neuronal structures segmentation method using a dual channel bidirectional LSTM (DC-BLSTM) and successfully converted the 3D image segmentation task into multiple 1D ray/sequence segmentation tasks.

1. Select **Neuron Segmentation > DC-BLSTM** (shown in Fig. 3).
2. Specify the expected weight factor for calculating local/global intensity features in the parameter **Lambda** (*see* **Note 7**).

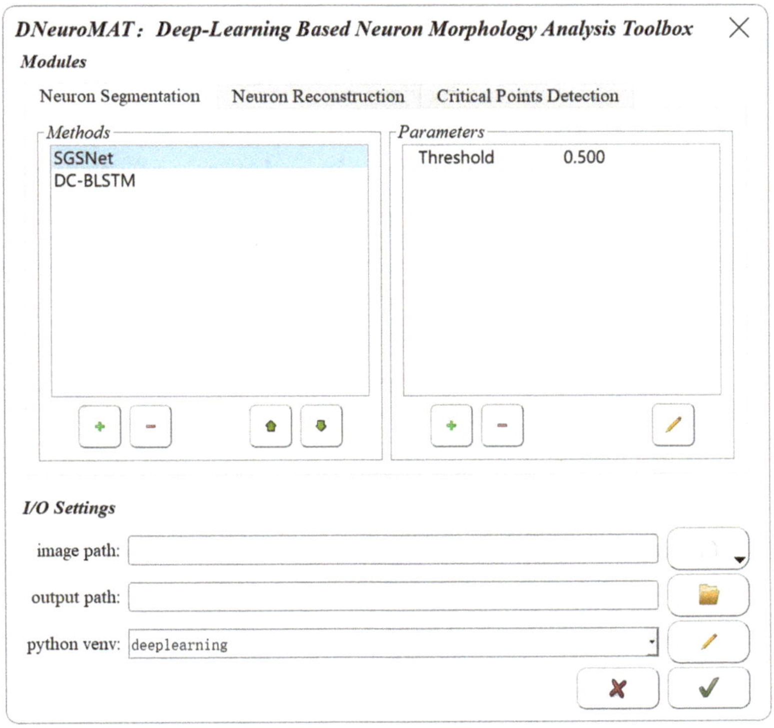

Fig. 2 Neuron segmentation using SGSNet

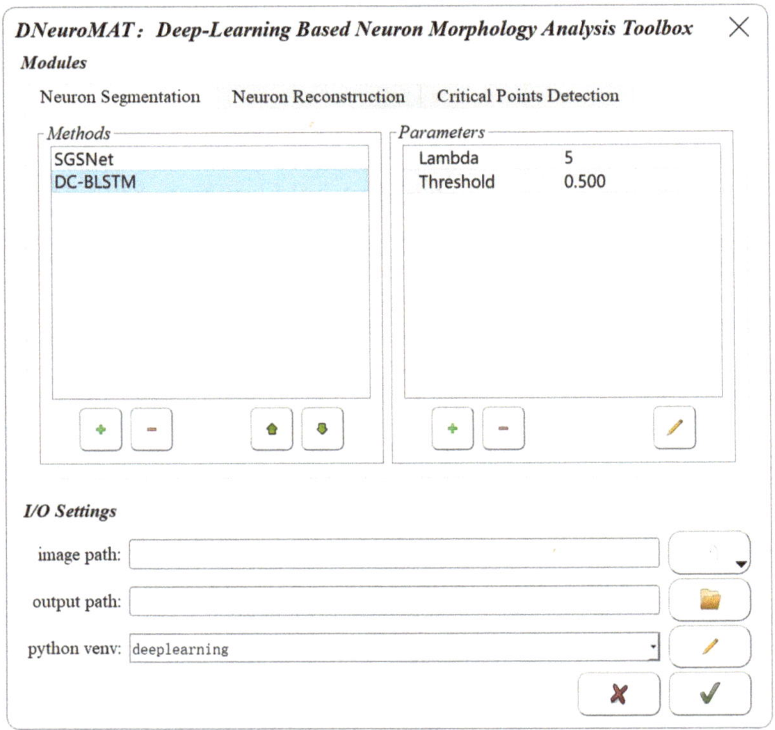

Fig. 3 Neuron segmentation using DC-BLSTM

Table 3
Summary of neuron reconstruction parameters

Parameter	Description
Ma	The number of azimuth angles for the SPE model (set to 32 by default).
Mp	The number of polar angles for the SPE model (set to 32 by default).
Md	The square root of the number of tracing directions (set to 32 by default).
N	The number of concentric spherical surfaces for the SPE model (set to 10 by default).
SphereRadiiStep	The radius growth step of concentric spherical surfaces for the SPE model (set to 1 by default).
Lambda	A scale factor to determine a local region (set to 4 by default).
AngleThreshold	The angle threshold to avoid tracing backward (set to $\pi/3$ by default).
MaxIteration	The maximum tracing iteration from a seed point (set to 100 by default).
StepSize	The minimum tracing step during the tracing process (set to 1 by default).
Z-ResampleRate	The resample rate along the *z*-axis (set to 1 by default).
GammaCorrection	The gamma correction factor for enhancing neuron images (set to 0.454 by default).
SaveSegmentation	Whether to save the segmentation results when running with neuron segmentation (set to True by default).
SaveSeed	Whether to save the seed points derived from skeletonized segmentation results (set to True by default).

3. Specify the expected threshold for classifying each node of the rays into foreground node or background node in the parameter **Threshold** (*see* **Note 8**).
4. Upon clicking the **Ok Button**, the rest of the procedure is the same as for SGSNet described above.

3.2 Neuron Reconstruction

Based on the spherical-patches extraction (SPE) model and CNNs, we proposed a deep-learning-based method named SPE-DNR [42], consisting of two functional heads (a neurite tracer and a classifier) for neuron reconstruction. The neurite tracer starts from a set of seed points, iteratively determining the tracing directions (seed points can be obtained from skeletonized neuron segmentation results). The classifier calculates the radii of the neuronal structures or stops the tracing process when the neurite tracer steps into background region. The following is the parameter setting process of the method, and all parameters have been summarized in Table 3.

1. Select **Neuron Reconstruction** > **SPE-DNR** (shown in Fig. 4).

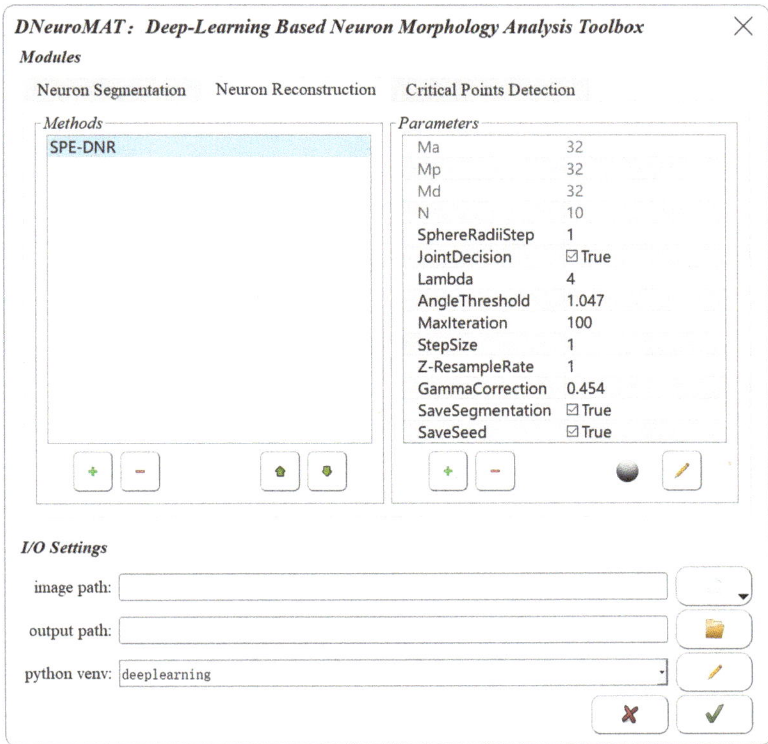

Fig. 4 Neuron reconstruction using SPE-DNR

2. For SPE-DNR, the parameters **Ma**, **Mp**, **Md**, and **N** are fixed, and thus can be ignored. Specify the expected radius growth step of concentric spherical surfaces for the SPE model in the parameter **SphereRadiiStep** (*see* **Notes 9** and **10**).
3. Specify whether to use the joint decision scheme to decide the tracing direction in the parameter **JointDecision** (*see* **Note 11**).
4. Specify the expected scale factor to determine a local region in the parameter **Lambda** (*see* **Note 12**).
5. Specify the expected angle threshold for tracing neurites in the parameter **AngleThreshold** (*see* **Note 13**).
6. Specify the expected maximum tracing iteration in the parameter **MaxIteration** (*see* **Note 14**).
7. Specify the expected minimum tracing step in the parameter **StepSize** (*see* **Note 15**).
8. Specify the expected resample rate along the *z*-axis in the parameter **Z-ResampleRate** (*see* **Note 16**).
9. Specify the expected gamma correction factor for enhancing the neuron image in the parameter **GammaCorrection** (*see* **Note 17**).

10. Specify whether to save the intermediate results in the parameters **SaveSegmentation** and **SaveSeed** (*see* **Note 18**).
11. Specify whether to run the neuron reconstruction module with the neuron segmentation module by checking the **Segmentation Checker** or not (*see* **Note 19**).
12. Click the **Ok Button**, all expected parameters are saved in a file named **reconstruction.yaml** in the output folder, and then the neuron reconstruction algorithm will start automatically. All results are stored in SWC format in a subfolder (named **reconstructions**) of the output folder after running. If only one TIFF image is taken as input, the result will automatically display.

3.3 Critical Points Detection

In ref. [43], we proposed a method named SPE-Net, consisting of the SPE model and a 2D multistream CNN for detecting critical points from 3D neuron images. The detected critical points can be fed to existing neuron reconstruction methods, which rely on the availability and quality of initial seed points. In our method, critical points detection is a classification task classifying all initial seed points (derived from skeletonized neuron segmentation results) into four classes (terminations, branch points, crossover points, and noncritical points). The following is the parameter setting process of the method, and all parameters have been summarized in Table 4.

Table 4
Summary of critical points detection parameters

Parameter	Description
Ma	The number of azimuth angles for the SPE model (set to 64 by default).
Mp	The number of polar angles for the SPE model (set to 64 by default).
N	The number of concentric spherical surfaces for the SPE model (set to 20 by default).
SphereRadiiStep	Same as neuron reconstruction (*see* Table 3).
Branch	Whether to save detected branch points (set to True by default).
Termination	Whether to save detected terminations (set to True by default).
Crossover	Whether to save detected crossover points (set to True by default).
Z-ResampleRate	Same as neuron reconstruction (*see* Table 3).
GammaCorrection	Same as neuron reconstruction (*see* Table 3).
SaveSegmentation	Similar to neuron reconstruction (*see* Table 3).
SaveSeed	Similar to neuron reconstruction (*see* Table 3).

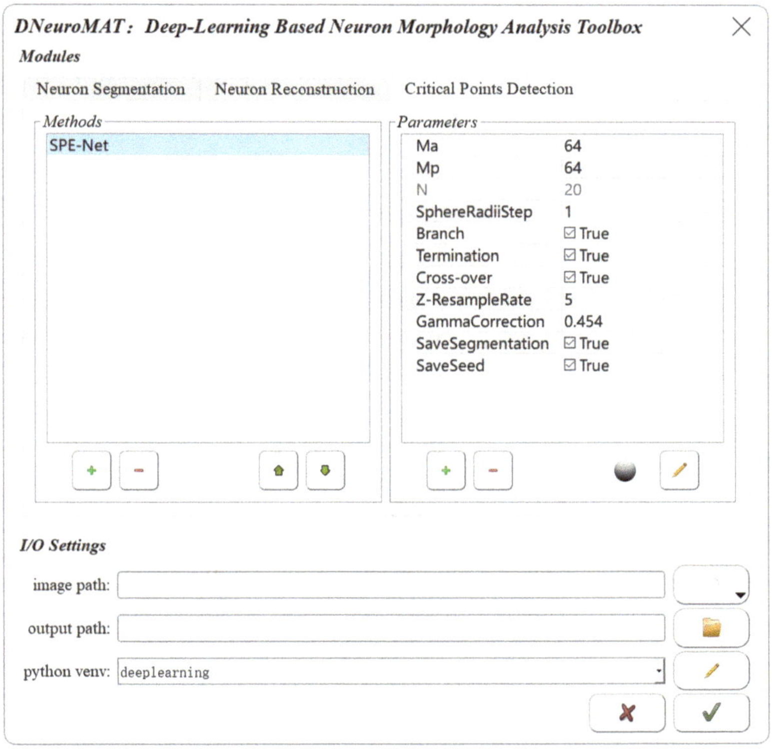

Fig. 5 Critical points detection using SPE-Net

1. Select **Critical Points Detection** > **SPE-Net** (shown in Fig. 5).
2. Specify the expected number of azimuth angles for the SPE model in the parameter **Ma** (*see* **Note 20**).
3. Specify the expected number of polar angles for the SPE model in the parameter **Mp** (*see* **Note 20**).
4. For SPE-Net, the parameter **N** is fixed and can be ignored.
5. Specify the expected **SphereRadiiStep**, **Z-ResampleRate**, and **GammaCorrection**, which are same parameters as for neuron reconstruction described above.
6. Specify whether to save the detected branch points, terminations, and crossover points in the parameters **Branch**, **Termination**, and **Crossover**, respectively (*see* **Note 21**).
7. Specify whether to save the intermediate results in the parameters **SaveSegmentation** and **SaveSeed**.
8. Specify whether to run the critical points detection module with the neuron segmentation module by checking the **Segmentation Checker** or not.
9. Click the **Ok Button**, all expected parameters are saved in a file named **detection.yaml** in the output folder, and then the critical points detection algorithm will start automatically. All

results are stored in Marker format in a subfolder (named **detections**) of the output folder after running. If only one TIFF image is taken as input, the result will automatically display.

3.4 How to Add New Method

In DNeuroMAT, users are allowed to add their own methods for better performance. Here, we describe how to add a new method.

1. Launch Vaa3D and activate DNeuroMAT, then select the corresponding module tab (Neuron Segmentation, Neuron Reconstruction, or Critical Points Detection) according to your method.
2. Click the **Add Method Button** in the ***Methods*** box, a dialog for naming method is displayed, then type the expected name of your method (shown in Fig. 6a).
3. Click the **Add Parameter Button** in the ***Parameters*** box, three dialogs are displayed successively for naming the parameter, specifying the parameter type, and setting the parameter default value (shown in Fig. 6b), then add all necessary parameters for your method (*see* **Note 22**).

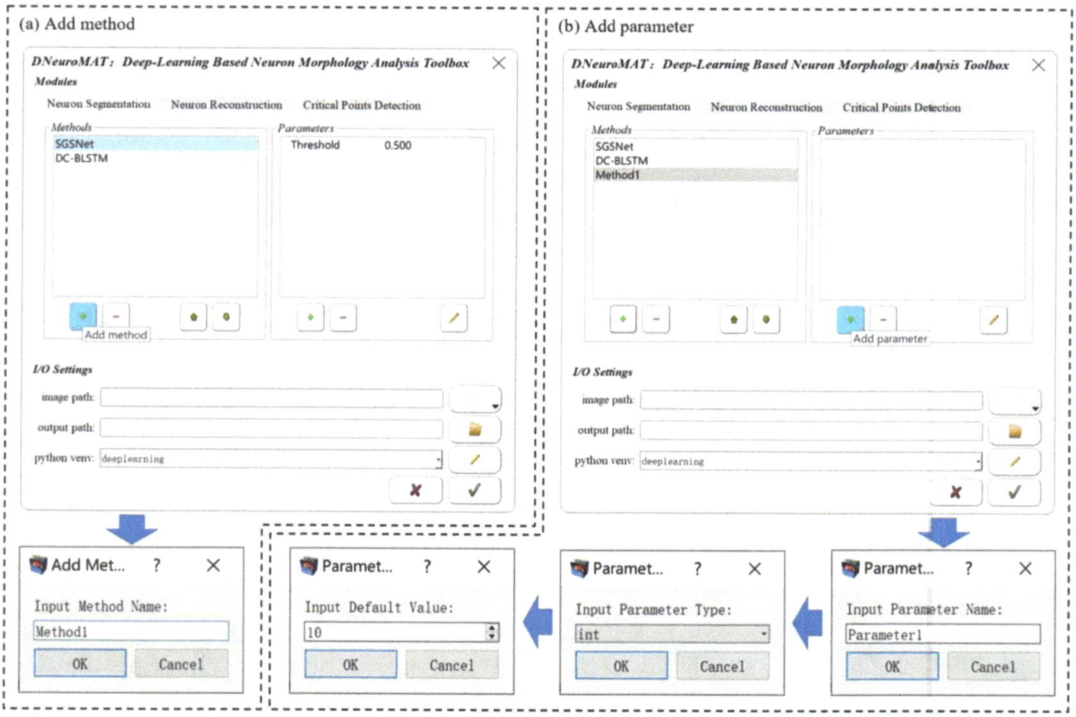

Fig. 6 Add a new method and corresponding parameter

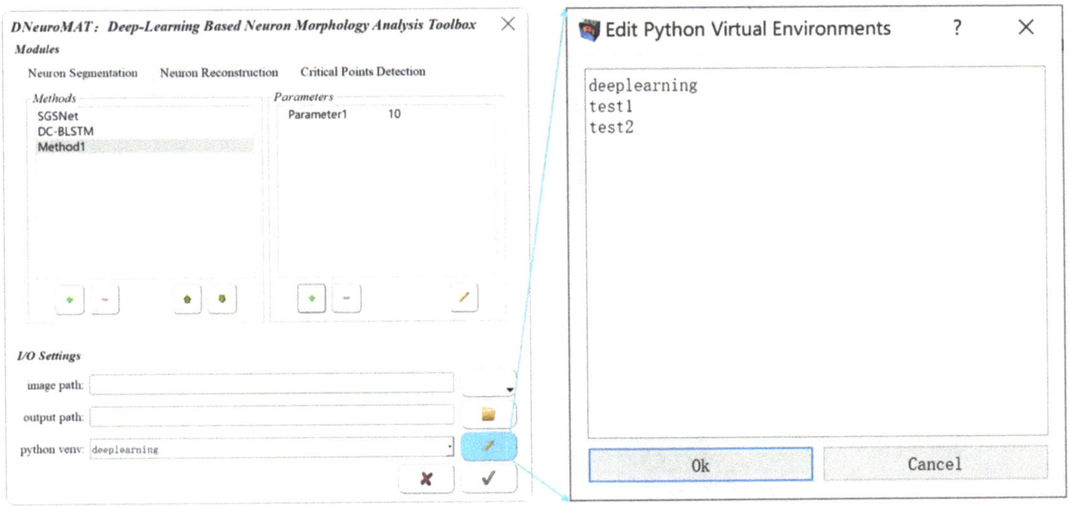

Fig. 7 Edit Python Environments

4. Click the **Edit Python Environment Button** (shown in Fig. 7) and add the Python environment satisfying your method to DNeuroMAT (*see* **Note 23**).
5. Put the source code of your method into "**{Vaa3DPath}**/vaa3d_tools/released_plugins/v3d_plugins/DNeuroMAT/src/python/" (*see* **Note 24**).

3.5 Example Use Case of DNeuroMAT

Here, we describe an example case of DNeuroMAT (shown in Fig. 8). Taking a 3D neuron microscopy image from the Whole Mouse Brain Sub-image (WMBSub) dataset [44] as input (Fig. 8a), SGSNet is firstly adopted to segment neuronal structures for producing initial seed points (Fig. 8b), and then according to the extracted seed points, SPE-DNR and SPE-Net can be used for neuron reconstruction and critical points detection, respectively (Fig. 8c).

4 Notes

1. The presented toolbox is built on Vaa3D with Qt version 4.8.6. Compatibility problems may occur when built on other versions. Besides, the current DNeuroMAT cannot be used on the latest Vaa3D (Qt6 version) as Qt6 is incompatible with Qt4. Here, we strongly recommend to use Vaa3D with Qt version 4.8.6 for avoiding unexpected issues when using DNeuroMAT.
2. Follow the steps below to compile DNeuroMAT with CMake:
   (a) Download and install CMake from https://cmake.org/ (make sure version ≥3.22), then add "**{CMakePath}**/bin" to the **PATH** environment variable of your system, where **{CMakePath}** is the installation path of CMake.

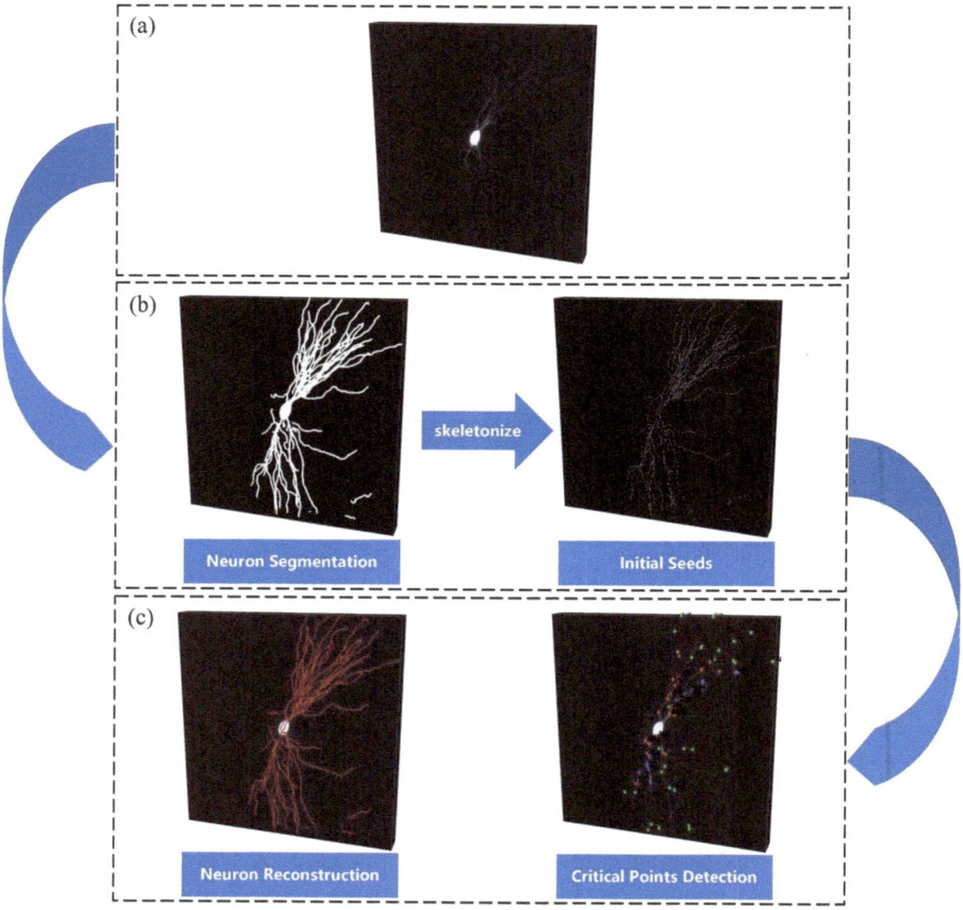

Fig. 8 Example use case of DNeuroMAT. (**a**) A 3D neuron microscopy image from the Whole Mouse Brain Sub-image (WMBSub) dataset; (**b**) segmenting neuronal structures using SGSNet and then producing initial seed points by skeletonization; (**c**) according to the extracted seed points, reconstructing neuron morphology using SPE-DNR and detecting neuronal critical points using SPE-Net

(b) Set the variable **VAA3DPATH** to **{Vaa3DPath}** in Line 5 of **CMakelists.txt** in **{Vaa3DPath}**/vaa3d_tools/released_plugins/v3d_plugins/DNeuroMAT/ (on Linux, rename CMakelists.txt to CMakeLists.txt).

(c) For Windows users, run the following commands in the command prompt, then open **DNeuroMAT.vcxproj** in **{Vaa3DPath}**/vaa3d_tools/released_plugins/v3d_plugins/DNeuroMAT/build using Visual Studio 2013 and build DNeuroMAT in "Release x64" mode:

① cd **{Vaa3DPath}**/vaa3d_tools/released_plugins/v3d_plugins/DNeuroMAT

② cmake . -G "Visual Studio 12 2013" -A x64 -B build

(d) For Linux users, run the following commands in the terminal:

① cd **{Vaa3DPath}**/vaa3d_tools/released_plugins/v3d_plugins/DNeuroMAT

② cmake . -B build

③ cd build && make

3. In addition to directly typing the path of TIFF image, there are other ways to specify the image to be processed for DNeuroMAT.

   (a) Click the **Select Image Path Button** and select a TIFF image or a folder, and DNeuroMAT will search for all TIFF images in the selected folder (nonrecursive) if the current image path is a folder after clicking the **Ok Button**.

   (b) Before activating DNeuroMAT, load TIFF images in Vaa3D, and DNeuroMAT will process the image in current window.

4. Make sure that the output path is not empty, or the outputs will be saved to **{Vaa3DPath}**/v3d_external/bin.

5. A suitable Python virtual environment is necessary for running neuron analysis modules of DNeuroMAT because our methods are implemented using Python. Follow the steps below to create a Python environment satisfying running requirements:

   (a) Download and install Anaconda (agree to **Add Anaconda3 to the system PATH environment variable** when installing) from https://www.anaconda.com/.

   (b) Create a new Python virtual environment using the command "conda create -n env_name python=3.8" (env_name is the name defined by users).

   (c) Activate the created environment using the command "conda activate env_name".

   (d) Run the following commands to install the required Python packages:

   ① cd **{Vaa3DPath}**/vaa3d_tools/released_plugins/v3d_plugins/DNeuroMAT/src/python

   ② pip install -r requirements.txt

6. SGSNet generates a probability map by processing the last convolutional layer with the Sigmoid function, and a suitable threshold is required to distinguish foreground and background.

7. Four features (the local voxel-intensity feature **LI**, the global voxel-intensity feature **GI**, the local boundary-response feature **LB**, and the global boundary-response feature **GB**) are calculated using the ray-shooting model, and then DC-BLSTM

takes **[LI, LB]** and **[GI, GB]** as inputs. The larger the **Lambda**, the larger the **LI** and **GI**. Modifying the **Lambda** will affect the performance.

8. DC-BLSTM generates a probability vector indicating the foreground probability of the nodes of a ray, and a threshold is used to classify the nodes into foreground node or background node.
9. In the spherical coordinate system, any point p_m belonging to a set of spherical surfaces can be determined by point p and (α_{m1}, φ_{m2}, r_i), where α_{m1} is the azimuth angle and $\alpha_{m1} = 2\pi\, m_1/\mathbf{Ma}$ ($m_1 = 1, 2, \ldots,$ **Ma**), φ_{m2} is the polar angle and $\varphi_{m2} = \arccos(2\, m_2/(\mathbf{Mp} + 1) - 1) - \pi/2$ ($m_2 = 1, 2, \ldots,$ **Mp**), and r_i is the radius of the ith spherical surface. The larger the **Ma** and **Mp**, the denser the distribution of azimuth angles and polar angles. The **Md** is the square root of the number of tracing directions, indicating that the neurite tracer outputs a ($\mathbf{Md}^2$)-dimensional vector to represent different directions on a unit sphere. For SPE-DNR, **Ma**, **Mp**, and **Md** are fixed because they are bound to the model during training.
10. Taking a seed point s at location (x, y, z) as a start, the first step of SPE is generating **N** (set to 10 for SPE-DNR and 20 for SPE-Net) concentric spherical surfaces centered at s with growing radii (starting from 1 with a fixed step defined by **SphereRadiiStep**) for extracting intensity distribution features around s. For both SPE-DNR and SPE-Net, **N** is fixed because it is bound to the models during training.
11. The joint decision scheme is used to prevent the tracing process from deviating from the centerline of the neurite when the neurite tracer outputs tracing directions with low confidences. In this case, seed points can be used to help the SPE-DNR to make a better decision since they are located on the neuronal centerlines.
12. The **Lambda** parameter is used to determine a local region with size **Lambda** · r, where r is the radius of the neuronal structure estimated by the classifier. The seed points in the local region will be used to calculate a new tracing direction. This parameter is valid only when using the joint decision scheme.
13. To prevent the SPE-DNR from moving backward, the angle between the previous tracing direction and the next tracing direction should be smaller than a specific threshold (namely the **AngleThreshold**).
14. The **MaxIteration** makes sure that the tracing process from one seed point is eventually stopped when all special stopping criteria (reach background regions, explored regions, or soma regions) are not met.

15. The **StepSize** indicates the minimum tracing step during the tracing process. It is used to calculate the next position when the radius of the neuronal structure estimated by the classifier is smaller than **StepSize**.

16. Considering the resolution in the *z* dimension of neuron images, which is usually lower than in the *x* and *y* dimensions, the **Z-ResampleRate** is used to scale the size of the *z*-axis (using cubic interpolation) before processing the images. For making neuronal structures sparse, keep the **Z-ResampleRate** > 1 to expand the *z*-axis. On the contrary, for reducing memory usage, keep the **Z-ResampleRate** < 1 to compress the *z*-axis.

17. As neuron microscopy images usually contain weak signals, gamma correction can be utilized to make the images brighter in order to facilitate tracing such weak structures. For this purpose, make sure the **GammaCorrection** < 1.

18. The initial seed points used for tracing neurites are derived from the skeletonized neuron segmentation image. To avoid segmenting the same image and skeletonizing the same segmentation result repeatedly when modifying reconstruction parameters, it is helpful to save the intermediate results after a reconstruction process.

19. For cases requiring neuron segmentation (e.g., an image that has not been segmented yet, using other segmentation methods, and so on), the **Segmentation Checker** makes sure that the neuron segmentation will be run during this process (green means ON, gray means OFF).

20. For SPE-Net, the **Ma** and **Mp** parameters are allowed to be modified, but since they were both set to 64 when training the network, modifying them may affect the performance, and thus we recommend keeping them as is.

21. By specifying the **Branch**, **Termination**, and **Crossover**, the users can choose what kinds of critical points they want to save.

22. Four types of parameters are supported in DNeuroMAT, namely ***Boolean***, ***Int***, ***Double***, and ***String***. Specify one of them from the drop-down box after naming (empty name, name containing space, and duplicate name are prohibited).

23. A text editor containing the added environments will be displayed after clicking the **Edit Python Environment Button**. Make sure that the name of the new environment to be added is the same as the name when creating it.

24. The core functions of DNeuroMAT are implemented using Python. According to the example we provided, write a Python file and name it "run_**{method_name}**.py" where **{method_name}** must be the same as the name when adding it.

Acknowledgments

This work was supported in part by the National Natural Science Foundation of China under Grant 62073126, in part by the Key Research and Development Program of Hunan Province under Grant 2022WK2011, and in part by International Scientific and Technological Innovation Cooperation Base for Biomedical Image Processing of Hunan Province under Grant 2021CB1009.

References

1. Meijering E (2010) Neuron tracing in perspective. Cytometry A 77:693–704
2. Donohue DE, Ascoli GA (2011) Automated reconstruction of neuronal morphology: an overview. Brain Res Rev 67:94–102
3. Liu M, Chen W, Wang C, Peng H (2019) A multiscale ray-shooting model for termination detection of tree-like structures in biomedical images. IEEE Trans Med Imaging 38:1923–1934
4. Peng H et al (2014) Virtual finger boosts three-dimensional imaging and microsurgery as well as terabyte volume image visualization and analysis. Nat Commun 5:1–13
5. Wang Y et al (2019) TeraVR empowers precise reconstruction of complete 3-D neuronal morphology in the whole brain. Nat Commun 10: 1–9
6. Xie J, Zhao T, Lee T, Myers E, Peng H (2011) Anisotropic path searching for automatic neuron reconstruction. Med Image Anal 15:680–689
7. De J et al (2016) A graph-theoretical approach for tracing filamentary structures in neuronal and retinal images. IEEE Trans Med Imaging 35:257–272
8. Peng H, Long F, Myers G (2011) Automatic 3D neuron tracing using all-path pruning. Bioinformatics 27:i239–i247
9. Xiao H, Peng H (2013) APP2: automatic tracing of 3D neuron morphology based on hierarchical pruning of a gray-weighted image distance tree. Bioinformatics 29:1448–1454
10. Liu S, Zhang D, Liu S, Feng D, Peng H, Cai W (2016) Rivulet: 3D neuron morphology tracing with iterative back-tracking. Neuroinformatics 14:387–401
11. Liu S, Zhang D, Somg Y, Peng H, Cai W (2018) Automated 3-D neuron tracing with precise branch erasing and confidence controlled back tracking. IEEE Trans Med Imaging 37:2441–2452
12. Radojević M, Meijering E (2017) Automated neuron tracing using probability hypothesis density filtering. Bioinformatics 33:1073–1080
13. Radojević M, Meijering E (2019) Automated neuron reconstruction from 3D fluorescence microscopy images using sequential Monte Carlo estimation. Neuroinformatics 17:423–442
14. Basu S, Ooi WT, Racoceanu D (2016) Neurite tracing with object process. IEEE Trans Med Imaging 35:1443–1451
15. Zhou Z, Liu X, Long B, Peng H (2016) TReMAP: automatic 3D neuron reconstruction based on tracing, reverse mapping and assembling of 2D projections. Neuroinformatics 14: 41–50
16. Wang Y, Narayanaswamy A, Tsai CL et al (2011) A broadly applicable 3-D neuron tracing method based on open-curve snake. Neuroinformatics 9:193–217
17. Peng H, Ruan Z, Long F, Simpson JH, Myers EW (2010) V3D enables real-time 3D visualization and quantitative analysis of large-scale biological image data sets. Nat Biotechnol 28: 348–353
18. Mukherjee S, Condron B, Acton ST (2015) Tubularity flow field—a technique for automatic neuron segmentation. IEEE Trans Image Process 24:374–389
19. Radojević M, Smal I, Meijering E (2016) Fuzzy-logic based detection and characterization of junctions and terminations in fluorescence microscopy images of neurons. Neuroinformatics 14:201–219
20. Brown KM et al (2011) The DIADEM data sets: representative light microscopy images of neuronal morphology to advance automation of digital reconstructions. Neuroinformatics 9:143–157
21. Peng H et al (2015) BigNeuron: large-scale 3D neuron reconstruction from optical microscopy images. Neuron 87:252–256

22. Cai H, Xu X, Lu J et al (2006) Shape-constrained repulsive snake method to segment and track neurons in 3D microscopy images. Paper presented at the 3rd IEEE International Symposium on Biomedical Imaging: Nano to Macro, 2006, Arlington, VA, April 2006
23. Basu S, Aksel A, Condron B, Acton ST (2010) Tree2Tree: neuron segmentation for generation of neuronal morphology. Paper presented at the 2010 IEEE International Symposium on Biomedical Imaging: From Nano to Macro, Rotterdam, Netherlands, April 2010
24. Santamaría-Pang A, Hernandez-Herrera P, Papadakis M et al (2015) Automatic morphological reconstruction of neurons from multiphoton and confocal microscopy images using 3D tubular models. Neuroinformatics 13:297–320
25. Gu L, Zhang X, Zhao H, Li H, Cheng L (2017) Segment 2D and 3D filaments by learning structured and contextual features. IEEE Trans Med Imaging 36:596–606
26. Gu L and Cheng L (2015) Learning to boost filamentary structure segmentation. Paper presented at the 2015 IEEE International Conference Computer Vision (ICCV), Santiago, Chile, December 2015
27. Li Q, Shen L (2020) 3D neuron reconstruction in tangled neuronal image with deep networks. IEEE Trans Med Imaging 39:425–435
28. Li R, Zeng T, Peng H, Ji S (2017) Deep learning segmentation of optical microscopy images improves 3-D neuron reconstruction. IEEE Trans Med Imaging 36:1533–1541
29. Yang B, Chen W, Luo H, Tan Y, Liu M, Wang Y (2021) Neuron image segmentation via learning deep features and enhancing weak neuronal structures. IEEE J Biomed Health Inform 25:1634–1645
30. Dai T et al (2019) Deep reinforcement learning for subpixel neural tracking. Paper presented at the 2nd International Conference on Medical Imaging with Deep Learning (MIDL), London, United Kingdom, July 2019
31. Huang Q et al (2020) Weakly supervised learning of 3D deep network for neuron reconstruction. Front Neuroanat 14:1–15
32. Zhao J et al (2020) Neuronal population reconstruction from ultra-scale optical microscopy images via progressive learning. IEEE Trans Med Imaging 39:4034–4046
33. Liu M et al (2011) 3D neuron tip detection in volumetric microscopy images. Paper presented at the 2011 IEEE International Conference on Bioinformatics and Biomedical Engineering, Atlanta, GA, USA, November 2011
34. Liu M, Gong R, Chen W et al (2018) 3D neuron tip detection in volumetric microscopy images using an adaptive ray-shooting model. Pattern Recogn 75:263–271
35. Liu M, Wang C, Chen W (2018) 3D neuron branch points detection in microscopy images. Paper presented at the 2018 IEEE International Conference on Bioinformatics and Biomedical Engineering, Madrid, Spain, December 2018
36. Tan Y, Liu M, Chen W, Wang X, Peng H, Wang Y (2020) DeepBranch: deep neural networks for branch point detection in biomedical images. IEEE Trans Med Imaging 39:1195–1205
37. Yu F, Liu M, Chen W et al (2020) Automatic repair of 3-D neuron reconstruction based on topological feature points and an MOST-based repairer. IEEE Trans Instrum Meas 70:1–13
38. Guo C, Liu M, Guan T et al (2021) Crossover structure separation with application to neuron tracing in volumetric images. IEEE Trans Instrum Meas 70:1–13
39. Cannon RC, Turner DA, Pyapali GK et al (1998) An on-line archive of reconstructed hippocampal neurons. J Neurosci Methods 84:49–54
40. Yang B, Liu M, Wang Y et al (2021) Structure-guided segmentation for 3D neuron reconstruction. IEEE Trans Med Imaging 41:903–914
41. Jiang Y, Chen W, Liu M, Wang Y, Meijering E (2021) 3D neuron microscopy image segmentation via the ray-shooting model and a DCBLSTM network. IEEE Trans Med Imaging 40:26–37
42. Chen W, Liu M, Du H, Radojević M, Wang Y, Meijering E (2022) Deep-learning-based automated neuron reconstruction from 3D microscopy images using synthetic training images. IEEE Trans Med Imaging 41:1031–1042
43. Chen W et al (2021) Spherical-patches extraction for deep-learning-based critical points detection in 3D neuron microscopy images. IEEE Trans Med Imaging 40:527–538
44. Jiang Y, Chen W, Liu M et al (2021) DeepRayburst for automatic shape analysis of tree-like structures in biomedical images. IEEE J Biomed Health Inform 26:2204–2215

Chapter 13

Ex Vivo Live Imaging in Brain Slices for Analysis of Neurite Formation in Combination with In Utero Electroporation

Xiaonan Liu and Kazuhito Toyo-oka

Abstract

The use of time-lapse live imaging enables us to track the dynamic changes in neurites during their formation. Ex vivo live imaging with acute brain slices provides a more physiological environment than cultured cells. To accomplish this, a certain method of labeling is necessary to visualize and identify neurite morphology. To understand the dynamics of neurite structure at early stages of neurite formation, we describe in this chapter ex vivo live imaging using a confocal microscope at P0 in combination with in utero electroporation (IUE).

Key words Ex vivo live imaging, In utero electroporation, Neurite initiation, Neurite outgrowth

1 Introduction

With the development of advanced microscopes, detailed observation of neuronal morphology had been feasible for us in lab. In neuroscience, live imaging under a physiological condition is difficult because the brain is under a protection of the scull and is highly sensitive after removal from the scull. Implanting a microscope in a live animal is hindered by high failure rate and high morbidity rate in the operated animals, and requires expertise in lens implantation as well as the advanced equipment. While ex vivo imaging with acute brain slices is more available throughout the labs and yet is a physiological-relevant condition as compared to in vitro [1]. It enables us to peek into the "inside" of brain structures and also manipulate with drug treatments or gene expressions.

In layer 2/3 cortical pyramidal neurons, neurite formation occurs immediately after terminal translocation, the last step in neuronal migration, around postnatal day 0 (P0). A thick neurite is seen growing up toward the pial surface, which is considered to be the precursor of the apical dendrite. Multiple thin neurites grow laterally and are likely the precursors of basal dendrites. As a result

Kazuhito Toyooka (ed.), *Neuronal Morphogenesis: Methods and Protocols*, Methods in Molecular Biology, vol. 2831,
https://doi.org/10.1007/978-1-0716-3969-6_13,

of the trailing process of migrating neurons, the axon emerges from the opposite side of the apical dendrite [2, 3]. The formation of neurites is the first step in neuronal morphogenesis. This is also the structural foundation for neuronal functions. When tracking neurite outgrowth using live imaging, it is possible to observe the dynamic changes occurring during this process, which is not possible when using fixed samples. By obtaining more detailed information about this key process, such as the rate of neurite growth and retraction and their frequency, we will be able to better understand the functions of molecules and the effects of treatments [4, 5].

The labeling of target neurons is important for the successful performance of live imaging. While there are several methods for marking neurons, we use in utero electroporation (IUE) to express fluorescence proteins in certain cell populations. A plasmid that encodes a fluorescent protein, such as green fluorescent protein (GFP), was transfected via IUE at embryonic day 15.5 (E15.5) in order to label pyramidal neurons in layer 2/3 of the cortex [6]. With different timing of IUE and different angles of the electrodes during IUE, neurons at a different layer and different brain regions can be labeled, respectively. For example, IUE at E13.5 allows us to label layer 5 cortical pyramidal neurons [7], and the use of different electrode angles allows us to label neurons within the hippocampus [8].

2 Materials

2.1 Glass Micropipette

1. A glass capillary with a length of 100 mm, an outside diameter (OD) of 1 mm, and an inside diameter (ID) of 0.58 mm with an inner filament. We use glass capillaries (1B100F-4) purchased from World Precision Instruments.
2. Micropipette puller: The P-97 (Sutter Instrument) is used in our laboratory.

2.2 In Utero Electroporation (IUE)

1. Analgesia.

   Meloxicam: The concentration is 5 mg/mL, and the dose is 1 μL/weight (g).

   Bupivacaine: Stock solution is 0.5%, while working solution is 0.125% diluted with saline. The dose is 2 μL/weight (g).

2. Anesthesia.

   Avertin: The stock solution (40×, 500μg/μL) is prepared by dissolving 2.5 g of 2,2,2-tribromoethanol in tert-amyl alcohol. It has been sterilized by filtration with a 0.22 μm pore size filter and is stored at −20 °C. Before using, dilute the stock solution with distilled water to create a 1× working solution, which should be filtered through a 0.22 μm filter. The dose is 20 μL/weight (g) (250μg/weight (g)).

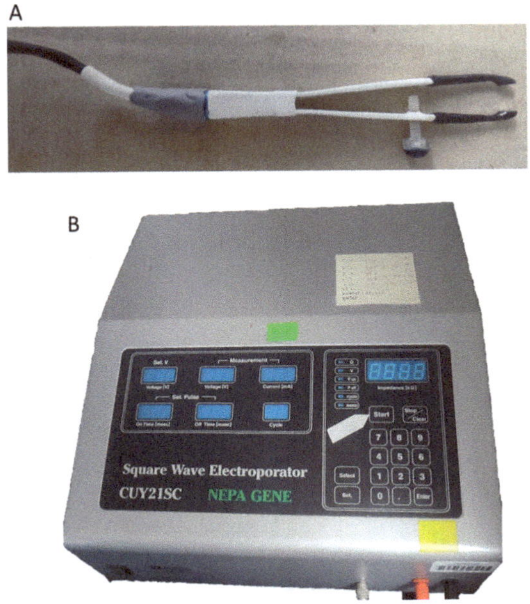

Fig. 1 Equipment used for in utero electroporation. (**a**) Electrode: Tweezer-type electrode, CUY650P5 (**b**) Electroporator: CUY21SC

3. Betadine wipe.
4. 70% Ethanol.
5. Gauze that has been autoclaved and sterilized (7 cm × 15 cm).
6. Surgical forceps and scissors: These devices are sterilized by autoclaving.
7. Ring forceps: This device is sterilized by autoclaving.
8. Saline: Autoclaved and warmed at 37 °C.
9. Fast Green: The stock solution is 0.1%. The final concentration is 0.01% in the plasmid solution.
10. Microliter syringe (Hamilton) for plasmid injection.
11. Electroporator to output square wave pulses (Fig. 1).
12. Electrode: Tweezer-type electrode, 5 mm in diameter (Fig. 1).
13. Surgical clips.
14. Heat pad.

2.3 Obtain Acute Brain Slices

1. Artificial cerebrospinal fluid (ACSF).

   10× stock solution: 2.5 mM KCL, 25 mM $NaHCO_3$, 1.25 mM NaH_2PO_4, 12 mM glucose.

   1× working solution (1000 mL): 100 mL 10× stock solution, 23.8 mM sucrose, 1 mM $CaCl_2{\cdot}2H_2O$, 1 mM $CaCl_2{\cdot}2H_2O$, 5 mM $MgCl_2{\cdot}6H_2O$.

   10× stock solution is stored at 4 °C. The 1× working solution is prepared fresh every time it is used.

2. A 4% solution of low melt agarose dissolved in 1× ACSF working solution. Microwaves can be used to accomplish this. It is then kept warm in a water bath at 40 °C until it is used.
3. A small cube-shaped plastic container for embedding the brain. We use a Peel-A-Way Mold (TED PELLA).
4. Vibratome.
5. Glass bottom dish, 35 mm.

2.4 Incubation

1. Dulbecco's Modified Eagle Medium (DMEM)/F12 cell culture media supplemented with 10% fetal bovine serum (FBS).
2. Fetal bovine serum (FBS).
3. Collagen type I from rat tail.
4. NaOH.
5. Mineral oil.

3 Methods

3.1 Preparation of a Glass Needle for DNA Injection

1. Pull a glass micropipette with the micropipette puller. For the P-97 puller, the following settings can be used: Pressure: 500, Heart: 800, Pull: 30, Velocity: 40, Time: 1.
2. Cut off the pulled pipette with forceps at a distance of 1–1.5 cm from the shoulder of the pipette (*see* **Note 1**).
3. The pipettes should be sterilized with a UV lamp.

3.2 In Utero Electroporation for Expressing Genes of Interest in Layer 2/3 Cortical Neurons

An in utero electroporation is conducted at E15.5 to target pyramidal neurons in layer 2/3. To perform in utero electroporation, plasmids encoding genes of interest are prepared in advance.

By expressing shRNA or fluorescently tagged molecules, such as GFP, we will be able to reveal neuronal morphology (Fig. 2).

1. Meloxicam (approximately 5 mg/kg) will be administered subcutaneously as a preemptive analgesic (*see* **Note 2**).
2. Avertin will be used as anesthesia. A 2.5% solution of avertin will be injected intraperitoneally (0.02 mL/gram body weight) (*see* **Note 3**).
3. The mouse will be placed on a working warm plate with the abdomen facing upward. The abdomen will be wiped with three alternate scrubs with betadine, followed by a wipe with 70% ethanol.
4. The abdomen will be covered with folded gauze with a 3 cm long slit in the center.
5. Gauze will be cleaned with 70% ethanol.

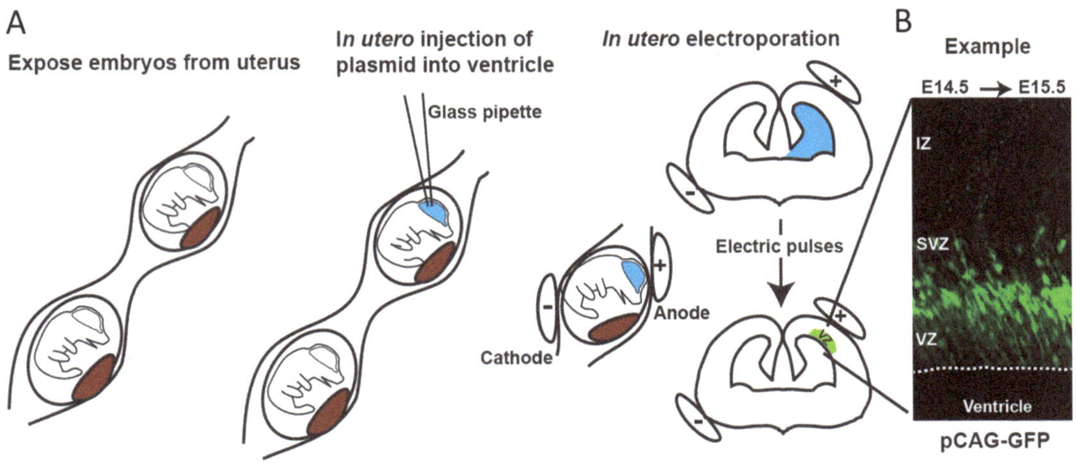

Fig. 2 An illustration and example of in utero electroporation (IUE). (**a**) The procedure for IUE. (**b**) An example of IUE. The plasmid coding for Green Fluorescent Protein (GFP) was injected into the lateral ventricle of the brain at E14.5, and IUE was performed. The brain was dissected at E15.5 and sliced using a cryostat. Following the fixation of the brain section, the image was captured using a confocal microscope. VZ Ventricular zone of the cortex, SVZ Subventricular zone of the cortex, IZ Intermediate zone of the cortex

6. The abdominal wall and skin will be cut with scissors. As soon as the skin incision is made, a drop of bupivacaine, a local analgesic, will be applied to the abdominal wall.
7. Evisceration of the uterus will be performed carefully by pinching gaps between embryos using ring forceps (*see* **Note 4**). To keep the uterus moist during surgery, warm saline will be dripped (the saline will be pre-heated to 37 °C).
8. A microliter syringe with glass needle is used to inject 1–2 μL of DNA solution mixed with Fast Green dye into the lateral ventricle of the brain.
9. A DNA-injected embryo will be held in parallel along its anteroposterior axis through the uterus using forceps-type electrodes, and electric pulses will be delivered to it (*see* **Note 5**).
10. We use the following settings for the CUY21SC electroporator:

    Voltage: 32 V, Pulse-On: 50 ms, Pulse-Off: 950 ms, Cycle: 3.
11. The uterus will be carefully repositioned into the abdominal cavity.
12. Warm saline will be injected into the abdominal cavity.
13. Using sutures, the abdominal wall will be closed separately from the skin, and the skin will be closed with clips (*see* **Note 6**).
14. Until the mouse recovers from the anesthetic, it will be placed on a heated pad maintained at 37 °C.

3.3 Preparation of Coronal Brain Slices at P0 (see Note 7)

1. The brain is dissected out in ice-cold ACSF at postnatal day 0 (P0). For live imaging of neurite initiation in cortical neurons, remove the caudal part of the brain containing the cerebellum.
2. Under a fluorescence microscope or fluorescence stereo microscope, confirm that fluorescent protein is expressed.
3. The brain is immediately placed in a plastic mold (Fig. 3).
4. Embed the brain sample in the low melt agarose solution (*see* **Note 8**).
5. For agarose to solidify, place the mold on ice for 5–10 min.
6. Set up the vibratome (Fig. 4). During the cutting process, the tray should remain cold by surrounding it with ice.
7. After the agarose has solidified, remove the mold and trim the excess agarose around the brain (Fig. 4).

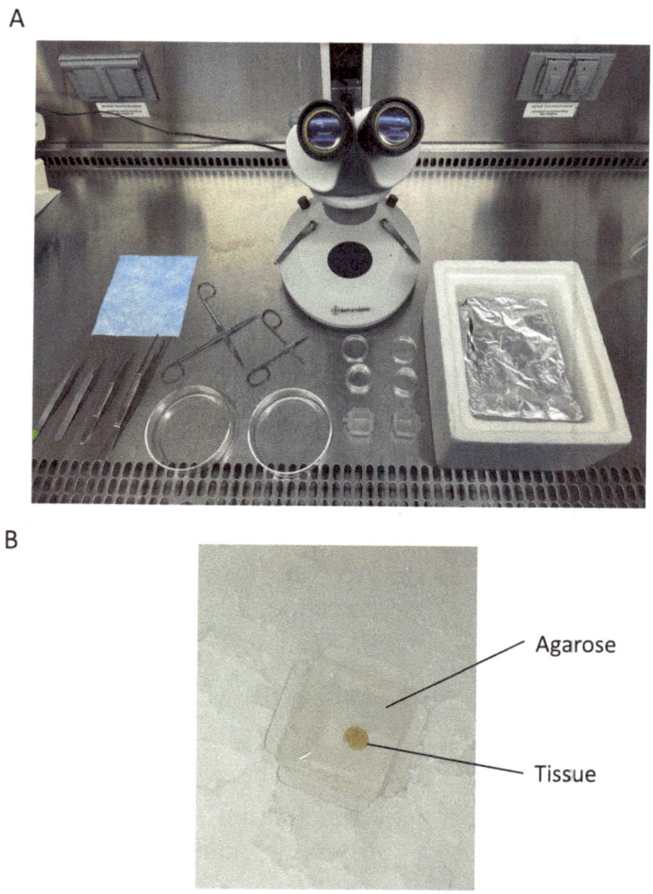

Fig. 3 Dissection of the brain at P0 for live imaging. (**a**) Setup for brain dissection in a hood. (**b**) Once the brain has been dissected, it is embedded in 4% low melt agarose in a plastic container, and the agarose gel is solidified on ice

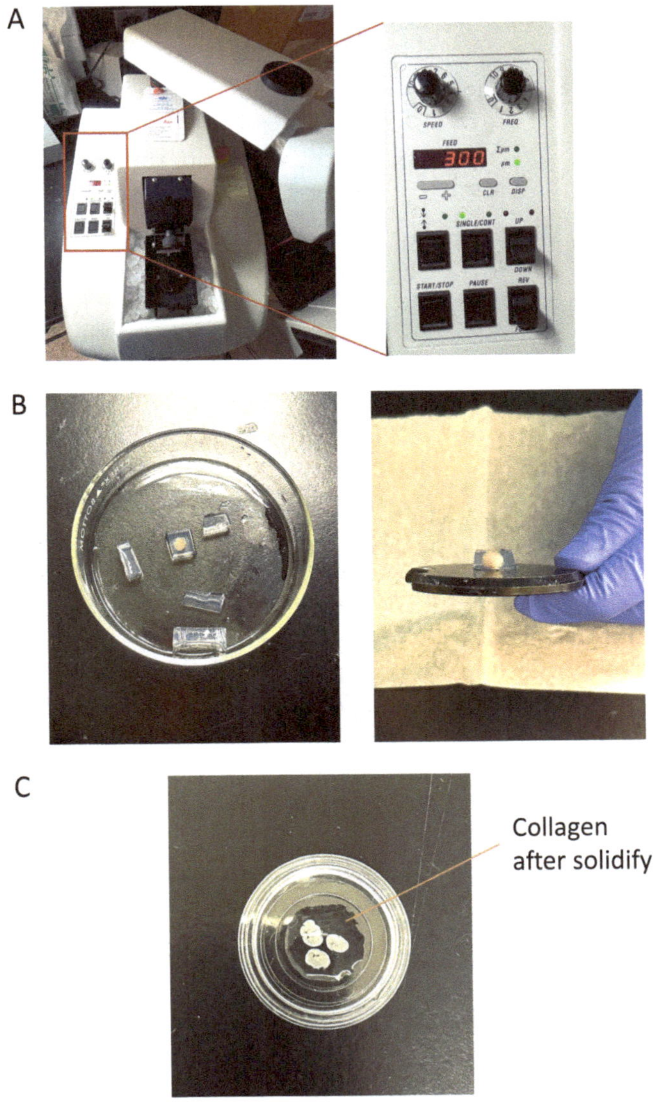

Fig. 4 Slicing brain tissue with a vibratome for imaging. (**a**) The setup of the vibratome. Note that the tray is surrounded by ice to maintain a cold temperature. The following condition is used: Speed = 4, Frequency = 5, and Feed (slice thickness) = 300 μm. (**b**) Preparation of the brain samples embedded in low melt agarose. Low melt agarose cube edges are trimmed. Superglue is used to adhere the agarose cube to the sample disc. (**c**) Mounting the sample on a glass bottom dish and solidifying the collagen. Brain samples should be placed in the center of a glass bottom covered with collagen

8. The agarose cube containing the brain should be glued to the specimen disc using superglue and allowed to dry for a few minutes.
9. Set up the vibratome with the disc.
10. Fill the tray with ice-cold ACSF to cover the agarose cube.

11. Set up the vibratome's slicing parameters. The following conditions are used: Speed: 4 (0.175 mm/s), Frequency: 5 (50 Hz).
12. Position the blade and agarose cube. To start cutting the agarose cube top, the edge of the blade should be at the same height as the agarose cube.
13. Slices of brain containing agarose are released into a tray filled with ice-cold ACSF. Pick and take the first few slices to a fluorescence microscope to check whether fluorescent proteins are expressed. For ex vivo live imaging, select slices with a decent fluorescence intensity level (*see* **Note 9**).

3.4 Incubation (see Note 10)

Following the preparation of acute brain coronal slices, the slices should be incubated in DMEM/F12 medium supplemented with 10% FBS in order to recover from damage caused by the slicing process. To avoid the initiation of neurite initiation, place them in 35 mm dishes in a CO_2 cell culture incubator for at least 1 h, but no more than 4 h.

3.5 Setup of the Confocal Microscope for Live Imaging

Set up the confocal microscope and temperature controller during incubation. Until live imaging begins, leave the unit on for at least 30 min until it reaches 37 °C. Neurite outgrowth is best observed with a 10× or 20× objective lens with zooming capability. We use a 10× objective lens with a 2.5× zoom.

3.6 Ex Vivo Time-Lapse Live Imaging

1. After incubation, select several brain slices with a decent level of fluorescent intensity under a fluorescence microscope.
2. Prepare the collagen solution and keep it on ice to prevent it from solidifying (*see* **Note 11**).
3. The brain slices should be placed on the glass bottom of a 35 mm glass bottom dish and excess medium should be removed.
4. To cover the brain slices, drop collagen solution over them (Fig. 4).
5. To solidify collagen, incubate the brain slices without medium for 30 min at 37 °C.
6. Place the dish into the holder of the confocal microscope and fill it with DMEM/F12 media with 10% FBS to cover the collagen. Place a thin layer of mineral oil over the culture media (*see* **Note 12**).
7. Set up the acquisition conditions in the confocal microscope, including the range of the *z*-axis. For the initiation and outgrowth of neurites, we use a 10-min interval.
8. Begin the acquisition of images (*see* **Note 13**).

4 Notes

1. A stereo microscope can be used to select a glass pipette with a sharp tip-end for easy injection of DNA solution.
2. It is also possible to use other analgesics.
3. It is possible to use other types of anesthesia. The level of anesthesia will be determined by physical signs such as the corneal reflex, paw pad pinch test, respiratory rate, and movement.
4. To avoid damaging the embryos and causing their death, do not pinch either the placenta or the embryos.
5. There is also an electroporator from BTX that can be used.
6. A surgical needle and thread may also be used to sew the skin.
7. Throughout the entire process, it is imperative to keep everything on ice or cold.
8. To allow the agarose to cool down to 37–40 °C before use, it should be prepared before the brain dissection. Ensure that you do not use hot agarose, which can cause serious damage to the brain. Also, ensure that the brain sample is completely covered with agarose. If necessary, remove any air bubbles.
9. In order to easily identify the target neurons on the slices, it is important to keep the brain slice as intact as possible.
10. The recovery incubation is very important. Shorter incubation times result in slices drifting during imaging. However, it is important to note that an incubation period that is too long would miss the stages of neurite initiation and outgrowth.
11. It is recommended that collagen solution be prepared right before application and used immediately after preparation. However, it may be possible to keep it on ice in order to prevent solidification.
12. To prevent the evaporation of culture media, mineral oil should completely cover the surface of the media. During live imaging, the sample would drift and the focus would be affected by the evaporation of the media.
13. Confirm that the fluorescence signals are within the range of the *z*-axis 2–3 h later. If the recovery incubation is not long enough, the fluorescence signals are outside the range of the *z*-axis. Set up the range of the *z*-axis and begin the acquisition process again. It is possible to perform the acquisition overnight (~16 h).

Acknowledgments

This work is supported by R01NS096098 to K.T and Dean's fellowship from Drexel University College of Medicine to X.L.

References

1. Ghoochani A et al (2016) A versatile ex vivo technique for assaying tumor angiogenesis and microglia in the brain. Oncotarget 7(2): 1838–1853
2. Polleux F, Snider W (2010) Initiating and growing an axon. Cold Spring Harb Perspect Biol 2(4):a001925
3. Hand RA et al (2015) Axon dynamics during neocortical laminar innervation. Cell Rep 12(2): 172–182
4. Liu X et al (2021) Glutathione S-transferase Pi (Gstp) proteins regulate neuritogenesis in the developing cerebral cortex. Hum Mol Genet 30(1):30–45
5. Bennison SA et al (2023) The cytoplasmic localization of ADNP through 14-3-3 promotes sex-dependent neuronal morphogenesis, cortical connectivity, and calcium signaling. Mol Psychiatry 28(5):1946–1959
6. Nagai T et al (2002) A variant of yellow fluorescent protein with fast and efficient maturation for cell-biological applications. Nat Biotechnol 20(1):87–90
7. Meyer-Dilhet G, Courchet J (2020) In utero cortical electroporation of plasmids in the mouse embryo. STAR Protoc 1(1):100027
8. dal Maschio M et al (2012) High-performance and site-directed in utero electroporation by a triple-electrode probe. Nat Commun 3(1):960

Chapter 14

Imaging of Structural Plasticity of Dendritic Spines with Two-Photon Microscopy

Takeo Saneyoshi and Yasunori Hayashi

Abstract

Plasticity of synaptic transmission underlies learning and memory. It is accompanied by changes in the density and size of synapses, collectively called structural plasticity. Therefore, understanding the mechanism of structural plasticity is critical for understanding the mechanism of synaptic plasticity. In this chapter, we describe the procedures and equipment required to image structural plasticity of a single dendritic spine, which hosts excitatory synapses in the central nervous system, and underlying molecular interactions/ biochemical reactions using two-photon fluorescence lifetime microscopy (2P-FLIM) in combination with Förster resonance energy transfer (FRET)-based biosensors.

Key words Two-photon fluorescence lifetime measurement microscopy, Dendritic spine, Synapse, Actin cytoskeleton, LTP, CaMKII

1 Introduction

In the neuronal network, synapses formed between neurons mediate information transmission. Synapses are functionally and structurally plastic to support higher order brain functions such as learning and memory. In representative forms of synaptic plasticity, long-term potentiation (LTP) and long-term depression (LTD), changes in spine density and size have been observed [1–5]. During the induction of Long-term potentiation (LTP), Ca^{2+} influx through N-methyl-D-aspartate (NMDA)-type glutamate receptors and the resulting activation of Ca^{2+}/calmodulin-dependent protein kinase II (CaMKII) triggers activation of signal transduction that eventually leads to the regulation of actin cytoskeleton [6, 7] and enlarges dendritic spines, a process called structural LTP (sLTP). This may act to translocate α-amino-3-hydroxy-5-methyl-4-isoxazolepropionate (AMPA)-type glutamate receptors to the enlarged synaptic surface and to maintain synaptic strength by providing new binding sites for synaptic molecules [6, 8–12].

Kazuhito Toyooka (ed.), *Neuronal Morphogenesis: Methods and Protocols*, Methods in Molecular Biology, vol. 2831, https://doi.org/10.1007/978-1-0716-3969-6_14,

Therefore, visualization of structural changes and underlying regulatory pathways during synaptic plasticity is crucial for understanding the mechanisms of learning and memory.

Traditionally, the biochemical signaling of these processes has been studied using homogenates of cells. However, this approach has major drawbacks in studying spatiotemporal dynamics. In this regard, the introduction of several key technologies has significantly advanced the field. First, the uncaging of caged-glutamate using a two-photon laser allows us to reliably induce LTP or LTD in a single spine under observation of a living neuron [13, 14]. Second, Förster resonance energy transfer (FRET), a non-radiative transfer of energy from an excited donor fluorophore to an acceptor fluorophore, can be used to monitor an interaction or a conformational change of proteins fused to different fluorophores [1, 15, 16]. It can be used as a molecular ruler to monitor the biochemical signaling that occurs within single dendritic spines under observation [17]. Finally, fluorescence lifetime imaging microscopy (FLIM), a microscopic method that detects the decay of fluorescence emission after excitation, is used to detect FRET. It has several advantages over conventional ratio-metric imaging. In particular, FLIM only uses the signal from the donor, and is therefore not affected by volume changes or the localization of proteins. This is particularly advantageous in spines where molecules are dynamically moving [18].

In this chapter, we describe the protocols for imaging structural plasticity of a dendritic spine, including sLTP induction and FLIM-FRET measurements using rat hippocampal slice cultures under two-photon laser scanning microscopy [19]. We use the TIAM1—CaMKII interaction as an example [19]. We will not discuss the design of FLIM-FRET probes; readers can refer to other reviews for further information [16].

2 Materials

2.1 Hippocampal Slice Cultures

1. Hippocampal culture media: 80% minimum essential medium (MEM) (Hanks' salts with L-glutamine, Gibco, 11575-032), 20% horse serum (Biowest, S0910), supplemented with 30 mM HEPES, 26.6 mM D-glucose, 5.8 mM $NaHCO_3$, 1 mM $CaCl_2$, 5 mM $MgSO_4$, 0.5 μg/mL insulin (Nacalai Tesque, 19251-24), 0.0012% ascorbic acid, pH 7.4, osmolality 320 ± 10 mOsm/L (Wescor 5600, vapor pressure osmometer, Elitech Group Inc.).
2. Dissection solution (2.5 mM KCl, 26 mM $NaHCO_3$, 1 mM NaH_2PO_4, 11 mM glucose, 238 mM sucrose, 1 mM $CaCl_2$, and 5 mM $MgCl_2$). Bubble with mixed gas containing 95% O_2 and 5% CO_2 for 20–30 min.

3. Artificial cerebrospinal fluid (ACSF): 119 mM NaCl, 2.5 mM KCl, 26.2 mM $NaHCO_3$, 1 mM NaH_2PO_4, 11 mM glucose, 4 mM $CaCl_2$, 1 μM tetrodotoxin (TTX), and 50 μM picrotoxin. Bubble with mixed gas containing 95% O_2 and 5% CO_2 for 20–30 min. Osmolality should be in the range of 320 ± 10 mOsm/L.
4. Uncaging ACSF: ACSF contains 2.5 mM 4-methoxy-7-nitroindolil (MNI)-L-glutamate (caged-glutamate. Tocris, No. 1490).

2.2 Bullet for Gene-Gun

1. DNA constructs are in a pCAGGS vector [20]. Since original enhanced green fluorescent protein (EGFP) tends to form dimers, we routinely use monomeric GFP with a A206K mutation, especially for fusion proteins.
2. 100% ethanol (desiccated with molecular sieves).
3. 0.05 M spermidine (Sigma, S-0266).
4. 1.0 M $CaCl_2$.
5. 20 mg/mL polyvinylpyrrolidone (Sigma, PVP360).
6. Gold powder (0.3–3 μm, Chempur, 009150-5).
7. Helios® Gene Gun system (Bio-Rad).
8. Helium tank.

2.3 Two-Photon Microscopy Setting

1. Two-photon microscope (FV1000ME, Olympus), 10× and 60× water immersion objective lens.
2. Two Ti-sapphire tunable pulsed lasers (Mai-tai DeepSee, Spectra-Physics), set at 910 nm for imaging, 720 nm for uncaging.
3. Fluorescent beads (0.5 μm diameter, Fluoresbrite YG, Polyscience, Inc.) covalently linked onto collagen-coated glass bottom dish.
4. Imaging chamber for circulation of ACSF with caged-glutamate (Fig. 1).

2.4 Fluorescent Lifetime Imaging Microscopy

1. A separate FLIM PC with time-correlated single photon counting (TCSPC) board (SPC-150, Becker-Hickl) and data acquisition software (Becker-Hickl).
2. A part of the excitation laser light is reflected on the surface of a cover glass placed in the light path and introduced into a photodiode module (PHD-400 PIN, Becker and Hickl). Then the pulsing is fed into the TCSPC board for synchronizing photon counting with excitation.
3. The PC controlling the two-photon microscope sends the frame start and sweep start timing signal to the FLIM PC.

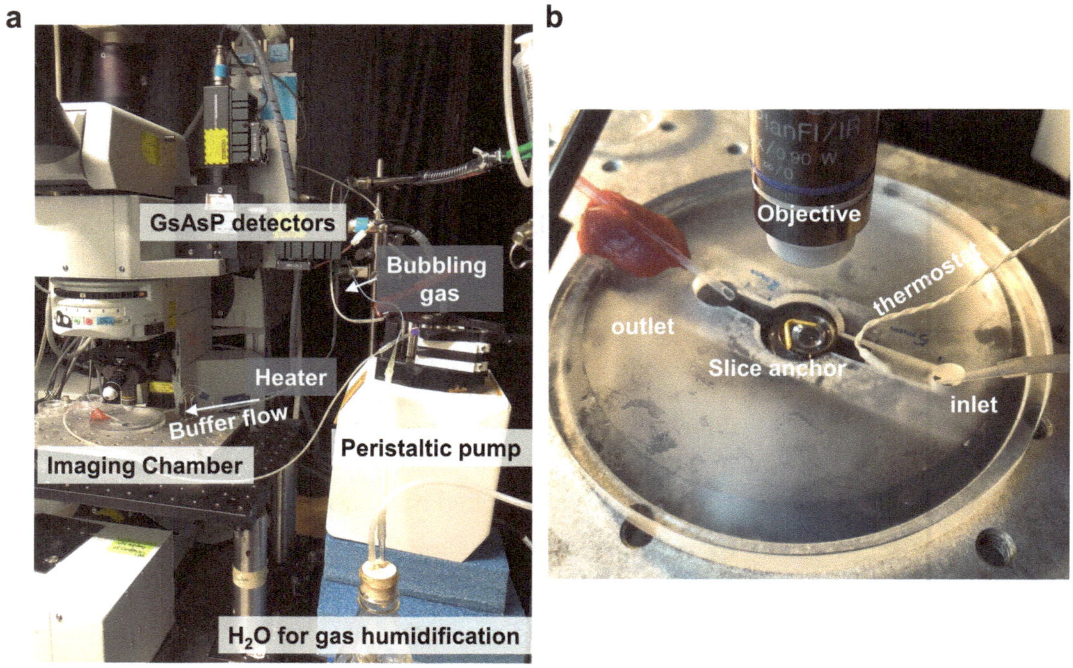

Fig. 1 Two-photon microscope with FLIM capability. (**a**) A FLIM detector is outfitted to a commercially available two-photon microscope (Olympus). ACSF is circulated through a peristaltic pump. The mixed gas is humidified with distilled water before bubbling the ACSF. (**b**) Imaging chamber. The slice is placed in the center of the chamber and held in place by the slice anchor. The temperature of the ACSF is controlled by the temperature controller via a thermostat

4. Gallium arsenide phosphide (GaAsP) detector (H7422-40, Hamamatsu Photonics) with custom-made power supply and amplifier (Olympus).

2.5 Image Analysis

FIJI (ImageJ) analysis for volume of dendritic spines.

Custom-made program to analyze FLIM images.

3 Methods

3.1 Hippocampal Slice Culture

It is generally perceived that synaptic plasticity is difficult to induce in dissociated culture. Therefore, we use hippocampal organotypic slice culture. Hippocampal slice cultures are prepared from postnatal day 6–9 rat pups of both sexes [19, 21]. The hippocampi are dissected out in ice-cold dissection solution. Slices (350 μm thick) are prepared using a McIlwain Tissue Chopper (TC752, Cavey Laboratory Engineering Co. Ltd.) and plated on a cell culture insert (Millicell CM, Millipore, PICM0RG50) in a six-well plate containing 1 mL of culture medium. After removing the excess dissection solution around the slice, the hippocampal slice cultures are placed in a CO_2 incubator (35 °C, 5% CO_2, 95% air) and the culture medium is changed every 2–3 days.

3.2 Gene-Gun Bullet Preparation and Gene Transfer into Slices

Prepare gene-gun bullets using 12.5 mg of gold powder and 50 μg of plasmid DNA according to the manufacturer's instructions (Bio-Rad). Bullets are stored in a plastic scintillation vial together with a desiccator (e.g., silica gel or Drierite) at −25 °C. Plasmids are transfected at 5–7 DIV using the gene-gun at a gas pressure of 180–200 psi. Expression of exogenous fluorescent proteins is imaged 2–7 days after the transfection.

To analyze the structural changes of the spine during LTP, GFP or another fluorescent protein can be used as a volume filler. For molecular dynamics analysis in a spine during LTP, GFP-fusion protein and DsRed2 as volume filler are co-expressed from a single gene-gun bullet. For gene knockdown experiments with shRNA, a knockdown vector construct such as pSuper and volume filler can be co-expressed as a transfection marker. The ratio of plasmid DNAs should be determined empirically.

3.3 Two-Photon Imaging and the Glutamate Uncaging Stimulation

1. For two-photon imaging and stimulation, one laser is tuned at 910 nm to image GFP, and the other laser at 720 nm to uncage the caged-glutamate. Before starting the experiment, the two lasers are adjusted using fluorescent beads (0.5 μm diameter). The images of fluorescent beads from the imaging laser and the stimulating laser must be perfectly overlapped within an error of <50 nm at the center of the field of view. The power of the stimulation laser is adjusted at 5 mW under the 60× objective using the power meter (LaserCheck, Coherent. Inc.).
2. Prepare perfusion. To perfuse ACSF, we use a peristaltic pump (MINIPULS 3, Gilson), and aerate the ACSF with a mixed gas (95% O_2, 5% CO_2), pre-humidified by passing distilled water through once. Turn on the temperature controller (TC324B, Warner Instruments) set to 30 °C. We typically use 6 mL of ACSF for circulation to save the caged glutamate.
3. Select an appropriate slice containing healthy GFP-positive neurons in area CA1, cut the membrane containing the slice from the culture-insert (~5 × 5 mm), and place it in the chamber. Immobilize it with a slice anchor (Warner Instruments) or platinum wire (Fig. 1).
4. Search for neurons expressing fluorescent proteins under illumination with a mercury lamp and an appropriate filter set through a 10× objective. Then use a 60× objective to find an isolated dendritic spine. Add the caged-glutamate to the ACSF to a final concentration of 2.5 mM. Incubate the slice for 30–40 min to stabilize the slice.
5. Find a spine separated from the dendritic shaft. The spines are imaged through a 60× objective and digitally zoomed at 10×. Using a Z-series of 512 × 512-pixel X-Y scans taken every 0.5–1 μm in depth, acquire images for 15 min as a baseline,

every 5 min, and then 35 min after the uncaging stimulation. The GFP signal is imaged through a 495–540 nm band-pass filter, and the red flureoscent protein (RFP) signal through a 575–630 nm band-pass filter with a 510 nm short-pass or long-pass dichroic mirror.

6. Stimulate a single spine by uncaging the caged-glutamate. The uncaging laser pulse should be given ~0.5 μm away from the tip of the dendritic spine at 2 msec for 30 pulses at 1 Hz or 4 msec for 30 pulses at 0.5 Hz [19, 22].
7. When an LTP measurement is completed, if there are multiple neurons expressing fluorescent protein, we select another neuron and repeat the recording of two or three neurons from one slice. When changing to another slice, we measure the osmolality and adjust it to 320 ± 10 mOsm/L with distilled water.
8. For image analysis, we subtract the background from each image. The change in spine volume is expressed as the difference of the averaged intensities from the baseline. Synaptic trafficking of protein can be calculated as the ratio between fluorescence of a protein of interest and volume filler [22].

3.4 2P-FLIM During Structural LTP

1. We typically use monomeric GFP as a donor and mCherry as an acceptor. For the TIAM1-CaMKII interaction, we used TIAM1 tagged with GFP at the carboxy tail and CaMKII tagged with mCherry at both ends (Fig. 2). The fluorescence lifetime is measured using the TCSPC system as described above at 910 nm excitation.
2. GFP emission light is filtered through a 510 ± 35 nm band-pass filter and a 680 nm short-pass blocking filter (Omega Optical). Images are binned at 64 × 64 pixels to increase the number of photons per pixel for accurate lifetime calculations.

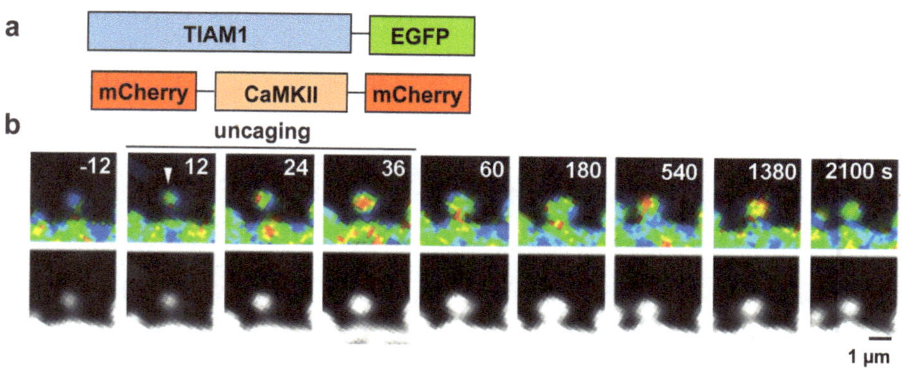

Fig. 2 FLIM-FRET imaging. (**a**) FLIM-FRET probe used to detect the interaction between TIAM1 and CaMKII. (**b**) Interaction between TIAM1 and CaMKII as visualized by FLIM-FRET before and during structural LTP (top). Warmer hues indicate shorter lifetime and more interaction. The distribution of TIAM1-GFP, as a proxy of structure, is shown (bottom). (Adapted from Saneyoshi et al. (2019) [19])

It is possible that the 720 nm uncaging laser bleeds through into the detector even with a blocking filter. This is readily noticeable in the images as bright horizontal lines at regular intervals. If this occurs, avoid including such images in the analyses.

3. The data from TCSPC are converted to the ASCII format. The analysis of the lifetime images is done with a custom-written macro in Igor Pro (Wavemetrics) [19, 22]. Due to the limited number of photons that can be recorded from a single spine, exponential fitting is difficult. We therefore calculate the average arrival time of all photons, which theoretically equals the decay time of a single exponential decay. We express the results as the difference from the lifetime during the baseline before the glutamate uncaging.

4 Tips

1. If the structural enlargement of the dendritic spine does not last long, it is often because the quality of the slice culture is not optimal. Ensure that the dissection is completed in the shortest time possible. Remove excess medium around the slice during culture. It is also advisable to perform a lot test of the horse serum. A good slice culture will maintain its thickness, while a poor slice culture will flatten during culture.

   The flow rate and temperature of the ACSF in the imaging chamber should be stable.

   If transfection efficiency is poor, the quality of the gene gun bullets might be compromised. Ensure that there is fresh desiccant in the bullet container. Before opening the container, leave it at room temperature for 20–30 min to prevent the accumulation of moisture.

2. When GFP or another fluorescent protein is used as a volume filler, the fine structure of dendritic spines near the dendritic shaft can be blurred as the shaft is brighter. In such case, use of membrane targeted GFP (such as myristoylated GFP, farnesylated GFP, or GFP with GPI-linker) is helpful. However, the spine volume will not be proportional to the GFP signal intensity.

3. Optimization of the FLIM-FRET sensors needs to be done empirically, such as the location of the fluorescent protein (N-, C-, or middle of the molecule), and the linker between the molecule and the fluorescent proteins.

4. Validation of the biosensor must be performed. For TIAM1-CaMKII, we first identified the interaction interface with alanine scanning and immunoprecipitation from HEK 293 cells

[19]. Then we used an identified mutant as a negative control in FLIM-FRET experiments.

5. Expression of the biosensor itself may cause a phenotype due to overexpression. Although it is difficult to totally avoid such an effect, it is advisable to monitor the density and structure of dendritic spines as well as the efficiency of sLTP in comparison with neurons expressing GFP only.

Acknowledgment

We thank E. Agnello for comments on the manuscript. This work was supported in part by KAKENHI (JP18H05434, JP22H04981, JP22K21353, JP21H05692, JP21H02595, 22H00434, 22H04981, 22K21353) from the Japan Society for the Promotion of Science (JSPS) and the Ministry of Education, Culture, Sports, Science, and Technology (MEXT), ISHIZUE 2022 of Kyoto University (TS), the Takeda Science Foundation (TS, YH), and Human Frontier Science Program RGP0022/2013 and RGP0020/2019, and JST CREST JPMJCR20E4.

References

1. Okamoto K et al (2004) Rapid and persistent modulation of actin dynamics regulates postsynaptic reorganization underlying bidirectional plasticity. Nat Neurosci 7(10): 1104–1112
2. Matsuzaki M et al (2004) Structural basis of long-term potentiation in single dendritic spines. Nature 429(6993):761–766
3. Nägerl UV et al (2004) Bidirectional activity-dependent morphological plasticity in hippocampal neurons. Neuron 44(5):759–767
4. Bosch M, Hayashi Y (2012) Structural plasticity of dendritic spines. Curr Opin Neurobiol 22(3):383–388
5. Wiegert JS, Oertner TG (2013) Long-term depression triggers the selective elimination of weakly integrated synapses. Proc Natl Acad Sci U S A 110(47):E4510–E4519
6. Saneyoshi T, Hayashi Y (2012) The Ca2+ and Rho GTPase signaling pathways underlying activity-dependent actin remodeling at dendritic spines. Cytoskeleton (Hoboken) 69(8): 545–554
7. Saneyoshi T (2021) Reciprocal activation within a kinase effector complex: a mechanism for the persistence of molecular memory. Brain Res Bull 170:58–64
8. Okamoto K, Bosch M, Hayashi Y (2009) The roles of CaMKII and F-actin in the structural plasticity of dendritic spines: a potential molecular identity of a synaptic tag? Physiology (Bethesda) 24:357–366
9. Goto A et al (2021) Stepwise synaptic plasticity events drive the early phase of memory consolidation. Science 374(6569):857–863
10. Gu J et al (2010) ADF/cofilin-mediated actin dynamics regulate AMPA receptor trafficking during synaptic plasticity. Nat Neurosci 13(10):1208–1215
11. Fukazawa Y et al (2003) Hippocampal LTP is accompanied by enhanced F-actin content within the dendritic spine that is essential for late LTP maintenance in vivo. Neuron 38(3): 447–460
12. Herring BE, Nicoll RA (2016) Long-term potentiation: from CaMKII to AMPA receptor trafficking. Annu Rev Physiol 78:351–365
13. Matsuzaki M et al (2001) Dendritic spine geometry is critical for AMPA receptor expression in hippocampal CA1 pyramidal neurons. Nat Neurosci 4(11):1086–1092
14. Oh WC, Hill TC, Zito K (2013) Synapse-specific and size-dependent mechanisms of spine structural plasticity accompanying synaptic weakening. Proc Natl Acad Sci U S A 110(4):E305–E312

15. Takao K et al (2005) Visualization of synaptic Ca2+/calmodulin-dependent protein kinase II activity in living neurons. J Neurosci 25(12): 3107–3112
16. Ueda Y, Kwok S, Hayashi Y (2013) Application of FRET probes in the analysis of neuronal plasticity. Front Neural Circuits 7:163
17. Miyawaki A (2011) Development of probes for cellular functions using fluorescent proteins and fluorescence resonance energy transfer. Annu Rev Biochem 80:357–373
18. Yasuda, R., Studying signal transduction in single dendritic spines. Cold Spring Harb Perspect Biol, 2012. 4(10).
19. Saneyoshi T et al (2019) Reciprocal activation within a kinase-effector complex underlying persistence of structural LTP. Neuron 102(6): 1199–1210.e6
20. Niwa H, Yamamura K, Miyazaki J (1991) Efficient selection for high-expression transfectants with a novel eukaryotic vector. Gene 108(2):193–199
21. Stoppini L, Buchs PA, Muller D (1991) A simple method for organotypic cultures of nervous tissue. J Neurosci Methods 37(2):173–182
22. Bosch M et al (2014) Structural and molecular remodeling of dendritic spine substructures during long-term potentiation. Neuron 82(2):444–459

Chapter 15

Visualization and Quantification of Organelle Axonal Transport in Cultured Neurons

Jayne Aiken and Erika L. F. Holzbaur

Abstract

The specialized function and extreme geometry of neurons necessitates a unique reliance upon long-distance microtubule-based transport. Appropriate trafficking of axonal cargos by motor proteins is essential for establishing circuitry during development and continuing function throughout a lifespan. Visualizing and quantifying cargo movement provides valuable insight into how axonal organelles are replenished, recycled, and degraded during the dynamic dance of outgoing and incoming axonal traffic. Long-distance axonal trafficking is of particular importance as it encompasses a pathway commonly disrupted in developmental and degenerative disease states. Here, we describe neuronal organelles and outline methods for live imaging and quantifying their movement throughout the axon via transient expression of fluorescently labeled organelle markers. This resource provides recommendations for target proteins/domains and appropriate acquisition time scales for visualizing distinct neuronal cargos in cultured neurons derived from human induced pluripotent stem cells (iPSCs) and primary rat neurons.

Key words Axonal transport, Trafficking, Organelle movement, Live imaging, iPSC-derived neuron, Primary neuron, Transfection

1 Introduction

Neurons are uniquely dependent on long-distance microtubule-based transport. The polarization of neurons requires components synthesized in the soma to be properly sorted and trafficked through long neuronal processes, called axons and dendrites. Most neurons contain multiple dendrites and one long axon, which can arborize and reach meters in total length [1]. Specialized targeting of cargos within cellular compartments is also essential, such as synaptic vesicle precursor (SVP) capture at axonal en passant presynapses [2] and distal axon accumulation of autophagy components for autophagosome initiation [3]. Thus, microtubule motor-based trafficking of neuronal cargos represents a highly complex system that requires deliberate and specialized trafficking and delivery strategies. Disruption to axonal transport is an early

Kazuhito Toyooka (ed.), *Neuronal Morphogenesis: Methods and Protocols*, Methods in Molecular Biology, vol. 2831, https://doi.org/10.1007/978-1-0716-3969-6_15,

and perhaps causative event in many neurodegenerative diseases [4], and mutations to the trafficking machinery can manifest as neurodevelopmental and degenerative disorders [5–8]. Understanding how axonal trafficking is spatially and temporally tuned for different tasks will provide invaluable insight into a keystone of neuronal function.

The cytoskeletal architecture of neuronal processes directly influences microtubule-based trafficking. In mammalian axons, microtubules are uniformly oriented with minus ends proximal to the soma and plus ends facing the distal axon tip [9]. Accordingly, plus-end directed kinesin motor proteins are responsible for transporting anterograde cargos outward, and the minus-end directed motor cytosolic dynein traffics retrograde cargos inward. Dendritic trafficking is more complex, with both classes of motor proteins involved in anterograde and retrograde movement. For more information of microtubule motors, their cargos, and cytoskeletal regulation within the neuron, we refer the reader to the following review: [10].

Here, we describe live-imaging of axonal trafficking in cultured human induced pluripotent stem cell (iPSC)-derived i^{3}Neurons [11] and primary rat neurons [12] utilizing lipofectamine-mediated vector transfection. For detailed methods on initiating these cultures, we suggest the following resources: i^{3}Neurons [11]; primary rat neurons [13, 14]. Two-dimensional neuronal cultures provide advantages for live imaging, including being amenable to transfection and existing primarily in a single z-plane. The low transfection efficiency of neuronal cultures permits identification and imaging of ectopically expressed, fluorescently tagged proteins/domains within a single axon. Special consideration should be given to promoter and expressed protein choice (*see* Tables 1 and 2, **Note 2**), as it is possible for protein overexpression to cause undesired cellular perturbation. The following study provides a comparison of constitutive mammalian promoters: [15]. Other neuronal transfection methods are available and may be advantageous depending on experiment purpose, but lipofection is quick, simple, cost-effective, and provides experiment flexibility [14]. Using commercially available dyes, alternatively, has the advantage of facilitating visualization of endogenous organelles. However, it is difficult to distinguish whether trafficking is occurring in one or multiple adjacent axons due to cell density and the propensity of neuronal processes to grow alongside one another.

Visualizing Neuronal Organelles

Neurons are post-mitotic cells specialized to transmit electrochemical signals that rapidly relay information across the nervous system. They rely on many of the same organelles present in other cells to carry out basic activities, with the added challenge of appropriate spatial distribution. Additionally, neurons contain unique

Table 1
Commonly used promoters for expression in mammalian neurons

Promoter	Description	Notes
CAG [53]	CMV early enhancer fused to modified chicken β-actin promoter	Strong ubiquitous promoter
CMV [54]	Human cytomegalovirus immediate early enhancer/promoter	Strong ubiquitous promoter
EF1α [55]	Human eukaryotic translation elongation factor 1 α1 promoter	Strong ubiquitous promoter, recommended for i^{3}Neurons
PGK [56]	Phosphoglycerate kinase 1 promoter	Medium-strength ubiquitous promoter
SV40 [57]	Simian virus 40 enhancer/early promoter	Medium-strength ubiquitous promoter
SYN1 [58]	Human synapsin I promoter	Expressed in mature neurons
Thy1 [59]	Mouse thymus cell antigen 1 promoter	Expressed in neurons
TUBA1A [60]	Rat α1A tubulin promoter	Expressed in developing neurons

Table 2
Recommendations for target protein/domains for visualizing distinct neuronal cargos

Organelle	Suggested protein/domain target(s)
Endoplasmic reticulum	Sec61β [20], KDEL [21, 27]
RNA granules	G3BP1 [23], TDP-43 [24]
Golgi apparatus	SialT2 [27], St3 [28], GM130* [61], GalT [27], St3 [28], manII [26]
Mitochondria	Mito [31], Cox8a [31]
Early endosomes	Rab5** [32, 35]
Late endosomes	Rab7** [32, 35]
Recycling endosomes	Rab11** [32, 35]
Autophagosomes	LC3 [3]
Lysosomes	TMEM192*** [39], LAMP1*** [32], SEP-LAMP1-RFP [41]
Peroxisomes	PTS1/SKL [21], PEX3 [62]
Dense core vesicles	NPY [42–44]
Synaptic vesicle precursors	SYP1 [2, 46], VAMP2, VMAT2 [46]
Piccolo-bassoon transport vesicles	Bassoon, Munc13, ELKS2 [47]

Provided reference includes example of live imaging in primary murine or iPSC-derived neurons
*GM130 overexpression may lead to ectopic multicompartment Golgi outposts in dendrites and axons [61]
**Rab5/7/11 overexpression can lead to inappropriate incorporation into additional endosomal compartments
*** TMEM192 and LAMP1 may mark early and/or late endosomes [39, 40]

organelles important for the formation and maintenance of synaptic machinery. The following section outlines neuronal organelles with suggestions on how to visualize their movement within live axons.

Endoplasmic Reticulum (ER) As in other cell types, neuronal ER comes in two forms, smooth ER (SER) and ribosome-containing rough ER (RER). While classical ultrastructural studies did not reveal obvious rough ER in axons [16], it is now believed that both SER and RER are present and may impact local protein translation, folding, and export [17], lipid biosynthesis, glucose homeostasis, Ca^{2+} storage, and organelle contact/regulation [18]. Axonal ER is tethered to the microtubule cytoskeleton at multiple sites [19], which limits its motility. In mammalian neurons, ER has been visualized using fluorescent labels on the ER translocon subunit Sec61β [20] or the target peptide sequence KDEL, which acts as an ER retention signal [21]. However, given that axons are filled with ER, it is impossible to resolve distinct structures by confocal microscopy using these targets. In contrast, specific ER regions, such as ER exit sites revealed by the Coat Protein Complex II (COPII) coat assembly protein Sec16L, can be visualized in the axon as discrete puncta [20].

RNA Granules In neurons, long-distance mRNA transport enables local translation at sites distant from the nucleus. mRNA, RNA binding proteins (RBPs), and ribosomes form heterologous, membraneless condensates that are trafficked bidirectionally in axons and dendrites, often piggybacking onto other organelles such as endosomes [22] or lysosomes [23]. RNA granules can be visualized by the expression of fluorescently tagged RBPs. Ras-GTPase-activating protein (GAP)-binding protein 1 (G3BP1, visualized as mCherry-G3BP1) and TAR-DNA binding protein 43 (TDP-43, visualized with enhanced green fluorescent protein as EGFP-TDP-43) are RBPs that co-traffic with RNAs in mammalian axons [23, 24]. Labeling specific RNA is also possible via fluorescence in situ hybridization (FISH), MS2-MCP (MS2 coat protein), or SunTag methods, described in the following resource: [25].

Golgi Apparatus The Golgi apparatus represents a continuum of membranous cellular compartments involved in the secretion and sorting of proteins. As such, there is no single marker that captures all Golgi states. Cis-golgi marker sialyltransferase 2 (SialT2) or 3 (St3), cis-golgi matrix protein GM130, medial-golgi marker mannosidase II (ManII), and trans-golgi markers β1,4-galactosyltransferase 1 (GalT) and β-1, 4-N-acetylgalactosaminyltransferase (GalNAcT) have all been tagged to visualize neuronal Golgi populations. In neurons, Golgi is found not only perinuclear but also as satellite structures in axons and dendrites. Dendritic Golgi outposts in particular have gained attention for their role in microtubule nucleation and protein secretion, trafficking, and modification [17, 26]. In the axon, satellite Golgi organelles are found in close proximity to ER and/or lysosomes and may play a role in local translation [17]. Golgi has been visualized in live mammalian neurons using SialT2-YFP [27], St3-GFP [28], and GalT-mCherry [29].

Mitochondria As the powerhouse of the cell, mitochondria are crucial in energetically expensive neurons and can be found stationed throughout the somatodendritic and axonal compartments. Mitochondria undergo both anterograde and retrograde trafficking in neurons, but 90% of mitochondria in mature axons are non-motile [30]. The mitochondrial targeting sequence (Mito) of the cytochrome *c* oxidase polypeptide VIII (COX8) is widely used to localize fluorescent proteins to the mitochondrial matrix [31]. Tagged mitochondria outer membrane proteins, such as TOM20, can be useful for distinguishing mitochondria as well as mitochondria-derived vesicles. Examples of fluorescent axonal mitochondria imaging in mammalian neurons: mito-mEmerald [32], Cox8a-BFP [33], and TOM20-GFP [34].

Endosomes Endosomes are membranous cellular sorting organelles that are generally differentiated by two factors: the time it takes for endocytosed material to reach them and the Ras-associated binding (Rab) proteins that decorate them. Early endosomes marked by Rab5 mature into Rab7-positive late endosomes primarily destined for lysosome fusion [35]. Recycling endosomes marked by Rab11 return internalized molecules back to the plasma membrane. Rab11 is also present on tropomyosin receptor kinase A (TrkA)-containing retrograde carriers in sympathetic neurons [35]. In axons, endosomes undergo bidirectional transport characterized primarily by short back-and-forth movement. N-terminal EGFP fusions of Rab5, Rab7, and Rab11 have been used in mammalian axonal imaging [32]. However, it is important to note that endosomal markers may incorporate inappropriately into other endosomal compartments when overexpressed.

Autophagosomes The autophagosome is a double-membrane vesicle formed around organelles and protein aggregates to promote their clearance via delivery to lysosomes. The major marker of autophagosomes, light chain 3 (LC3), is incorporated into the autophagosome membrane during formation and remains stably associated with the organelle during transport and acidification [36]. Axonal autophagosomes initiate at distal axon tips and mature during retrograde transport to the cell soma [3]. Live axonal autophagosome trafficking can be visualized using mCherry- or EGFP-LC3 [3, 32, 37]. The maturation state of axonal autophagosomes can be assessed using the dual-color LC3 reporter mCh-EGFP-LC3 [38]. GFP fluoresces when the pH is greater than 6 but is quenched within an acidic environment. Therefore, upon autophagosome fusion with acidic lysosomes, EGFP quenches and the autophagosome shifts from dually labeled green/red puncta to red only.

Lysosomes Lysosomes are acidified vesicles hosting hydrolytic enzymes that degrade engulfed biomolecules. Because lysosomes exist within the continuum of the autophagy and endolysosomal systems, it is challenging to definitively draw a line where one state ends and another begins using protein labels. Conventional lysosomal tags such as lysosomal-associated membrane protein 1 (LAMP1) or transmembrane protein 192 (TMEM192) can sometimes be found on late or even early endosomes [39, 40]. Chimeric constructs such as SEP-LAMP1-RFP, where superecliptic pHluorin (SEP) is fused to the luminal domain of LAMP1-RFP, may provide a more definitive label for acidified lysosomes at low levels of expression; as with endosomes there can be a loss of compartment specificity at higher expression levels. SEP is a GFP variant that is similarly quenched in at low pH, so loss of SEP signal when using the SEP-LAMP1-RFP construct (i.e., only RPF detected) indicates a "true" acidic lysosome [41].

Peroxisomes Peroxisomes are small, single-membrane vesicles important for metabolism of lipids and reactive oxygen species, and they are often found in close proximity to other organelles such as ER, mitochondria, or lysosomes [21]. Neuronal peroxisomes exhibit comparable motility to mitochondria and have been live imaged in mammalian neurons using EGFP-SKL [21] and PEX-mRFP [42].

Dense Core Vesicles (DCV) Special neuronal organelles that store and release neuropeptides and neurotrophins, DCVs are filled with cargo at the Golgi and transported through dendrites and axons to reach synapses where they can be released via exocytosis. DCVs are distributed equally throughout axons and dendrites, but are mainly exocytosed within axons [43]. Fluorescently tagged neuropeptide Y (NPY) is a reliable DCV reporter that is targeted to the vast majority of DCVs in cultured primary or iPSC-derived neurons irrespective of which neuropeptide is endogenously expressed [43]. NPY-pHluorin [43] and NYP-mCherry [44] have been used in mammalian neurons to detect DCV exocytosis or intracellular trafficking, respectively.

Synaptic Vesicle Precursors (SVP) Another neuron-specific organelle, SVPs carry synaptic vesicle proteins synthesized and preassembled in the soma, such as synaptophysin (SYP), synaptotagmin I (SYT1), vesicular monoamine transporter 2 (VMAT2), and vesicle-associated membrane protein 2 (VAMP2), along the axon to distant presynaptic boutons [45]. Trafficked anterogradely

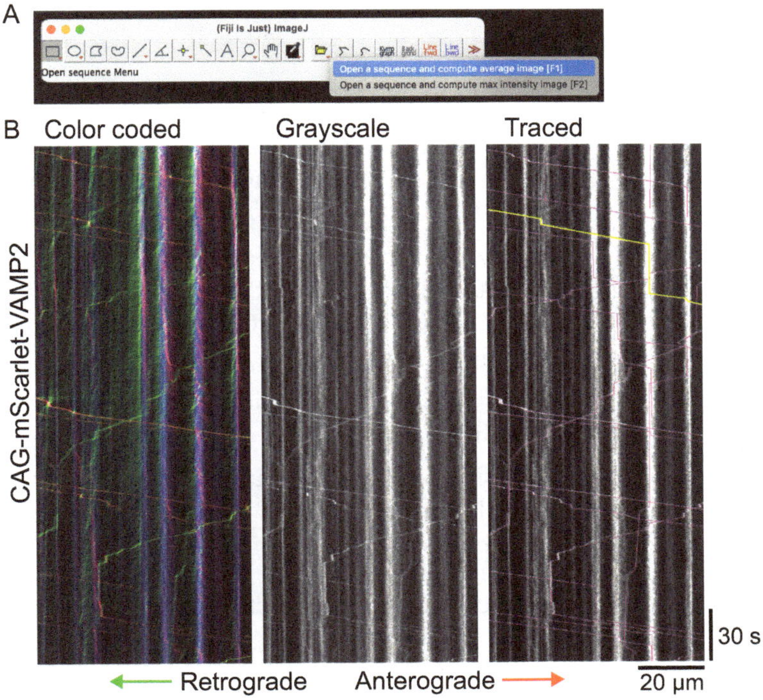

C

x pixel	y pixel	(x) distance (um)	(y) time (s)	Xf-Xi	Abs Xf-Xi	Yf-Yi	Positive InstVel_NoPauses	Pause count	Pause time (s)	Pauses	Inst Velocty (μm/s)
5	151	0.64725	37.75								
127	173	16.44015	43.25	15.7929	15.793	5.5	2.871436364	0		FALSE	2.87
126	186	16.3107	46.5	-0.12945	0.129	3.25		1	3.25	TRUE	0.04
380	233	49.191	58.25	32.8803	32.880	11.75	2.798323404	0		FALSE	2.80
380	347	49.191	86.75	0	0.000	28.5		1	28.5	TRUE	0.00
464	358	60.0648	89.5	10.8738	10.874	2.75	3.954109091	0		FALSE	3.95
470	365	60.8415	91.25	0.7767	0.777	1.75	0.443828571	0		FALSE	0.44
502	372	64.9839	93	4.1424	4.142	1.75	2.367085714	0		FALSE	2.37

Run Length (μm)	Absolute Run Length (μm)	Time of Residency (s)	Average Velocity (μm/s)	#paused timepoints	total pause duration (s)	Average Pause duration/pause
64.33665	64.596	55.25	2.486956629	2	31.75	15.875

Fig. 1 Example of axonal imaging analysis of SVPs in DIV21 i^{3}Neurons. SVPs are visualized by mScarlet-Syp under Cytomegalovirus (CMV) promoter and imaged at a frame rate of 250 ms for 5 min. (**a**) After installing KymographClear plugin in Fiji, open TIF image sequence and compute average intensity image using the folder icon. (**b**) KymographClear generates multiple kymograph versions, including color coded directions (left), forward movement filtered, backward movement filtered, static filtered, and grayscale (middle). Using the segmented line tool, traces can be tracked along the kymograph (left). (**c**) *x* and *y* coordinates from the yellow trace are displayed in an example Excel sheet. Calculations of various run parameters are shown

by the "marathon runner" kinesin-3 KIF1A, SVPs have been visualized speeding through mammalian axons using fluorescently tagging synaptic proteins such as mScarlet-Syp ([2]; Fig. 1) and VMAT2-GFP [46].

Piccolo-Bassoon Transport Vesicles (PTVs) Sometimes called active zone precursor vesicles, PTVs exit the trans-Golgi network and axonally transport proteins necessary to set up the Cytoskeletal matrix assembled at the active zone (CAZ), including piccolo, bassoon, Munc13-1, RIM1α, and ELKS2 (CAST1/ERC2) [47]. Axonal trafficking of PTVs has been visualized using GFP-bassoon [47–49], EGFP-Munc13, and ELKS2-iEGFP [47] in mammalian neurons.

2 Materials

2.1a Primary neuron culture and transfection:

1. Primary murine neurons plated at 200–220,000 cells per 35 mm glass-bottom imaging dish (MatTek) coated in 200 μg/mL poly-L-lysine.
2. Primary neuron media: Neurobasal media supplemented with 2% B-27 (Gibco), 58 mM glucose (Sigma), 2 mM GlutaMAX (Gibco), 100 U/mL penicillin, and 100 mg/mL streptomycin (Sigma).
3. Primary neuron lipofection reagent: Lipofectamine™ 2000 (Invitrogen).
4. Expression plasmids (*see* Table 1 for suggested organelle tags).
5. Opti-MEM (Gibco).

2.1b i^{3}Neuron culture and transfection:

1. i^{3}Neurons plated at 300,000 cells (*see* **Note 1**) per 35 mm glass-bottom imaging dish (MatTek) coated in 100 μg/mL poly-L-ornithine.
2. i^{3}Neuron media: BrainPhys Neuronal Medium (STEMCELL Technologies) supplemented with 2% N21-MAX (R&D Systems), 10 ng/mL Brain-derived neurotrophic factor (BDNF) (PeproTech), 10 ng/mL Glial cell line-derived neurotrophic factor (GDNF) (PeproTech), 10 ng/mL NT-3 (PeproTech), and 1 μg/mL Laminin (Corning).
3. i^{3}Neuron lipofection reagent: Lipofectamine™ Stem (Invitrogen).
4. Expression plasmids (*see* Table 1 for suggested organelle tags).
5. Opti-MEM (Gibco).

2.2 Live cell microscopy:

1. i^{3}Neuron imaging media: Hibernate A media (BrainBits) supplemented with 2% N21-MAX (R&D Systems), 10 ng/mL BDNF (PeproTech), 10 ng/mL GDNF (PeproTech), 10 ng/mL NT-3 (PeproTech), and 1 μg/mL Laminin (Corning).
2. Primary neuron imaging media: Hibernate E media (BrainBits), supplemented with 2% B27 and 2 mM GlutaMAX.

3. Spinning-disk confocal microscope equipped with temperature-controlled chamber and 60x or 100x 1.40 NA oil-immersion objectives.

2.3 Image Analysis:

1. Computer equipped with Fiji [50] with KymographClear plugin [51] and Microsoft Excel.

3 Methods

3.1 Culture neurons prior to transfection:

1. Culture primary rat neurons [12] or iPSC-derived i^3Neurons [11] plated at 200–220,000 or 300,000 cells per 35 mm glass-bottom imaging dish (*see* **Note 1**), respectively, until desired day in vitro (DIV). For axonal imaging, neurons must be allowed to polarize and mature. Recommended day for live axonal imaging: i^3Neurons DIV21–28, rat primary neurons DIV8–14.

3.2a Primary neuron transfection (steps performed in a laminar flow hood):

1. Equilibrate 2.5 mL primary neuron media per dish for >30 min in a CO_2 incubator at 37 °C.
2. For each imaging dish, mix 0.3–2 μg plasmid DNA in 150 μL Neurobasal medium without supplements. Prepare Lipofectamine mixture by adding 4–8 μL Lipofectamine 2000 to 150 μL Neurobasal medium without supplements (*see* **Note 2**). Incubate for 5 min at room temperature.
3. Mix the DNA and Lipofectamine solutions together. Incubate at room temperature for 20 min.
4. Remove conditioned media from the cultured neurons and save. For each plate, add 0.5 mL fresh media to the conditioned media. Reserve at 37 °C.
5. Add 2 mL fresh, pre-equilibrated primary neuron media to cultured neurons and add Lipofectamine-DNA mixture dropwise to cells.
6. Place dishes in 5% CO_2 incubator at 37 °C for 45 min.
7. After incubation, remove transfection media and replace with reserved, conditioned media collected in **step 3**.
8. Return cells to 37°C, 5% CO_2 incubator for 24–48 h (*see* **Note 3**).

3.2b i³Neuron transfection (steps performed in a laminar flow hood):

1. Equilibrate 2.5 mL i³Neuron media per dish for >30 min in a CO_2 incubator at 37 °C.
2. For each imaging dish, mix 0.5–2 μg plasmid DNA with 3–6 μL Lipofectamine Stem (*see* **Note 2**) in 200 μL Opti-MEM medium. Incubate at room temperature for 10 min.
3. Remove conditioned media from the cultured neurons and save. For each plate, add 0.5 mL fresh media to the conditioned media. Reserve at 37 °C.
4. Gently and slowly add 2 mL fresh, pre-equilibrated i³Neuron media to cultured neurons (*see* **Note 4**) and add Lipofectamine-DNA mix dropwise to cells.
5. Place dishes in 5% CO_2 incubator at 37 °C for 1–2 h.
6. After incubation, remove transfection media and replace with reserved media collected in **step 3**.
7. Return cells to 37°C, 5% CO_2 incubator for 48–72 h (*see* **Note 3**)

3.3 Image transfected neurons:

1. Turn on spinning-disk confocal microscope and set temperature control to 37 °C. Depending on the environmental control, it typically takes 30–45 min to achieve stabilized temperature. Warm objective oil to 37 °C.
2. Warm imaging media to 37 °C.
3. Transfer transfected neuron cultures to laminar flow hood. Remove culture media from dish. Gently and slowly add warmed imaging media to side of dish (*see* **Note 4**).
4. Add a small drop of pre-warmed objective oil to 60x or 100x objective.
5. Place cultured neuron dish in the microscope chamber and focus on neurons expressing vector of interest.
6. Based on microscope setup and neuronal vector expression, select optimal laser power and exposure for each channel to be imaged (*see* **Notes 5**–7).
7. Identify axons based on morphological parameters: long processes at least ~300 μm in length with uniform diameter. Avoid imaging axons with noticeable swellings as this indicates an unhealthy state.
8. Start live imaging organelle of interest at appropriate framerate (use Table 3 as a guide). Typical imaging duration ranges from 2–10 min depending on experiment purpose (*see* **Note 8**). Each dish can be imaged in Hibernate™-based media for up to 1 h.

Table 3
Axonal organelle transport parameters and recommendations for acquisition time scale in mammalian neurons (*see* Note 8)

Organelle	Velocity range	Primary direction	Frame rate
Endoplasmic reticulum	More research needed	Bidirectional, primarily oscillatory* [20]	0.5–1 s
RNA granules	0.01–0.8 μm/s [23]	Bidirectional [63]	1 s
Golgi apparatus	~0.8 μm/s* [26]	Bidirectional, primarily stationary** [64]	0.5–1 s
Mitochondria	0.01–2 μm/s [32]	Bidirectional, primarily stationary [32]	1 s
Early endosomes	0.01–4 μm/s [32]	Bidirectional, [32]	500 ms
Late endosomes	0.01–4 μm/s [32]	Bidirectional [32]	500 ms
Recycling endosomes	0.01–6 μm/s [32]	Bidirectional [32]	200–400 ms
Autophagosomes	0.01–2 μm/s [32]	Retrograde [32, 37]	1 s
Lysosomes	0.01–5 μm/s [32]	Bidirectional [32]	200–400 ms
Peroxisomes	~0.1 μm/s [21]	Bidirectional, primarily stationary [21]	1 s
Dense core vesicles	0.01–5 μm/s [2, 42]	Bidirectional [42]	200–400 ms
Synaptic vesicle precursors	0.01–6 μm/s [2, 46]	Anterograde [2, 46]	200–400 ms
Piccolo-bassoon transport vesicles	0.01–0.5 μm/s [47, 49]	Anterograde [47]	0.5–1 s

*As determined in primary mouse DRG axons using GFP-Sec16L
**As determined in *Drosophila* dendrites using ManII:GFP and GalT:TagRFPt

3.4 Analysis of live axonal imaging using kymographs:

1. Open image series in Fiji [50] installed with KymographClear plugin [51].
2. If multiple channels were acquired, split and save each channel as individual TIF for separate kymograph processing.
3. Organize TIFs into individual labeled folders.
4. Open desired TIF using the Fiji plugin KymographClear by clicking the folder icon and selecting "Open a sequence and compute average image" (Fig. 1a).

5. Enter number of images in sequence. Click "OK."
6. In the projection image, draw a line along the axon of interest and click "Kymograph" button.
7. Enter line width of ~3 pixels (depending on thickness of axon) and click "OK" to generate kymographs. Kymographs will be automatically saved in a folder labeled "kymograph" within the host folder. KymographClear generates multiple kymographs with color-coded directional tracks. We recommend performing analysis on the grayscale "kymograph1.tif" image (Fig. 1b).
8. Save the "kymograph1.tif" image as new file for tracing movement and convert image type by selecting "Image → Type → RGB Color." Change background color to help distinguish traces.
9. Draw segmented line along a single organelle's track (*see* **Note 9**). Press "Delete" to change the color of underlying pixels. This helps keep track of the traces drawn (Fig. 1b).
10. Use "y" hotkey to bring up the Properties window. Select "List coordinates" and press "OK." Copy and paste coordinates into Excel worksheet.
11. Using Excel equations, convert *x* values from pixels to μm using microscope and objective-specific conversion factor. Convert *y* values from pixels to seconds using frame per second conversion.
12. The distance (*x*) and duration (*y*) values can be used to calculate run lengths, time of residence, instantaneous velocity, average velocity, pause frequency/duration, etc. Example Excel calculations for an individual trace are provided (Fig. 1c). Anterograde and retrograde runs are defined as tracks with run length greater than 10 μm or less than −10 μm, respectively. Stationary runs are defined as tracks with run length between ±10 μm. A pause is called when the instantaneous velocity is between ±0.083 μm.

4 Notes

1. i^{3}Neuron seeding density vs. actual number of cells that adhere and mature can vary across neuronal inductions and with iPSC genetic background. Low density cultures can experience Lipofectamine toxicity, so we recommend seeding a higher number of i^{3}Neurons (up to 750,000 per dish) if cell death following transfection is observed.
2. Neuron transfection efficiency is notoriously low. For each vector of interest, time should be spent troubleshooting optimal DNA:Lipofectine amounts/ratios within the

recommended ranges to achieve desired transfection outcome. Lower amounts of plasmid DNA should be used for proteins that cause unwanted cellular perturbation and/or lose compartment specificity at high levels.

3. Certain promoters take longer to express adequate levels of tagged proteins. Consequently, there can be differences in recommended time between transfection and imaging. For primary neurons, 24 h post-transfection is the default imaging time. For i^3Neurons, 48 h post-transfection is sufficient for CAG promoter vectors, but PGK promoter constructs may take 72 h before reaching optimal expression.
4. i^3Neurons weakly adhere to the culture surface. Special care should be given when changing media to prevent cells from lifting off. We recommend adding media slowly and gently to the side of the dish.
5. Laser power, exposure, and imaging frame rate should be carefully considered depending on organelle/tag to be visualized and experiment purpose. Increasing laser power and exposure can improve signal-to-noise ratio, but comes at the expense of faster signal bleaching over time and potential phototoxicity.
6. Two-channel imaging can provide invaluable information (i.e., how two organelles interact with one another in the same axon across time), but collecting two channels per time frame reduces frame rate.
7. Some organelles densely populate the axon. It can therefore be challenging to distinguish movement amid the crowded environment. Using fluorescence recovery after photobleaching (FRAP) can be useful in these instances [52].
8. Acquisition frame rate and imaging duration should be thoughtfully chosen and optimized depending on experiment purpose. Some questions may require longer observation time, which may necessitate imaging at a slower frame rate than recommended in Table 2.
9. Cross-referencing the kymograph with the image sequence can help distinguish individual traces.

References

1. Guedes-Dias P, Holzbaur ELF (2019) Axonal transport: driving synaptic function. Science 366. https://doi.org/10.1126/science.aaw9997
2. Guedes-Dias P, Nirschl JJ, Abreu N et al (2019) Kinesin-3 responds to local microtubule dynamics to target synaptic cargo delivery to the presynapse. Curr Biol 29:268–282.e8. https://doi.org/10.1016/j.cub.2018.11.065
3. Maday S, Wallace KE, Holzbaur ELF (2012) Autophagosomes initiate distally and mature during transport toward the cell soma in primary neurons. J Cell Biol 196:407–417. https://doi.org/10.1083/jcb.201106120

4. De Vos KJ, Grierson AJ, Ackerley S, Miller CCJ (2008) Role of axonal transport in neurodegenerative diseases. Annu Rev Neurosci 31: 151–173. https://doi.org/10.1146/annurev.neuro.31.061307.090711
5. Tischfield MA, Cederquist GY, Gupta ML, Engle EC (2011) Phenotypic spectrum of the tubulin-related disorders and functional implications of disease-causing mutations. Curr Opin Genet Dev 21:286–294. https://doi.org/10.1016/j.gde.2011.01.003
6. Aiken J, Buscaglia G, Bates EA, Moore JK (2017) The α-Tubulin gene TUBA1A in brain development: a key ingredient in the neuronal isotype blend. J Dev Biol 5:8. https://doi.org/10.3390/jdb5030008
7. Boyle L, Rao L, Kaur S et al (2021) Genotype and defects in microtubule-based motility correlate with clinical severity in KIF1A associated neurological disorder. Hum Genet Genomics Adv 2:100026. https://doi.org/10.1016/j.xhgg.2021.100026
8. Chen XJ, Xu H, Cooper HM, Liu Y (2014) Cytoplasmic dynein: a key player in neurodegenerative and neurodevelopmental diseases. Sci China Life Sci 57:372–377. https://doi.org/10.1007/s11427-014-4639-9
9. Baas PW, Deitch JS, Black MM, Banker GA (1988) Polarity orientation of microtubules in hippocampal neurons: uniformity in the axon and nonuniformity in the dendrite. Proc Natl Acad Sci 85:8335–8339. https://doi.org/10.1073/pnas.85.21.8335
10. Aiken J, Holzbaur ELF (2021) Cytoskeletal regulation guides neuronal trafficking to effectively supply the synapse. Curr Biol 31:R633–R650. https://doi.org/10.1016/j.cub.2021.02.024
11. Fernandopulle MS, Prestil R, Grunseich C, et al (2018) Transcription factor–mediated differentiation of human iPSCs into neurons. Curr Protoc Cell Biol 79:1–48. https://doi.org/10.1002/cpcb.51
12. Kaech S, Banker G (2006) Culturing hippocampal neurons. Nat Protoc 1:2406–2415. https://doi.org/10.1038/nprot.2006.356
13. Sahu MP, Nikkilä O, Lågas S et al (2019) Culturing primary neurons from rat hippocampus and cortex. Neuronal Signal 3
14. Datta PK (2013) Neuronal cell culture. Neuronal Cell Cult Methods Protoc 1078:35–44
15. Qin JY, Zhang L, Clift KL et al (2010) Systematic comparison of constitutive promoters and the doxycycline-inducible promoter. PLoS One 5:3–6. https://doi.org/10.1371/journal.pone.0010611
16. Merianda TT, Lin AC, Lam JSY et al (2009) A functional equivalent of endoplasmic reticulum and golgi in axons for secretion of locally synthesized proteins. Mol Cell Neurosci 40: 128–142. https://doi.org/10.1016/j.mcn.2008.09.008
17. Kemal S, Richardson HS, Dyne ED, Fu M meng (2022) ER and Golgi trafficking in axons, dendrites, and glial processes. Curr Opin Cell Biol 78:102119. https://doi.org/10.1016/j.ceb.2022.102119
18. Öztürk Z, O'Kane CJ, Pérez-Moreno JJ (2020) Axonal endoplasmic reticulum dynamics and its roles in neurodegeneration. Front Neurosci 14:1–33. https://doi.org/10.3389/fnins.2020.00048
19. Foster HE, Ventura Santos C, Carter AP (2021) A cryo-ET study of microtubules in axons. bioRxiv 2021.03.29.437471
20. Maday S, Holzbaur ELF (2014) Autophagosome biogenesis in primary neurons follows an ordered and spatially regulated pathway. Dev Cell 30:71–85. https://doi.org/10.1016/j.devcel.2014.06.001
21. Wang Y, Metz J, Costello JL et al (2018) Intracellular redistribution of neuronal peroxisomes in response to ACBD5 expression. PLoS One 13:1–22. https://doi.org/10.1371/journal.pone.0209507
22. Cioni JM, Lin JQ, Holtermann AV et al (2019) Late endosomes act as mRNA translation platforms and sustain mitochondria in axons. Cell 176:56–72.e15. https://doi.org/10.1016/j.cell.2018.11.030
23. Liao YC, Fernandopulle MS, Wang G et al (2019) RNA granules hitchhike on lysosomes for long-distance transport, using annexin A11 as a molecular tether. Cell 179:147–164.e20. https://doi.org/10.1016/j.cell.2019.08.050
24. Gopal PP, Nirschl JJ, Klinman E, Holzbaurb ELF (2017) Amyotrophic lateral sclerosis-linked mutations increase the viscosity of liquid-like TDP-43 RNP granules in neurons. Proc Natl Acad Sci U S A 114:E2466–E2475. https://doi.org/10.1073/pnas.1614462114
25. Bauer KE, Kiebler MA, Segura I (2017) Visualizing RNA granule transport and translation in living neurons. Methods 126:177–185. https://doi.org/10.1016/j.ymeth.2017.06.013
26. Ori-McKenney KM, Jan LY, Jan Y-N (2012) Golgi outposts shape dendrite morphology by functioning as sites of acentrosomal microtubule nucleation in neurons. Neuron 42:115–125. https://doi.org/10.1016/j.neuron.2012.10.008.Golgi

27. Buchanan SK, Walker JM (2016) The golgi complex. Springer New York, New York
28. Govind AP, Jeyifous O, Russell TA et al (2021) Activity-dependent Golgi satellite formation in dendrites reshapes the neuronal surface glycoproteome. Elife 10:1–24. https://doi.org/10.7554/eLife.68910
29. Jensen CS, Misonou H (2015) Live-cell imaging of post-golgi transport vesicles in cultured hippocampal neurons. Cell Mol Neurobiol 35: 123–135. https://doi.org/10.1007/s10571-014-0103-4
30. Lee A, Hirabayashi Y, Kwon S-K et al (2018) Emerging roles of mitochondria in synaptic transmission and neurodegeneration. Curr Opin Physiol 3:82–93. https://doi.org/10.1016/j.cophys.2018.03.009
31. Liu X, Yang L, Long Q et al (2017) Choosing proper fluorescent dyes, proteins, and imaging techniques to study mitochondrial dynamics in mammalian cells. Biophys Reports 3:64–72. https://doi.org/10.1007/s41048-017-0037-8
32. Boecker CA, Olenick MA, Gallagher ER et al (2019) ToolBox: live imaging of intracellular organelle transport in induced pluripotent stem cell-derived neurons. Traffic 21:138–155. https://doi.org/10.1111/tra.12701
33. Goldsmith J, Ordureau A, Harper JW, Holzbaur ELF (2022) Brain-derived autophagosome profiling reveals the engulfment of nucleoid-enriched mitochondrial fragments by basal autophagy in neurons. Neuron 110:967–976.e8. https://doi.org/10.1016/j.neuron.2021.12.029
34. Cho B, Cho HM, Jo Y et al (2017) Constriction of the mitochondrial inner compartment is a priming event for mitochondrial division. Nat Commun 8:1–17. https://doi.org/10.1038/ncomms15754
35. Villarroel-Campos D, Schiavo G, Lazo OM (2018) The many disguises of the signalling endosome. FEBS Lett 592:3615–3632
36. Wong YC, Holzbaur ELF (2015) Autophagosome dynamics in neurodegeneration at a glance. J Cell Sci 128:1259–1267. https://doi.org/10.1242/jcs.161216
37. Cason SE, Mogre SS, Holzbaur ELF, Koslover EF (2022) Spatiotemporal analysis of axonal autophagosome–lysosome dynamics reveals limited fusion events and slow maturation. Mol Biol Cell 33. https://doi.org/10.1091/mbc.E22-03-0111
38. Cason SE, Carman PJ, Van Duyne C et al (2021) Sequential dynein effectors regulate axonal autophagosome motility in a maturation-dependent pathway. J Cell Biol 220. https://doi.org/10.1083/jcb.202010179
39. Nguyen TL, Schneppenheim J, Rudnik S et al (2017) Functional characterization of the lysosomal membrane protein TMEM192 in mice. Oncotarget 8:43635–43652. https://doi.org/10.18632/oncotarget.17514
40. Fermie J, Liv N, ten Brink C et al (2018) Single organelle dynamics linked to 3D structure by correlative live-cell imaging and 3D electron microscopy. Traffic 19:354–369. https://doi.org/10.1111/tra.12557
41. Farías GG, Guardia CM, De Pace R et al (2017) BORC/kinesin-1 ensemble drives polarized transport of lysosomes into the axon. Proc Natl Acad Sci U S A 114:E2955–E2964. https://doi.org/10.1073/pnas.1616363114
42. Gumy LF, Katrukha EA, Grigoriev I et al (2017) MAP2 defines a pre-axonal filtering zone to regulate KIF1- versus KIF5-dependent cargo transport in sensory neurons. Neuron 94:347–362.e7. https://doi.org/10.1016/j.neuron.2017.03.046
43. Persoon CM, Moro A, Nassal JP et al (2018) Pool size estimations for dense-core vesicles in mammalian CNS neurons. EMBO J 37:1–18. https://doi.org/10.15252/embj.201899672
44. Bharat V, Siebrecht M, Burk K et al (2017) Capture of dense core vesicles at synapses by JNK-dependent phosphorylation of synaptotagmin-4. Cell Rep 21:2118–2133. https://doi.org/10.1016/j.celrep.2017.10.084
45. Goldstein AY, Wang X, Schwarz TL (2008) Axonal transport and the delivery of pre-synaptic components. Curr Opin Neurobiol 18:495–503. https://doi.org/10.1016/j.conb.2008.10.003
46. De Pace R, Britt DJ, Mercurio J et al (2020) Synaptic vesicle precursors and lysosomes are transported by different mechanisms in the axon of mammalian neurons. Cell Rep 31: 107775. https://doi.org/10.1016/j.celrep.2020.107775
47. Maas C, Torres VI, Altrock WD et al (2012) Formation of Golgi-derived active zone precursor vesicles. J Neurosci 32:11095–11108. https://doi.org/10.1523/JNEUROSCI.0195-12.2012
48. Bury LAD, Sabo SL (2011) Coordinated trafficking of synaptic vesicle and active zone proteins prior to synapse formation. Neural Dev 6: 24. https://doi.org/10.1186/1749-8104-6-24
49. Shapira M, Zhai RG, Dresbach T et al (2003) Unitary assembly of presynaptic active zones

from Piccolo-Bassoon transport vesicles. Neuron 38:237–252. https://doi.org/10.1016/S0896-6273(03)00207-1

50. Schindelin J, Arganda-Carreras I, Frise E et al (2012) Fiji: an open-source platform for biological-image analysis. Nat Methods 9: 676–682. https://doi.org/10.1038/nmeth.2019

51. Mangeol P, Prevo B, Peterman EJG (2016) KymographClear and KymographDirect: two tools for the automated quantitative analysis of molecular and cellular dynamics using kymographs. Mol Biol Cell 27:1948–1957. https://doi.org/10.1091/mbc.E15-06-0404

52. Zhang XM, Cordelieres FP, Herzog E (2021) Investigate synaptic vesicles mobility in neuronal culture axons by FRAP imaging. Bio-protocol 11:1–11. https://doi.org/10.21769/BioProtoc.3962

53. Niwa H, Yamamura K, Miyazaki J (1991) Efficient selection for high-expression transfectants with a novel eukaryotic vector. Gene 108:193–199. https://doi.org/10.1016/0378-1119(91)90434-d

54. Boshart M, Weber F, Jahn G et al (1985) A very strong enhancer is located upstream of an immediate early gene of human cytomegalovirus. Cell 41:521–530. https://doi.org/10.1016/s0092-8674(85)80025-8

55. Mizushima S, Nagata S (1990) pEF-BOS, a powerful mammalian expression vector. Nucleic Acids Res 18:5322. https://doi.org/10.1093/nar/18.17.5322

56. Adra CN, Boer PH, McBurney MW (1987) Cloning and expression of the mouse pgk-1 gene and the nucleotide sequence of its promoter. Gene 60:65–74. https://doi.org/10.1016/0378-1119(87)90214-9

57. Benoist C, Chambon P (1981) In vivo sequence requirements of the SV40 early promotor region. Nature 290:304–310. https://doi.org/10.1038/290304a0

58. Glover CPJ, Bienemann AS, Heywood DJ et al (2002) Adenoviral-mediated, high-level, cell-specific transgene expression: a SYN1-WPRE cassette mediates increased transgene expression with no loss of neuron specificity. Mol Ther 5:509–516. https://doi.org/10.1006/mthe.2002.0588

59. Vidal M, Morris R, Grosveld F, Spanopoulou E (1990) Tissue-specific control elements of the Thy-1 gene. EMBO J 9:833–840. https://doi.org/10.1002/j.1460-2075.1990.tb08180.x

60. Gloster A, Wu W, Speelman A et al (1994) The T alpha 1 alpha-tubulin promoter specifies gene expression as a function of neuronal growth and regeneration in transgenic mice. J Neurosci 14:7319–7330

61. Yang SZ, Wildonger J (2020) Golgi outposts locally regulate microtubule orientation in neurons but are not required for the overall polarity of the dendritic cytoskeleton. Genetics 215: 435–447. https://doi.org/10.1534/genetics.119.302979

62. Lipka J, Kapitein LC, Jaworski J, Hoogenraad CC (2016) Microtubule-binding protein doublecortin-like kinase 1 (DCLK1) guides kinesin-3-mediated cargo transport to dendrites. EMBO J 35:302–318. https://doi.org/10.15252/embj.201592929

63. Wang B, Pan L, Wei M et al (2015) FMRP-mediated axonal delivery of miR-181d regulates axon elongation by locally targeting Map1b and Calm1. Cell Rep 13:2794–2807. https://doi.org/10.1016/j.celrep.2015.11.057

64. Du Q, Chang J, Cheng G et al (2021) Sunday driver mediates multi-compartment golgi outposts defects induced by amyloid precursor protein. Front Neurosci 15:1–12. https://doi.org/10.3389/fnins.2021.673684

Chapter 16

Analysis of Microtubule Polymerization During Axon Outgrowth Using Fluorescently Labeled Microtubule Plus-End Tracking Proteins

Isabel Pérez-Ferrer and Carlos Sánchez-Huertas

Abstract

The study of microtubules arrangements and dynamics during axon outgrowth and pathfinding has gained scientific interest during the last decade, and numerous technical resources for its visualization and analysis have been implemented. In this chapter, we describe the cell culture protocols of embryonic cortical and retinal neurons, the methods for transfecting them with fluorescent reporters of microtubule polymerization, and the procedures for time-lapse imaging and quantification in order to study microtubule dynamics during axon morphogenesis.

Key words Microtubule, Microtubule plus-end, Axon outgrowth, Growth cone, EB3, +TIPs, Microtubule dynamics, Cortical neuron, Retinal neuron

1 Introduction

Microtubules (MTs) are major composers of the neuronal cytoskeleton, providing shape, organelle arrangement, motility, and intracellular cargo transport. During maturation, neurons migrate toward their final destinations while simultaneously extending their axon to reach their synaptic partners. This axonal pathfinding is led by a small amoeboid-like protrusion at the tip called the growth cone (GC). Both axon extension and navigation require a dynamic MT network to propel growth and instruct steering movements. In the GCs, scattered MTs explore the structure, interacting with actin fibers and substrate adhesion points, getting eventually captured and stabilized. Then, the stabilized MTs allow the asymmetric entry of organelles and vesicles to the GC periphery, triggering the turning [1–3].

MTs are intrinsically polar polymers with a stable/slow-growing minus-end and a highly dynamic/fast-growing plus-end that alternate polymerization and shrinking phases [4]. This, so

Kazuhito Toyooka (ed.), *Neuronal Morphogenesis: Methods and Protocols*, Methods in Molecular Biology, vol. 2831,
https://doi.org/10.1007/978-1-0716-3969-6_16,

called, MT plus-end dynamic instability determines the growth and stability of MT networks and is controlled by an expanding family of MT-associated proteins denominated plus-tip interacting proteins (+TIPs). +TIPs exert collective effects on MTs and must follow certain hierarchies in their MT recruitment to achieve specific functions. For instance, microtubule end-binding proteins (EB1, EB2, EB3 in metazoans)—which are the most abundant +TIPs—directly associate with MT plus-ends and are autonomous regulators of plus-end dynamic instability. EBs targeting to the MT plus-end favor its continuous polymerization and reduce the number of catastrophes. Therefore, EBs tip-tracking mainly correlates with MT polymerization events. While EB1 protein is cell ubiquitous, EB3 is predominantly expressed by neurons, but EB1 and EB3 are both necessary for axon extension. EBs are also scaffold providers for other +TIPs, for this reason EBs are considered master +TIP controllers and widely employed as general reporters of MT dynamics [5, 6]. The time-lapse studies of MT plus-end polymerization using EBs tracking encompass in vitro systems, immortalized cell lines and different neuron types from several experimental organisms [7–11]. Our imaging analyses of the MT plus-end polymerization events, so called comets, using the EB3-GFP fluorescent reporter in two different mouse neuron subtypes—cortical and retinal ganglion cell (RGC) neurons—showed the following figures. In cortical neurons, the average comet density and MT growth velocity in the axonal shaft and GCs were ~10 EB3 comets/100 micron/min at ~0.16 micron/s (shaft) and ~3 EB3 comets/10 micron^2/min at ~0.267 micron/s (GC), respectively. In retinal neurons, the average comet density and MT growth velocity in the axonal shaft and GCs were ~13 EB3 comets/100 micron/min at ~0.19 micron/s (shaft) and ~2 EB3 comets/10 micron^2/min at ~0.18 micron/s (GC), respectively.

2 Materials

2.1 Neuron Cultures

Consumables: Use the transfection reagent Lipofectamine 2000 to transiently transfect DNA constructs in dissociated cortical cultures in suspension or in adhesion.

Equipment: Cabinets of laminar-flow sterile air, an illuminated stereo-microscope, and a 37 °C incubator with a humidified atmosphere containing 5% CO_2.

2.1.1 Cortical Culture Media

Prepare the plating medium supplementing Dulbecco's Modified Eagle Medium (DMEM) (carrying 4500 mg/L glucose, sodium pyruvate) with 10% (vol/vol) fetal bovine serum (heat-inactivated at 55 °C during 30 min), 1% GlutaMAX, 50 U/mL penicillin and 50 μg/mL streptomycin (Pen/Strep).

Prepare the maintenance medium supplementing Neurobasal with 2% B27, 0.6% glucose, 1% GlutaMAX, and Pen/Strep.

Prepare the transfection medium supplementing Neurobasal with 2% B27, 0.6% glucose, and 1% GlutaMAX.

2.1.2 Explants Growth Medium

To prepare the plating media supplement DMEM:F12 with 0.4% methylcellulose, supplemented with 2% B-27, 10% (vol/vol) bovine serum albumin (BSA), and Pen/Strep. Incubate methylcellulose under shaking at 4 °C with the culture medium until its complete dissolution. Then, sterilize the complete medium using a vacuum filter (0.2 μm pore).

2.2 In Utero Electroporation of the Retina of Mouse Embryos

Consumables: Borosilicate glass pipettes, aspirator tube for pipettes, syringes and needles (25G × 5/8″), braided silk suture 6/0 USP, steel tongs 11 cm, straight needle holder 12.5 cm, serrated iris forceps 10 cm, serrated dissection scissors 10 cm, isoflurane, ritodrine, sterile 0.9% saline and 70% ethanol (EtOH).

Equipment: Electroporator and a surgical setup for rodents, including O_2 concentrator, isoflurane vaporizer, induction chamber, warm pad and an illuminated stereo-microscope.

2.3 Borate Buffer 0.1 M pH 8.5

1. Dissolve 1.24 g of boric acid and 1.9 g of sodium tetraborate in 400 mL ddH_2O.
2. Add sodium tetraborate slowly and reduce constantly the pH of the solution dropping pure HCl. Maximal solubility takes place between pH 4–7.
3. Adjust pH to 8.5. Complete to 500 mL H_2O. Readjust pH 8.5 sterilize with a 0.22 μm-pore filter and store at 4 °C in an autoclaved bottle.

2.4 Hank's Buffer

1. 137 mM NaCl, 5.4 mM KCl, 0.25 mM Na_2HPO_4, 0.44 KH_2PO_4, 4.2 mM $NaHCO_3$, 5 mM glucose. Adjust pH to 7.
2. Sterilize by vacuum filtration using a 0.22 μm-pore filter, do not autoclave.

2.5 Image Acquisition and Analysis

For monitoring the MT plus-end growth events (comets) use a confocal spinning disk microscope equipped with incubation system, motorized stage, and an acquisition camera with the following requirements: 1024 × 1024 sensor, 13 μm pixel size, and 26 frames per second (full frame). Also, the following objectives will be needed: dry 20× and oil 40×, 60×, or 100×.

The Fiji software (National Institutes of Health, NIH) can be obtained at: https://imagej.net/software/fiji/downloads.

The Fiji-macro Velocity_Measurement_Tool is a modified version of code from kymograph macro tsp050706.txt (R. Basel and A. Seitz, EMBL, Heidelberg). It allows measuring velocities of moving spots or structures in image time series. It can be obtained at: https://dev.mri.cnrs.fr/projects/imagej-macros/wiki/Velocity_Measurement_Tool.

The Fiji macro Microtubule plus-end 2D tracking (CountComets_1_7) is intended to count persistent plus-end comets in discrete cell soma areas (S. Tosi and C. Sanchez-Huertas, ADM IRB Barcelona). It requires a Fiji version 1.50e (2015). Macro can be obtained at: https://adm.irbbarcelona.org/bioimage-analysis/image-j-fiji#h.p_ArC1561RjVLD.

3 Methods

3.1 Dissociated Neuron Cultures of Embryonic Mouse Brain Cortices

Dissociated cortical neurons grown as monolayer over a glass-bottom culture dish allow the delivery by transient transfection of DNA constructs encoding for fluorescent proteins fused to plus-end tracking proteins (EB3-GFP or EB3-mCherry), to subsequently monitor MT plus-end dynamics by time-lapse microscopy [7, 11–13].

1. Coat 35 mm diameter glass-bottom dishes with 0.1 mg/mL Poly-D-Lysine diluted in borate buffer and incubate 3 h or overnight in the incubator (*see* **Note 1**).
2. Wash abundantly with sterile ddH_2O, 3–4 washes during 60 min.
3. Coat dishes with 5 μg/mL of laminin diluted in Phosphate-Buffered Saline (PBS) and incubate 2–3 h at 37 °C.
4. Sacrifice the pregnant female mouse at embryonic stage from 14.5 to 18.5 (e14.5–e18.5), obtain the embryos, and dissect the brains in chilled Hank's buffer.
5. Dissect all the cortices in chilled Hank's buffer, removing the basal ganglia and the meninges, and collect the tissue in cold Neurobasal medium. To dissect and manipulate the tissue use straight N5 tweezers and microsurgical knives.
6. Aspirate Neurobasal carefully and digest the tissue with 0.25% trypsin in the presence of 1 mg/mL of DNase I during 12–15 min at 37 °C, diluted in sterile Hank's buffer. Invert carefully the tube 2–3 times to facilitate the extracellular matrix digestion.
7. Remove the trypsin-DNA solution by decantation and stop digestion by washing the tissue with plating medium, ×3 washes (2–3 mL each wash).
8. In the last wash medium, carefully dissociate the tissue, using a P1000 pipette with a blue tip. Avoid making bubbles and aggressive pipetting. Once obtained a homogenous turbid solution, add extra plating medium up to 1 mL per embryo.

9. Count individual cells using a Neubauer chamber, discarding the unviable dark blue-colored cells. Seed 600,000 cells per 35 mm diameter glass-bottom dish in 1 mL of plating medium (10^5 cells per cm^2, approximately) into dishes containing 1mL of pre-warmed plating medium (2 mL total plating medium volume). Use a master-mix cell solution for seeding multiple dishes.
10. Move the dishes gently to facilitate a homogeneous cell distribution and enter the dishes in the incubator.
11. 1–2 h later aspirate the plating medium avoiding touching the dish surface, add gently 2 mL of pre-warmed maintenance medium, and re-enter dishes to the incubator. At 3 days-in-vitro (3 DIV), add 1 μM of Cytosine arabinoside (AraC) to the cultures to avoid excessive proliferation of non-neuronal cells. Refresh 1/3 of volume of the maintenance medium every 4–5 DIV (*see* **Note 1**).

3.2 Transfection of Dissociated Neuron Cultures

Transfect the MT plus-end fluorescent reporters at 0 DIV or 2–3 DIV for immature neuron analysis, or at 14 DIV for mature neuron analysis (*see* **Note 2**).

3.2.1 Transfection in Suspension at 0 DIV

1. Mix 4 μL of Lipofectamine 2000 in 250 μL Neurobasal and 4 μg of DNA in 250 μL Neurobasal, using separated tubes for each solution. If other DNA constructs are co-transfected together with the MT plus-end fluorescent reporter (EB3-GFP or EB3-mCherry) keep a 3:1 or 2:1:1 proportions, where MT plus-end reporter represents the lowest amount.
2. Incubate 5 min at RT.
3. Add the Lipofectamine 2000 solution into the DNA solution and mix thoroughly by pipetting or vortexing.
4. Incubate 20 min at RT.
5. Following the **step 8** of Subheading 3.1 (trypsinization and cell counting), centrifuge 600,000 cells at 100 g during 5 min at RT and resuspend the cellular pellet in 500 μL of the Lipofectamine 2000/DNA mix.
6. Incubate 5 min at 37 °C in the incubator.
7. Seed neurons in a 35 mm diameter glass-bottom dish pre-coated with Poly-D-Lysine and laminin (Subheading 3.1, **steps 1–3**), containing 1 mL of pre-warmed transfection media.
8. Move the dish gently to facilitate a homogeneous cellular distribution and enter the cultures in the incubator.
9. 1 h 30 min later aspirate transfection media, add pre-warmed maintenance medium, and re-enter the dishes into the incubator.

3.2.2 Transfection in Suspension at 0 DIV

1. Use neuron cultures from Subheading 3.1 at either 2–3 DIV or 14 DIV. Collect and save the conditioned maintenance medium from cultures, add 1 mL of pre-warmed transfection medium and re-enter the dishes into the incubator.
2. Mix Lipofectamine 2000 and DNA as in Subheading 3.2.1, **steps 1–4**.
3. Drop 500 μL of Lipofectamine 2000/DNA mix all over the surface of a 35 mm diameter glass-bottom dish and move it gently to facilitate a homogeneous culture transfection. Re-enter dishes into the incubator.
4. 1 h 30 min later aspirate transfection media, add the pre-warmed conditioned medium previously saved in **step 1**, and re-enter dishes into the incubator.

3.3 In utero Electroporation (IUE) of the Retina of Mouse Embryos

Dissociated cultures of retinal ganglion cell (RGC) neurons frequently require the purification of the RGCs by immunopanning [14]. In contrast, culturing retina explants during 24 h allows the immediate identification of RGC axons projecting out of the explants. This technical approach to monitor RGC axons skips the expensive and time-consuming immunopanning protocol, although it requires the in utero electroporation of the MT plus-end fluorescent reporters in the embryo retinas 24 h before performing the explants procedure [15].

1. Anaesthetize timed-pregnant mouse females at embryonic day 13.5 (e13.5) in an induction chamber using isoflurane mixed with 0.8% oxygen.
2. Program the electroporator with the following settings: 5 pulses of 50 ms at 38 V, with 1 s of interval between pulses.
3. Place the anesthetized mouse on its back on an electric warm blanket in order to avoid loss of body temperature. Ensure that the nose and mouth are inside the anesthetic device and reduce the isoflurane dose to maintain the surgical plane.
4. Make an intraperitoneal injection of rytodrine (0.1 mL of a 14 mg/mL solution), a muscle relaxant to prevent uterine contractions during procedure.
5. Clean up surgery area with 70% EtOH and make an incision with scissors in the skin along the midline of approximately 2 cm. Then, make an incision in the abdominal wall.
6. Carefully, pull the uterus containing the embryos out, avoiding the placenta and direct contact with the embryos.
7. Aspirate the DNA mix preparation (1 μg/μL pCAG-EB3-GFP with 0.03% fast green in PBS) with a mouth aspirator tube and a pulled glass micropipette (*see* **Note 3**).

8. Using the pulled glass micropipette inject DNA into one eye of each embryo. To do so, punch through the uterus and yolk sac to enter the embryo's eye and blow 1 μL of DNA into the eye using the mouth aspirator tube.
9. Place the head of each embryo between the tweezer-type electrodes and electroporate the embryos. Put embryos back into the mother's abdomen after electroporation (*see* **Note 4**).
10. Close up the incision with braided silk suture from the bottom to the top of the abdominal incision. At the lower end of incision do a double knot and zig-zag up along the abdomen, then tie off at the top of the incision.

3.4 Explant Cultures of Embryonic Mouse Retinas

Twenty-four hours after the in utero electroporation of the MT plus-end fluorescent reporter (e14.5) perform the retina explant cultures [16, 17].

1. Coat 35 mm diameter glass-bottom dishes with 0.01% Poly-L-lysine and incubate 24 h at 37 °C.
2. Remove Poly-L-lysine and wash with sterile Hank's buffer 3–4 times, during 1 h. Then, coat with 20 μg/mL laminin and incubate 1 h at 37 °C.
3. During the coating time, sacrifice the e14.5 pregnant mouse to obtain the electroporated embryos and dissect their heads in chilled PBS.
4. Pin embryo heads with pins orienting the left or right side up in a dish coated with a silicone elastomer and filled with chilled PBS. Place one pin in the snout and another in the center of the head to avoid movements.
5. Remove tissue surrounding the eye using straight N5 tweezers and with a syringe needle (25G × 5/8″) drill the center of the lens.
6. Pulling the edges of the hole separate the pigmented epithelium from the retina, and extract the retina pressing with the forceps on both sides of the eye. Then, cut the optic nerve. Finally, collect the retinas in a multi-well-24 plate filled with cold PBS.
7. Check the electroporation of the retinas using a fluorescence microscope and select for further explants only those retinas that are fluorescent.
8. Transfer retinas one by one to a 35 mm culture dish lid containing DMEM:F12 and cut the retina in smaller square pieces using microsurgical knives. To do so, first cut retina edges, orient it in a ventral-side up position, and do straight/clean cuts.

9. Remove laminin from the culture glass-bottom dish, wash with Hank's buffer 2–3 times, and add 2 mL of pre-warmed explant growth medium to the dishes.
10. Aspirate the 6–10 pieces of retina (~2 retinas) using a pipette with a cut yellow tip and transfer them to the dish. Using the microsurgical knives, sink, flatten, and attach the retina pieces to the glass with the flat pigment side up, recognized by scattered black dots on surface. Attach the pieces to the center of the dish and separate fragments ~1 cm each other to avoid axon crossing.
11. Incubate 18–24 h in the incubator and proceed with the time-lapse imaging.

3.5 Time-Lapse Imaging of MT Plus-Ends Dynamics in Axons and Growth Cones

For monitoring the polymerization events of MT plus-ends (comets) use a confocal spinning disk microscope with the technical specifications described above (*see* Subheading 2, Materials).

To achieve optimal fluorescent signal at MT plus-ends in axons and CGs, image the neurons 18–24 h after retina explant culture or 24–48 h after cortical culture transfection (*see* **Note 5**).

Monolayer Cortical Cultures

1. Find the neuron's focal plane by using transmitted light and—if available—activate the Perfect Focus System (PFS) mode of the microscope to stay in-focus during the dish screening.
2. Place the image field at the periphery of the round glass surface of the dish and screen for transfected neurons by systematic scrolling of the dish surface using the live-mode of the microscope, the corresponding EB3 excitation laser, and the 60×/100× oil objective.
3. For an unbiased imaging of axons/GCs, select axons that randomly enter the image field and track them toward their proximal or distal regions in order to acquire the proximal axon segment or the growth cone, respectively (*see* **Notes 6** and **7**).
4. Acquire EB3 time-lapse image stacks with the corresponding excitation laser and the 60×/100× objective in a proximal axon segment, 100–200 micron away from the cell body, or in axonal GCs. Use the following time-lapse acquisition settings: time interval, 1 s; video duration, 2 min; z-step size, 0.5 micron; 2 z-planes for axons and 2-3 z-planes for GCs; binning 1 (*see* **Note 8**).
5. If other fluorescent reporters need to be imaged sequentially with different excitation lasers, adjust the camera exposure time for each channel and the number of z-planes in order to not exceed the time interval of 1 s during the time-lapse.

Retina Explants

1. Find the explants focal plane in the dish using transmitted light and the 20× objective, and switch to fluorescence in order to find the electroporated explants. Save the X, Y, Z positions of the electroporated explants with the multi-position mode of the microscope.
2. Switch to 60×/100× oil objective, move to the saved positions across the dish, and image sparse electroporated RGC axons/GCs growing out of the explants. Use the acquisition settings previously described in Subheading 3.5 (monolayer cortical cultures), **step 4**. For an unbiased imaging of axons/GCs, select axons that randomly enter the image field (*see* **Notes 6** and 7).

3.6 MT Plus-End Tracking Analysis in Axons

Process and analyze the time-lapse movies of the axons using Fiji software and the macro Velocity_Measurement_Tool described in the Subheading 2 Materials.

1. Load time-lapse movie and do maximum projection of the 2–3 z-planes for each time frame (Image > Stacks > Z Project > Maximum projection).
2. Convert time-lapse to 8-bit (Image > Type > 8-bit), set the scale, and save the projected time-lapse as new .tiff file to work with (Fig. 1a).
3. Trace the axon segment of interest using the straight line tool and add the tracing to the Region Of Interest (ROI) manager (Analyze > Tools > ROI manager) (*see* **Note 9**).
4. Install the macro Velocity_Measurement_Tool (Plugins > Macros > Install).
5. Select the axon tracing at the ROI manager and execute kymograph (Plugins > Macros > kymograph create kymogram) (Fig. 1b).
6. Contrast the kymograph to visualize the flow lines of the MT plus-end comets (Image > Adjust > Brightness/Contrast > Auto > Apply). Save the kymograph as a new .tiff file.
7. Trace a line on the trajectory of each comet of the kymograph using straight line tool (always in a top-bottom orientation). Add comet tracings to the ROI manager and visualize them over the kymograph image activating "Show all" in the ROI manager (Fig. 1c).
8. To calculate the comet's density in the axon shaft, divide the *n* number of comet tracings by the length of the axon segment analyzed and normalize to 100 micron of axon length and 1 min of video duration.

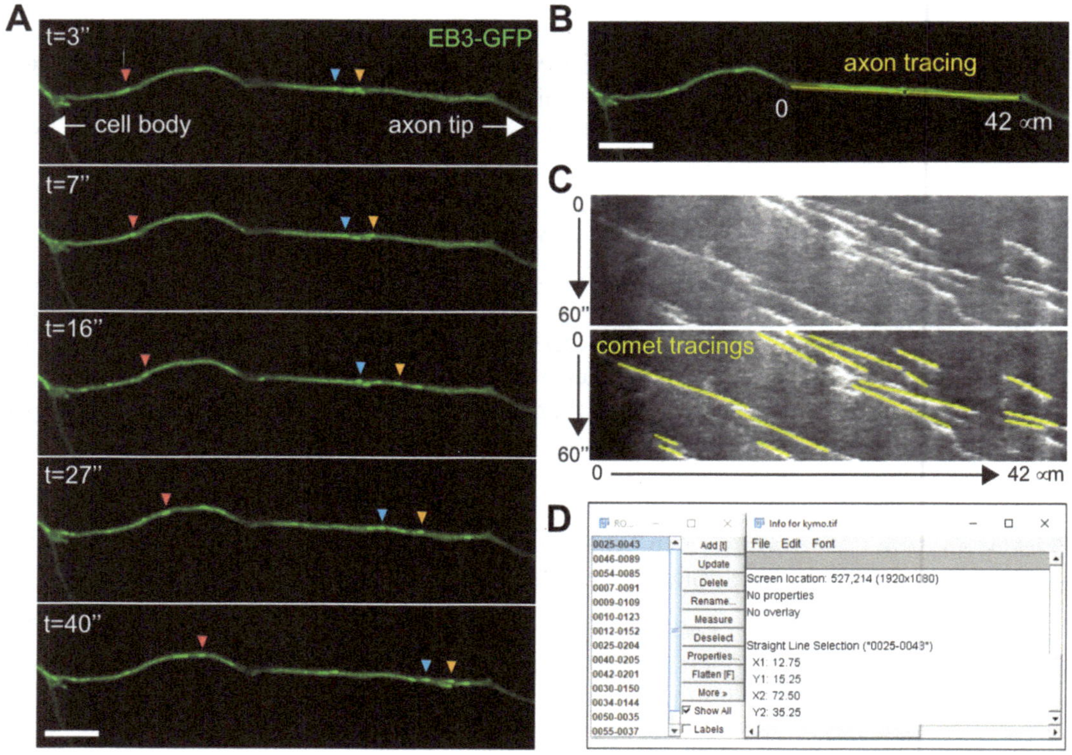

Fig. 1 MT polymerization events (comets) in an RGC axon from retina explant electroporated with EB3-GFP. (**a**) Time-lapse composition showing comets' movement. Colored arrowheads point individual comets at different time points. *t* means seconds. Scale bar, 10 μm. (**b**) Axon tracing to produce the kymograph of the segment. (**c**) Kymograph of 1 min time-lapse. Yellow lines mark the comet tracings used to calculate speed in D

9. To calculate the comet's speed in μm/s obtain the coordinates X1, Y1; X2, Y2 for each comet tracing of the ROI manager by selecting the tracing and typing "Ctrl" + "I" (Fig. 1d). Apply the following formula to the X1, Y1; X2, Y2 coordinates:

$$[(X2 - X1)^{*} \text{ pixel size value in } \mu\text{m}]/[(Y2 - Y1)^{*} \text{ time interval in s}].$$

10. Pixel size value (μm/pixel): physical length of a pixel on the camera/objective magnification.
11. Average the comet speed values of all the comets analyzed per axon segment to obtain an average speed value of comets per axon.
12. To calculate the proportion of comets moving in reverse direction along the axon shaft, divide the *n* number of comets with negative speed value by the total number of comets (reverse direction: comets growing toward the neuron cell body).

3.7 MT Plus-End Tracking Protein Analysis in Growth Cones

Process and analyze the time-lapse movies of GCs using Fiji software and, either the macro CountComets_1_7.ijm [11] or the Fiji plugin TrackMate [18] (*see* **Note 10**).

1. Load time-lapse movie and do maximum projection of the 2–3 z-planes for each time frame (Image > Stacks > Z Project > Maximum projection). Crop the image area of the GC, set the scale, and save the cropped time-lapse as a new .tiff file to work with (Fig. 2a).
2. Install the macro CountComets_1_7.ijm (Plugins > Macros > Install) and execute it.
3. Activate the option of "Manual ROI drawing" and trace the GC periphery using the polygon tool to restrict comet detection to the GC area, excluding the axon shaft.
4. Compare the resulting masks of automated comet-tracking (yellow, comets; green, cell periphery) with manual comet-tracking in the original time-lapse (Plugins > Tracking > MTrackJ) in several GCs to validate the automated tracking done by the macro (Fig. 2b–d) (*see* **Note 11**).
5. When automated macro tracking overestimates the comets number, increase the detection threshold value >775. When automated tracking underestimates, decrease the detection threshold value <775. When automated tracking counts fluorescent artifacts in the cellular edge, increase the border retraction value to 2–3 (Fig. 2c).
6. To calculate the comet density in the GC, normalize the detected number of objects (comets) by the GC area (μm^2) measured in the first frame of the time-lapse. For GC area measurement use an actin fluorescent reporter (fluorescent actin or Lifeact fusion proteins) or just the background EB3-fluorescent signal in a highly contrasted image. Normalize the obtained density values to 10 μm^2 of GC area and 1 min of video duration.
7. CountComets macro does not perform particle tracking, but just reliably counts spots persistent across time. For an automated tracking analysis of MT plus-ends polymerization in GCs, including the growth track speed, lifetime and length, use the Fiji plugin TrackMate (Plugins>Tracking>TrackMate).
8. Use the following TrackMate settings: LoG detector (estimated blob diameter, 0.8 micron; Threshold, 30; activate median filter and sub-pixel localization). Blob diameter and threshold values may be adjusted according to specific image scale and quality conditions. Set "quality" filter on spots (auto). Linear motion LAP tracker (initial search radius, 0.5 micron; search radius, 0.5 micron; Max frame gap, 2 frames). Set the following filters on tracks: "duration," 2 s; "displacement," 0.5 micron; "track index," 1 (Fig. 2e).

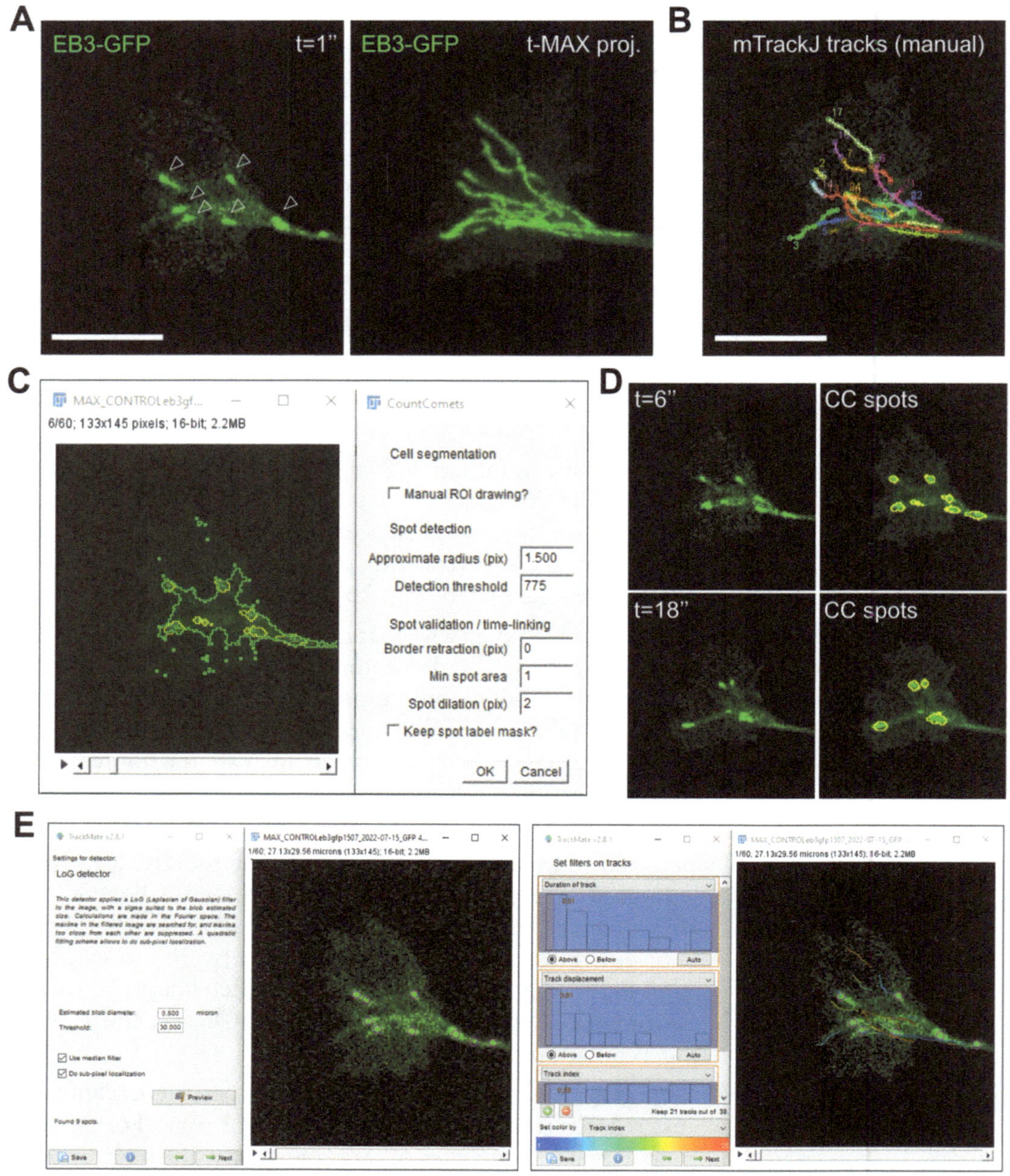

Fig. 2 MT polymerization events (comets) in a growth cone from an RGC axon in a retina explant electroporated with EB3-GFP. (**a**) Arrowheads mark comets at the t=1 sec time-lapse image. *t*-MAX proj is the maximum projection of all images of the EB3-GFP time-lapse. Scale bar, 10 μm. (**b**) Manual comet tracings in the growth cone using mTrackJ plugin of Fiji. (**c**) CountComets macro display. (**d**) Automated comet selection done by CountComets (CC) macro. Original time-lapse images (left) compared with comets selection (right) marked in yellow in the same second. (**e**) TrackMate plugin (Fiji) displays for automated comet tracking analysis

9. Compare the resulting tracks of automated comet analysis with the manual comet-tracking with mTrackJ in several GCs as in **step 4**.
10. Execute "Analysis" to obtain the data.
11. Number of comets is "number of tracks" and density can be calculated by normalizing to GC area and time; growth track speed is "track mean speed" in micron/s; lifetime is "track duration" in s; length is "track displacement" in micron.
12. Average the values of speed, lifetime, or duration of all comets per GC to obtain averaged comets dynamics values in each category per individual GC.

4 Notes

1. We discourage the use of glass-bottom dishes coated with Poly-L-lysine and laminin and stored for periods of time longer than 2 days. Also, when working with long-term 14 DIV cultures it is critical to minimize its manipulation out of the incubator during the first 3 DIV.
2. The transfection efficiency of dissociated neurons using Lipofectamine 2000 typically ranges between 1–5% of neurons in culture. Therefore, using an appropriate neuron density in cultures is essential. Plating too low neuron densities may lead to a deficient number of transfected neurons per dish, and using too high plating densities (especially in 14 DIV cultures) may cause axon overcrossing and hinder the tracking of individual axons.
3. To visualize MT plus-end comets by time-lapse imaging in the electroporated retina explants, use CAG promoter-driven DNA constructs and avoid CMV promoter-driven constructs.
4. During electroporation, keep moist with saline the portion of uterus out of the abdomen. Maintain only a few embryos out of the abdomen at a time to keep them warm and moist. Also, verify that the current amperage values are 0.08–0.15 Amp to prevent an electric shock in the mouse.
5. The ectopic expression of fluorescent EB3-reporters in neurons for longer than 48 h may eventually label the MT lattice, precluding an optimal visualization of the fluorescent MT plus-end particles (comets).
6. To avoid bias during image acquisition, acquire the proximal axon segments and GCs randomly found during the scrolling of the dish surface. Discard only the axons and GCs with no visible comets and the stalled GCs with no dynamic filopodia.

7. When tracking the axons in the microscope live-mode toward the proximal segment or the GC, keep in mind that axonal branches form the sharper angles toward the axon distal end.
8. To prevent photo toxicity during time-lapse imaging of axons and growth cones keep the laser intensity at ~2–4%, <= 100 ms of exposure time and a low number of z-planes. In addition, lower laser and camera settings can be employed during the dish screening. Verify that GCs and axonal filopodia remain dynamic after the imaging and avoid acquisition settings that produce net filopodia retraction or GC collapse.
9. To trace the axon segment of interest, select straight fragments of the axon and avoid abrupt trajectory changes and thickenings. Trace the axon shaft segment at least at 100 microns away from the growth cone to exclude MT comets entering the GC.
10. As an alternative to TrackMate, microtubule dynamics, by tracking fluorescently labeled particles, can be measured with the open source software plusTipTracker [19]. plusTipTracker, now incorporated into the package u-track 2.0 (K. Applegate and G. Danuser, Scripps Institute, La Jolla), is a Matlab-based software.
11. Comet density values in GCs obtained with CountComets macro in each video should be similar to the comet density values obtained by using TrackMate.

Acknowledgment

IPF is funded by a Severo Ochoa (SO)-PhD grant (PRE2021-098605). CS-H is funded by the SO-Program for Centers of Excellence in R&D (SEV-2013-0317) and by *Fundació "la Caixa"* (HR21-00824), awarded to Prof. Eloisa Herrera (*Instituto de Neurociencias, CSIC-UMH*).

References

1. Geraldo S, Gordon-Weeks PR (2009) Cytoskeletal dynamics in growth-cone steering. J Cell Sci 122(20) Art. no. 20. https://doi.org/10.1242/jcs.042309
2. Coles CH, Bradke F (2015) Coordinating neuronal actin–microtubule dynamics. Curr Biol 25(15) Art. no. 15. https://doi.org/10.1016/j.cub.2015.06.020
3. Sánchez-Huertas C, Herrera E (2021) With the permission of microtubules: an updated overview on microtubule function during axon pathfinding. Front Mol Neurosci 14: 759404. https://doi.org/10.3389/fnmol.2021.759404
4. Paz J, Lüders J (2018) Microtubule-organizing centers: towards a minimal parts list. Trends Cell Biol 28(3) Art. no. 3. https://doi.org/10.1016/j.tcb.2017.10.005
5. Akhmanova A, Steinmetz MO (2015) Control of microtubule organization and dynamics: two ends in the limelight. Nat Rev Mol Cell Biol 16(12) Art. no. 12. https://doi.org/10.1038/nrm4084
6. van de Willige D, Hoogenraad CC, Akhmanova A (2016) Microtubule plus-end tracking proteins in neuronal development. Cell Mol

Life Sci 73(10) Art. no. 10. https://doi.org/10.1007/s00018-016-2168-3

7. Stepanova T et al (2003) Visualization of microtubule growth in cultured neurons via the use of EB3-GFP (End-Binding Protein 3-Green Fluorescent Protein). J Neurosci 23(7) Art. no. 7. https://doi.org/10.1523/JNEUROSCI.23-07-02655.2003
8. Stout A, D'Amico S, Enzenbacher T, Ebbert P, Lowery LA (2014) Using plusTipTracker software to measure microtubule dynamics in Xenopus laevis growth cones. J Vis Exp 91: e52138. https://doi.org/10.3791/52138
9. Pecani K, Lieberman K, Tajima-Shirasaki N, Onishi M, Cross FR (2022) Control of division in Chlamydomonas by cyclin B/CDKB1 and the anaphase-promoting complex. PLoS Genet 18(8):e1009997. https://doi.org/10.1371/journal.pgen.1009997
10. Maurer SP, Cade NI, Bohner G, Gustafsson N, Boutant E, Surrey T (2014) EB1 accelerates two conformational transitions important for microtubule maturation and dynamics. Curr Biol 24(4) Art. no. 4. https://doi.org/10.1016/j.cub.2013.12.042
11. Sánchez-Huertas C, Freixo F, Viais R, Lacasa C, Soriano E, Lüders J (2016) Non-centrosomal nucleation mediated by augmin organizes microtubules in post-mitotic neurons and controls axonal microtubule polarity. Nat Commun 7:12187. https://doi.org/10.1038/ncomms12187
12. Freixo F et al (2018) NEK7 regulates dendrite morphogenesis in neurons via Eg5-dependent microtubule stabilization. Nat Commun 9(1) Art. no. 1. https://doi.org/10.1038/s41467-018-04706-7
13. Sánchez-Huertas C et al (2020) The +TIP Navigator-1 is an actin–microtubule crosslinker that regulates axonal growth cone motility. J Cell Biol 219(9) Art. no. 9. https://doi.org/10.1083/jcb.201905199
14. Winzeler A, Wang JT (2013) Purification and culture of retinal ganglion cells. Cold Spring Harb Protoc 2013(7):614–617. https://doi.org/10.1101/pdb.top070961
15. García-Frigola C, Carreres MI, Vegar C, Mason C, Herrera E (2008) Zic2 promotes axonal divergence at the optic chiasm midline by EphB1-dependent and -independent mechanisms. Development 135(10): 1833–1841. https://doi.org/10.1242/dev.020693
16. Murcia-Belmonte V et al (2019) A retino-retinal projection guided by Unc5c emerged in species with retinal waves. Curr Biol 29(7) Art. no. 7. https://doi.org/10.1016/j.cub.2019.02.052
17. Morenilla-Palao C et al (2020) A Zic2-regulated switch in a noncanonical Wnt/βcatenin pathway is essential for the formation of bilateral circuits. Sci Adv 6(46) Art. no. 46. https://doi.org/10.1126/sciadv.aaz8797
18. Tinevez J-Y et al (2017) TrackMate: An open and extensible platform for single-particle tracking. Methods 115:80–90. https://doi.org/10.1016/j.ymeth.2016.09.016
19. Vinopal S et al (2023) Centrosomal microtubule nucleation regulates radial migration of projection neurons independently of polarization in the developing brain. Neuron:S0896-6273(23)00070–3. https://doi.org/10.1016/j.neuron.2023.01.020

Chapter 17

Interrogating the Molecular Clutch in Neuronal Growth Cones: Measuring Traction Forces, F-actin Retrograde Flow, and Point Contact Demographics

Ketakee Ghate, Sampada P. Mutalik, and Aurnab Ghose

Abstract

Growth cone-dependent outgrowth of neuronal processes is essential for the development, plasticity, and regenerative capacity of the nervous system. This process involves the attachment of the growth cone to the substrate and the cyclical engagement/disengagement of the molecular clutch at the sites of adhesive contact. In this chapter, we describe protocols for traction force microscopy, measurement of F-actin retrograde flow velocities, and the assessment of adhesive point contacts by immunofluorescence. These complementary techniques collectively facilitate investigations into the regulation of the molecular clutch in neuronal growth cones.

Key words Growth cone, Motility, Molecular clutch, Extracellular matrix, Point contact, F-actin retrograde flow, Traction force microscopy, Mechanotransduction

1 Introduction

The directional outgrowth of neuronal growth cones is essential for the formation of functional connections during neuronal development [1–6] and regeneration following injury [6, 7]. During directed migration, the growth cone actively senses the environmental cues, remodels the cytoskeleton, and exerts propulsive traction forces onto the growth substrate. Point contacts are adhesion complexes that function as transient "molecular clutches," coupling the intracellular actomyosin machinery to the extracellular matrix via surface receptors [8–10] (Fig. 1). This coupling allows the transmission of the forces generated by the actomyosin cytoskeleton onto the substrate as traction forces. The engagement of F-actin filaments undergoing retrograde flow at the focal point

Authors Ketakee Ghate and Sampada P. Mutalik have equally contributed to this chapter.

Kazuhito Toyooka (ed.), *Neuronal Morphogenesis: Methods and Protocols*, Methods in Molecular Biology, vol. 2831,
https://doi.org/10.1007/978-1-0716-3969-6_17,

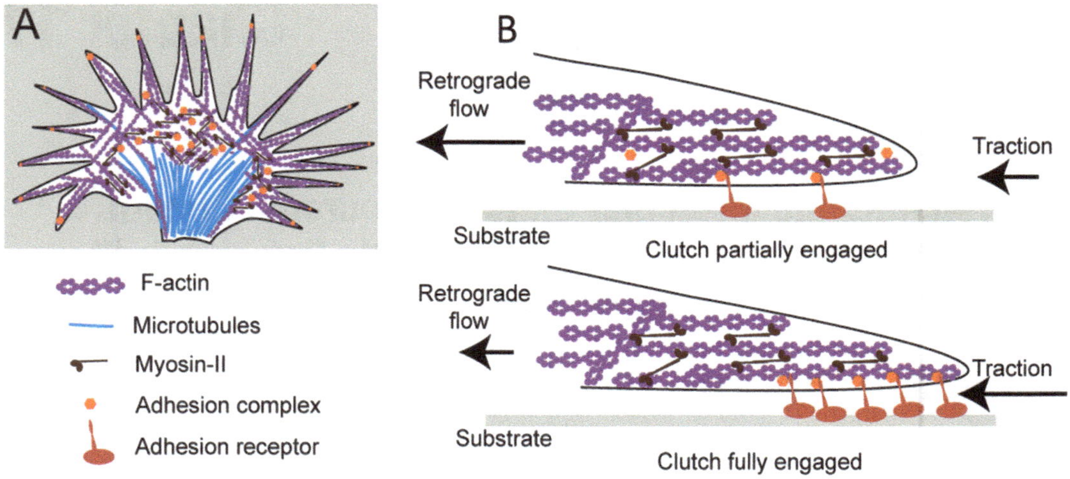

Fig. 1 Schematic representation showing the interaction of the actomyosin cytoskeleton with the adhesive point contacts to form the proposed molecular clutch. (**a**) Schematic of the top view of a neuronal growth cone depicting the underlying cytoskeleton and the adhesion complexes. (**b**) Side view of the growth cone peripheral region. Actomyosin networks drive the retrograde flow that generates transmissible traction forces when coupled to the substrate. When fewer clutches are engaged, the F-actin retrograde flow is higher, and transmitted traction forces are weak (top). When more F-actin filaments are engaged, the F-actin retrograde flow attenuates and the traction forces increase (bottom). Increased polymerization of F-actin (not shown) initiated by strong clutch engagement also facilitates protrusion. The length of the arrows indicates the magnitude of the F-actin retrograde flow or the cumulative traction force.

contacts leads to the development of traction forces while locally reducing the F-actin retrograde flow at these sites. Disengagement of F-actin filaments from the adhesion complexes results in reduced traction forces and an increase in the F-actin retrograde flow [9]. Further, the forces generated at the point contacts regulate the assembly-disassembly kinetics of the adhesion complexes [11, 12].

Measurement of traction forces, the velocity of the F-actin retrograde flow, and point contact dynamics enable the comprehensive evaluation of the mechanisms mediating growth cone mechanotransduction and motility [10, 13, 14].

In this chapter, we describe protocols to evaluate the traction forces generated by growth cones growing on deformable polyacrylamide (PAA) gels using traction force microscopy (TFM) [15, 16]. We further describe methods to evaluate growth cone point contacts and measure F-actin retrograde flow that complement TFM.

2 Materials

2.1 Traction Force Microscopy (TFM)

1. 28 mm (Fischer Scientific) glass coverslips, number 1.5.
2. 18 mm glass coverslips (Bluestar).

3. 35 mm plastic Petri dishes.
4. 1× Phosphate-buffered saline (PBS), pH 7.4.
5. (3-Aminopropyl)trimethoxysilane (APTMS; Sigma): Prepare 0.5% APTMS solution in distilled water in a glass bottle inside a fume hood. Prepare a fresh stock for each experiment.
6. Glutaraldehyde, 25% aqueous solution (Sigma): Prepare 0.5% of glutaraldehyde in 1× PBS from 25% glutaraldehyde stock. Add 500 μL of glutaraldehyde in 24.5 mL of 1× PBS. Store at 4 °C.
7. Fibronectin (Sigma): Reconstitute 1 mg Fibronectin in cold 1 mL of 1× PBS to make a 1 mg/mL stock. Aliquot 50 μL in low protein binding tubes. Use fresh aliquots to make 50 μg/mL stock for each experiment.
8. HEPES buffer, pH 8: Dissolve 23.83 g of HEPES salt in 80 mL Milli-Q water in an autoclaved bottle, adjust the pH to 8 using 10 N NaOH, and make the volume up to 100 mL (1M). Dilute 2 mL of this stock and in 98 mL of Milli-Q water to make a 20 mM stock, pH 8. Filter sterilize using a 0.2 μm filter.
9. Sulfo-SANPAH (Thermo Scientific or Covachem): Bring the vial (stored at −80 °C) to room temperature. Add 1 mL of HEPES buffer (20 mM, pH 8) to 50 mg of sulfo-SANPAH in the vial. Pipette up and down. Transfer to a polystyrene tube containing 24 mL HEPES buffer (20 mM, pH 8). Mix thoroughly. Make smaller aliquots of this 2 mg/mL stock based on how much you need per experiment. Store at −80 °C.
10. FluoSpheres™ (Thermo Fisher Scientific): Fluorescently labeled carboxylate-modified microspheres; 200 nm diameter; 508/605 nm emission/emission wavelength.
11. Acrylamide solution (30%): Weigh 30 g of acrylamide and dissolve in 60 mL of water. Heat to 37 °C to dissolve. Make the volume up to 100 mL once dissolved.
12. Bis-acrylamide (2.5%): Weigh 2.5 g of acrylamide and dissolve in 60 mL of water. Heat to 37 °C to dissolve. Make the volume up to 100 mL once dissolved.
13. 10% Ammonium persulphate (APS).
14. N,N,N′,N′-tetramethylethylenediamine (TEMED).
15. Distilled water.
16. Handheld UV source (365 nm).
17. 10× Trypsin (Sigma).
18. Spin coater.
19. 37 °C cell culture incubator.
20. Hot air oven.

21. Water bath.
22. Sterile cell culture hood.
23. Silicon sealant (Dow Corning silicone glue).
24. pCAG-mGFP-actin (Addgene #21948).
25. MATLAB (2007a/b).
26. Fiji (https://imagej.net/software/fiji/).
27. Fiji plugin 1: Template matching (https://sites.google.com/site/qingzongtseng/template-matching-ij-plugin).
28. Fiji plugin 2: TFM plugin [17] (https://sites.google.com/site/qingzongtseng/tfm).

2.2 Measurement of F-actin Retrograde Flow

1. 35mm coverslip-bottomed dishes (Fisher Scientific).
2. Poly-L-lysine (Sigma) diluted to 1 mg/mL in sterile PBS.
3. Fibronectin (Sigma): Reconstitute 1 mg Fibronectin in cold 1 mL of 1× PBS to make a 1 mg/mL stock. Aliquot 50 μL in low protein binding tubes. Use fresh aliquots to make 20 μg/mL stock for each experiment.
4. pCAG actin-mGFP (Addgene #21948).
5. Metamorph (https://www.moleculardevices.com/products/cellular-imaging-systems/acquisition-and-analysis-software/metamorph-microscopy).
6. Fiji (https://imagej.net/software/fiji/).
7. FlowTrack code [18] (https://oddelab.umn.edu/software).

2.3 Analysis of Growth Cone Point Contacts

All solutions are made fresh unless otherwise mentioned.

1. 1× Kreb's buffer, pH 7.4: Prepare the buffer with the indicated final concentrations in a final volume of 1 L using Milli-Q water. 145 mM NaCl, 5 mM KCl, 1.2 mM $CaCl_2$, 1.3 mM $MgCl_2$, 1.2 mM NaH_2PO_4, 10 mM glucose, 20 mM HEPES. Add HEPES and glucose after all the other ingredients have dissolved. Adjust the pH to 7.4 and filter sterilize using a 0.2 μm filter (*see* **Note 1**). Store at 4 °C [19].
2. Bouin's fixative: Combine 75 mL of saturated picric acid, 25 mL of formalin, and 5 mL of glacial acetic acid. Store at 4 °C.
3. Modified Kreb's fixative: Weigh and dissolve sucrose in 1× Kreb's buffer to a final concentration of 0.4 M. Mix this solution with Bouin's fixative in a 1:1 ratio.
4. 1× Phosphate-buffered saline (PBS) pH 7.4.
5. Permeabilization solution: Mix heat-inactivated normal goat serum (final concentration of 10%) and saponin (final concentration 0.01%) in 1× phosphate-buffered saline (PBS).

6. Wash buffer: Dissolve 0.5% bovine serum albumin (BSA; Sigma) in 1× Kreb's buffer.
7. Blocking solution: Dissolve 3% BSA in 1× PBS, pH 7.4.
8. Primary antibodies: Focal adhesion kinase (FAK) (Abcam ab33141), phospho-FAK(Y397) (Abcam ab4803), Vinculin (Abcam ab11194-200), Paxilin (Pax) (Abcam ab2264), phospho-Pax(Y31) (Abcam ab4832).
9. Secondary antibodies: Goat anti-mouse IgG (H+L) cross-adsorbed secondary antibody, Alexa Fluor™ 488 (Invitrogen), goat anti-rabbit IgG (H+L) cross-adsorbed secondary antibody, Alexa Fluor™ 568 (Invitrogen).
10. Mounting media: Mix 80% glycerol in Milli-Q water until homogeneous. This can be stored at room temperature. To preserve the fluorescence signal the following alternative mounting media can be used. Dissolve 0.05 g of n-propyl gallate in a solution of 90% glycerol and 20 mM Tris-HCl, pH 8. Heat for 10–15 min at 60 °C to dissolve. Store in the dark at 4 °C.
11. Fiji (https://imagej.net/software/fiji/).

3 Methods

3.1 Traction Force Microscopy

Traction force microscopy is performed on primary neurons cultured on polyacrylamide (PAA) gels. There are several advantages of using PAA gels [20]:

1. PAA gel deforms in proportion to a wide range of applied forces and is perfectly elastic.
2. The gel stiffness can be reproducibly tuned to specific values by choosing appropriate acrylamide/bis-acrylamide ratios.
3. The optical properties of PAA gels are well-suited for epifluorescence imaging.
4. The culture medium efficiently diffuses through the gel and facilitates the growth of cells.

3.1.1 Activation of Coverslips

PAA gels adhere poorly to glass and can detach over the incubation period. Hence, it is essential to covalently attach the gels to the glass surface by chemical activation as described below.

1. All coverslips need to be cleaned in advance. Immerse coverslips in 1 N HCl and incubate at 37 °C in a water bath with continuous shaking for 2 h. Subsequently, wash coverslips 4–5 times with an excess of double distilled water. The acid-washed coverslips can be stored in 70% ethanol for months at 4 °C.

2. Air-dry 28 mm coverslips to remove the ethanol. Coat coverslips with 250 μL of 0.5% APTMS (3-Aminopropyltrimethoxysilane; Sigma) for 10 min. Arrange them on a metal rack for the subsequent steps. Wash extensively with water three times followed by baking at 160 °C in a hot air oven for 1 h.
3. Drill a 22 mm diameter hole at the center of a 35 mm plastic petri dish and glue the activated coverslips to it using a silicon sealant. Ensure that the treated side faces the inside of the petri dish. Give enough time for the sealant to cure (overnight) before proceeding.
4. Once the culture dishes are ready, add 200 μL of 0.5% glutaraldehyde onto the coverslip and incubate for 1 h followed by washing with an excess of water four to five times.

3.1.2 Depositing Fluorescent Bead onto Coverslips

1. Take the required number of 18 mm acid-washed coverslips and coat them with 50 ug/mL fibronectin for 1 h at room temperature.
2. Dilute the fluorescent beads (FluoSpheres™) to a 1:1 ratio in water. This stock of beads can be further diluted to 1:20 or 1: 50 dilution based on experimental requirements* (*see* **Note 2**). Aspirate the excess fibronectin and place 100 μL of diluted bead solution onto the coverslip. Sonicate the bead solution for 15 min in a bath sonicator before use to avoid clumping. Spin coat the beads using a spin coater at 410 × *g* for 5 min.

* The dilution determines the density of the bead distribution.

3.1.3 Preparation of Gels

Neurons are weak force generators [21] necessitating the use of low-stiffness PAA gels to ensure adequate gel deformation. We use 0.438 Pa PAA gels prepared using 3% acrylamide and 0.1% bis-acrylamide [22]. Multiple techniques, like an atomic force microscope or the ball indentation method, can be used to accurately determine the stiffness of PAA gels [16, 23, 24]. These are beyond the scope of this chapter and are not discussed here.

1. Prepare the gel in a 50 mL polystyrene tube by adding the reagents listed in Table 1 below in the order indicated. Mix thoroughly.

   Stock solutions of acrylamide and bis-acrylamide can be stored at 4 °C for up to one year. Always use a fresh stock of APS. Use 1/100 volume of APS and 1/1000 volume of TEMED solutions [22].
2. Immediately add 15 μL of the prepared gel on the activated coverslip (attached to the bottom of the petri dish) (Fig. 2a).
3. Invert and place the bead-coated 18 mm coverslip on top of the gel immediately (Fig. 2b, c). Once gels are solidified, add water

Table 1
Composition of polyacrylamide (PAA) gel

Reagent	Volume (total 4000 μL)
Water	3396 μL
Acrylamide (30%)	400 μL
Bis-acrylamide (2.5%)	160 μL
Ammonium persulphate, APS (10%)	40 μL
N,N,N′,N′-tetramethylethylenediamine, TEMED	4 μL

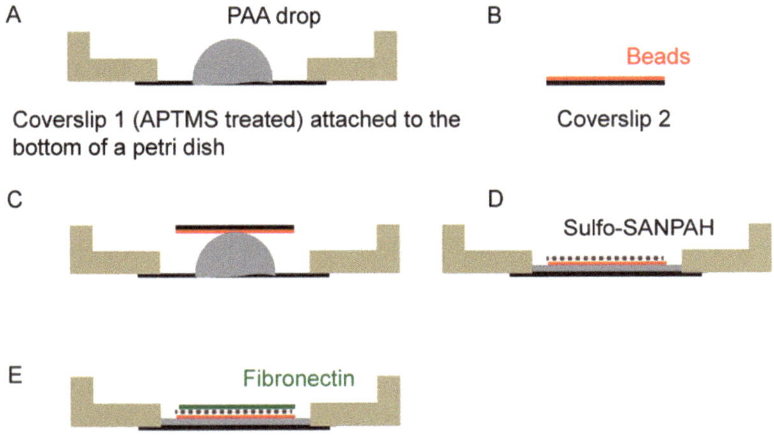

Fig. 2 Steps for making deformable PAA substrates for traction force microscopy. (**a**) PAA mix is placed on top of APTMS treated coverslips attached to the bottom petri dishes. (**b**) 18 mm coverslip are coated with fluorescent beads in advance. (**c**) The bead-coated coverslip is inverted onto the polymerizing PAA. (**d**) Solidified PAA gel is functionalized sulfo-SANPAH and (**e**) coated with fibronectin

to avoid drying. The layer of beads on the top coverslips enables maximum bead distribution on the surface of the gel.

4. After the gel has solidified, carefully remove the top 18 mm coverslips from the gel surface using a bent razor blade.

   The gel thickness in this method is approximately 60 μm. If gels are too thin, the deformation of the gel by the growth cone will be modified by the stiffness of the coverslip below [25].

3.1.4 Crosslinking Fibronectin to PAA Gel Surface

Cells attach poorly to PAA gels. It is necessary to conjugate the relevant extracellular matrix proteins to facilitate neuronal growth on PAA surfaces. A heterobifunctional reagent, sulfo-SANPAH, is used to conjugate proteins to the gel. Sulfo-SANPAH contains a succinimidyl ester group that interacts with the NH_2 group of lysines [26]. Simultaneously, the phenyl azide group of sulfo-SANPAH interacts with polyacrylamide to achieve crosslinking.

From this point, all procedures should be conducted under sterile conditions inside a cell culture hood.

1. Add 200 μL of 2 mg/mL sulfo-SANPAH on the gels and place the Petri dish under a hand-held UV source (365 nm) at a distance of 5 to 8 cm for one h (Fig. 2d).
2. Wash the plates with 50 mM HEPES (pH 8) buffer three times.
3. Add 250 μL of 50 μg/mL fibronectin (Sigma) on gels and incubate overnight at 4 °C (Fig. 2e).
4. Isolate neurons from the source (we used transfected primary neurons isolated from the spinal cord of the developing chick embryo; *see* **Note 3**) and plate on the prepared gels after equilibrating with the appropriate neuronal culture media. Place in cell culture incubator and incubate for 48 h.

3.1.5 Imaging

1. Replace the culture media containing 10% FBS with 1 mL of media without serum. Incubate for 30 min in the incubator.
2. Place the culture dish with PAA and neurons on the stage of an inverted epifluorescence microscope using an appropriate stage adapter. All imaging is performed using a 100×/1.4 NA oil immersion objective. Appropriate temperature and pH need to be maintained during imaging.
3. Locate a growth cone and image it using an appropriate set of filters (we transfect the neurons with an actin-GFP expression construct). This image provides the spread area of the growth cone (Fig. 3a).
4. Change filters to image the fluorescent beads (Ex. 508/ Em.605 nm). This image constitutes the stressed image (due to the deformation of the gel by the growth cone-generated traction forces; Fig. 3b).
5. Gently add 1 mL of 10× trypsin and let the growth cone completely detach or retract. Displacement of the dish in the x-y or the z plane should be avoided while adding trypsin. The use of hardware-based stable focusing systems can greatly aid image acquisition.
6. Acquire an image of beads of the same region following the detachment of the growth cone. This image is the relaxed image where the gel is no longer deformed by the growth cone (Fig. 3b).
7. Multilocation imaging of several regions of the gel, before and after trypsinization, significantly improves throughput.

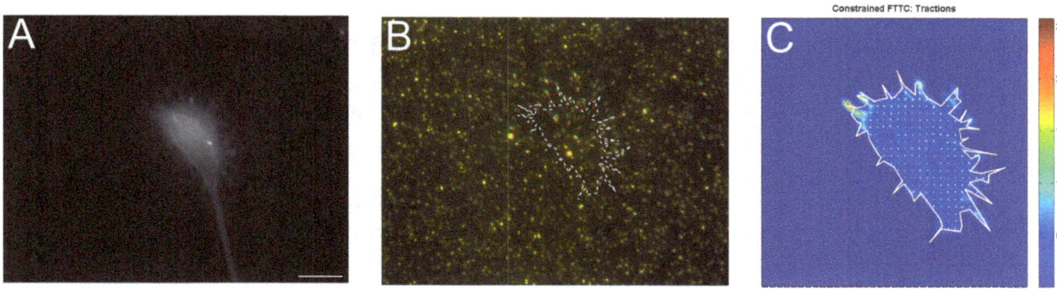

Fig. 3 Representative images for traction force microscopy. (**a**) Actin-GFP expressing growth cone on PAA gel coated with fluorescent beads (shown in **b**). (**b**) Overlayed image of the fluorescent beads before (pseudocolored red) and after (pseudocolored green) trypsin-mediated detachment of the growth cone (outline shown). Note the separation of the red and green signals within the demarcated area and its vicinity due to the traction forces generated by the growth cone. Yellow beads indicate no deformation. (**c**) Results obtained following the analysis by the TFM code; the image indicates traction forces constrained under growth cone area. The heat map indicates range of traction in Pascal (Pa) and the arrows within the demarcated growth cone boundary indicates direction and magnitude of the forces (Scale bar: 10 μm)

3.1.6 TFM Analysis

1. Import the stressed (before trypsin), relaxed (after trypsin), and the growth cone images to Image J. Correct for translational shifts in the images, if any, using the template matching plugin (https://sites.google.com/site/qingzongtseng/template-matching-ij-plugin).
2. Traction values for the growth cones are calculated using a MATLAB-based code developed by Butler et al, 2002, that uses displacement fields [25]. The main challenge in calculating traction using beads is to detect displacements in beads between stressed and relaxed images. The software divides the images into 64 × 64 pixel windows that are separated by 16 pixels from the center of each successive window. The cross correlation between the windows is then used to calculate displacement vectors. This is repeated for the entire image to obtain displacement fields. The code can calculate both the traction that is constrained under the growth cone area and also unconstrained traction across the entire field of view.
3. In the MATLAB code, import the stressed, relaxed, and growth cone (to obtain the cell boundary) images. Input the gel stiffness value (0.438 Pa), the Poisson ratio for the gel (0.48), and the thickness of the gel (60 μm). The code will generate displacement fields using cross-correlation between stressed and relaxed images after regularization and will calculate the traction fields (Fig. 3c).

   Alternatively, an Image J plugin that uses particle image velocimetry to generate the displacement fields and compute the tractions can be used (https://sites.google.com/site/qingzongtseng/tfm).

3.2 Evaluating F-actin Retrograde Flow in Growth Cones

Transfect neurons with a fluorophore-tagged actin expression vector (pCAG actin-mGFP; Addgene #21948) or a fluorophore-tagged F-actin probe. Grow the transfected neurons in appropriate culture media on coverslip-bottomed dishes coated with poly-L-lysine followed by fibronectin.

1. Image the actin-mGFP expressing growth cones after 24 h of culture in the same culture media (*see* **Note 4**).
2. Use 63×/1.4 NA oil-immersion objective to image the fluorescent growth cones at 1 min intervals for 30 min. Select growth cones with weak and speckled expression of GFP-actin for imaging. Excessive expression obscures flow patterns and can also affect F-actin dynamics.
3. Process the raw movies in the MetaMorph software. Use the segmented line tool in MetaMorph to draw a 5 pixel wide line from the growth cone center to the periphery, through the filopodia and lamellipodia (Fig. 4a), to generate kymographs from four to five different regions of the growth cone (Fig. 4b).
4. Process the resultant kymograph via the MATLAB code, FlowTrack, developed by Dr. David Odde, University of Minnesota to determine the velocity of actin retrograde flow (https://oddelab.umn.edu/software [18]).
5. Import the kymograph into FlowTrack and follow the instructions generated by the code.
6. The code generates a heat map to visualize the flow rates (Fig. 4c). This can also be used for quality control and the selection of kymographs for analysis.
7. Avoid regions close to the edge of the membrane in the kymograph to minimize artifacts due to membrane fluctuations.
8. Export the data generated by the FlowTracker software to Excel or GraphPad Prism, perform statistical analysis, and plot graphs.

3.3 Analysis of Growth Cone Point Contacts

All the steps are performed at room temperature without shaking unless otherwise mentioned. Ensure that the volume of all solutions is enough to cover the cultured neurons. All solutions should be added gently with the flow directed toward the walls of the culture dish/well to avoid detachment of neurons.

1. Culture primary neurons on coverslips placed in multiwell plates or on coverslip-bottomed chambers for the required duration under appropriate conditions. We grow primary neurons on growth surfaces coated with poly-L-lysine followed by fibronectin.
2. Fix cultured neurons using Kreb's fixative for 15 min at room temperature.
3. Replace the fixative with the permeabilization buffer and incubate for 20 min.

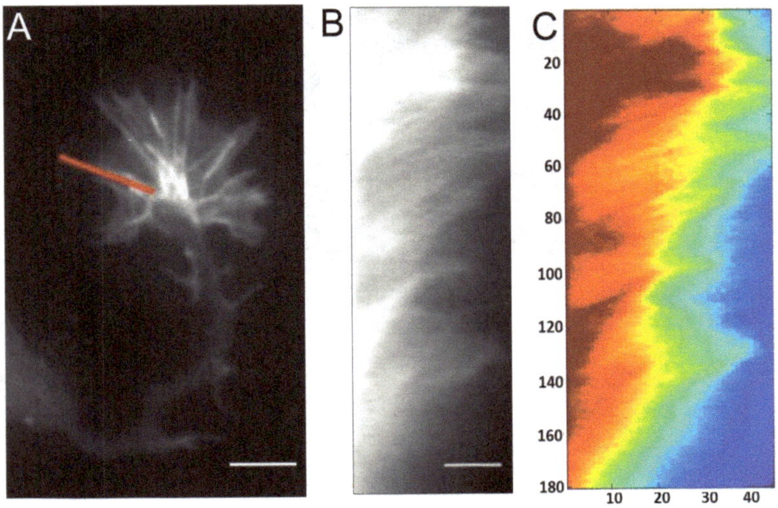

Fig. 4 Representative images for F-actin retrograde flow in growth cones. (**a**) Representative growth cone expressing pCAG actin-mGFP. The red line indicates the region selected to generate the kymograph. (**b**) Kymograph generated along the red line indicated in (**a**) by the MetaMorph software. (**c**) Heat map generated by the FlowTrack code used to analyze F-actin retrograde flow velocity. The displacement (in pixels) is indicated of the *x*-axis and time (in seconds) is on the *y*-axis (Scale bar: A—5 μm; B—2 μm)

4. Wash the cells twice using the wash buffer for 10 min each.
5. Block cells with blocking buffer for 1 h.
6. Replace the blocking solution with the primary antibodies in the blocking solution at 1:600 dilution and incubate at 4 °C overnight.
7. Remove the primary antibody solution and rinse twice with wash buffer for 10 min each.
8. Block cells with blocking solution for 45 min to 1 h.
9. Replace the blocking solution with secondary antibodies in the blocking solution at 1:300 dilution and incubate the cells in the dark for 1 h.
10. Remove the secondary antibody solution and wash the cells thrice with washing buffer for 10 min each.
11. Remove the washing solution and mount the samples in mounting media. If the neurons are cultured on coverslips in multiwell plate, gently lift the coverslip using forceps and invert it on a drop of mounting media (~80 μL) placed on a glass slide. Take precautions to avoid introducing air bubbles during mounting. If the neurons are cultured on a coverslip bottom chamber, simply replace the washing buffer with the mounting media ensuring that it covers the cells entirely (~100 μL to 150 μL depending on chamber size). Store the stained cells in dark at 4 °C until imaging.

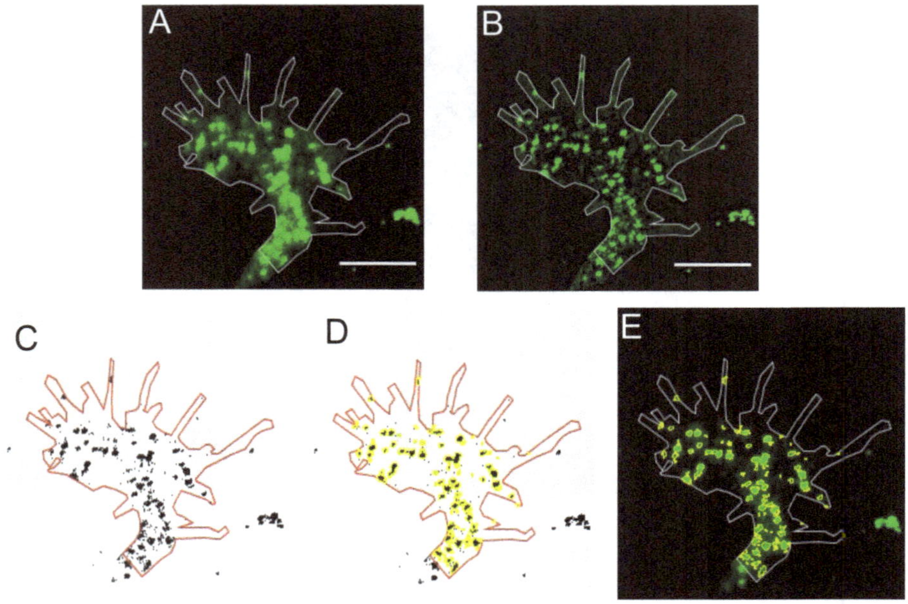

Fig. 5 Step-by-step representation of the analysis of growth cone point contacts. (**a**) Micrograph of the unprocessed immunofluorescence signal for immunolabeled focal point contacts (FAK in this example). (**b**) Point contacts following background subtraction and application of the unsharp mask. (**c**) Point contacts thresholded using the Otsu algorithm. (**d**) ROIs obtained following particle analysis. (e) Mask obtained by overlaying ROIs from (**d**) onto (**a**) (Scale bar: 5 μm)

12. Image cells using confocal or wide-field microscopy, preferably within a couple of days of immunolabeling (*see* **Note 5**).
13. For visualization of point contacts use 63×/1.4 NA or 100×/ 1.4 NA oil-immersion objectives. For fluorescence intensity analysis, ensure all imaging settings are constant across samples and treatments.
14. Use Fiji to analyze growth cone point contacts. Demarcate the boundary of the growth cone manually using the polygon selection tool. Duplicate this region of interest (ROI) and select it for all the subsequent steps (Fig. 5a).
15. Perform background subtraction to reduce noise with a rolling ball radius of 50 pixels and sliding paraboloid settings. Next, apply the unsharp mask with a Sigma radius set to 2.5 pixels and mask weight of 0.80 to sharpen point contact boundaries (Fig. 5b).
16. Threshold this image using the in-built Otsu algorithm. Set the threshold manually such that the point contacts are accurately selected and apply it to the demarcated area. Make sure the dark background option is selected in the thresholding window (Fig. 5c).
17. Prior to processing the resultant thresholded image for analysis, go to "set measurements" menu in analysis tab of Fiji. Select area and mean grey value options, which will allow

measurement of point contact area and intensity. To further perform the analysis, use the "analyze particles" feature from the analyze menu of Fiji. Here, set the size of particles to 1.5 pixel units to infinity and make sure the "summarize" and "add to manager" options are selected. Additionally, within this feature, set the dropdown menu "show" to outlines which will yield result in outlined point contact mask (Fig. 5d).

18. The above steps will generate a list of ROIs (outlined point contacts) added into the ROI manager along with a floating window result (overlay of counted point contacts in demarcated growth cone outline).
19. Go back to the original antibody channel and select all the ROIs obtained from the previous step such that they overlay on the demarcated area of the stained growth cone (Fig. 5e). Measure from the ROI manager to obtain intensities observed in the regions within the outlined mask.
20. Export the parameters generated by the particle analyzer, such as point contact number and area, along with the intensity measurements from Fiji to Microsoft Excel and/or GraphPad Prism for statistical analysis.

4 Notes

1. To prepare sterile Kreb's buffer, all constituents can be autoclaved separately except for HEPES and glucose. The latter items should be filter sterilized and added to the autoclaved solution inside a sterile hood.
2. For the TFM measurements, imaging neurons growing on a gel surface with a good distribution of beads is crucial. Poor bead distribution will not yield analyzable force measurements.
3. Protocols for chick spinal neuron cultures are discussed extensively in [10, 13]. Methods described in this chapter can be used for primary cultures of neurons from any source.
4. The media should be devoid of phenol-red to avoid non-specific, background signal during imaging of actin retrograde flow.
5. For immunofluorescence, perform control experiments with the secondary antibodies alone to ensure there is no background signal from the non-specific binding of the secondary antibodies.

Acknowledgment

The authors acknowledge the DBT-Wellcome Trust India Alliance (IA/TSG/20/1/600137) for supporting the study and the IISER Pune Microscopy Facility.

References

1. Pinto-Costa R, Sousa MM (2021) Microtubules, actin and cytolinkers: how to connect cytoskeletons in the neuronal growth cone. Neurosci Lett 747:135693
2. Dent EW, Gertler FB (2003) Cytoskeletal dynamics and transport in growth cone motility and axon guidance. Neuron 40(2):209–227
3. Omotade OF, Pollitt SL, Zheng JQ (2017) Actin-based growth cone motility and guidance. Mol Cell Neurosci 84:4–10
4. Gomez TM, Letourneau PC (2014) Actin dynamics in growth cone motility and navigation. J Neurochem 129(2):221–234
5. McCormick LE, Gupton SL (2020) Mechanistic advances in axon pathfinding. Curr Opin Cell Biol 63:11–19
6. Leite SC, Pinto-Costa R, Sousa MM (2021) Actin dynamics in the growth cone: a key player in axon regeneration. Curr Opin Neurobiol 69: 11–18
7. Stoll G, Müller HW (1999) Nerve injury, axonal degeneration and neural regeneration: basic insights. Brain Pathol 9(2):313–325
8. Abe K et al (2021) Mechanosensitive axon outgrowth mediated by L1-laminin clutch interface. Biophys J 120(17):3566–3576
9. Case LB, Waterman CM (2015) Integration of actin dynamics and cell adhesion by a three-dimensional, mechanosensitive molecular clutch. Nat Cell Biol 17(8):955–963
10. Ghate K et al (2020) Fmn2 regulates growth cone motility by mediating a molecular clutch to generate traction forces. Neuroscience 448: 160–171
11. Renaudin A et al (1999) Organization of point contacts in neuronal growth cones. J Neurosci Res 55(4):458–471
12. Suter DM, Forscher P (1998) An emerging link between cytoskeletal dynamics and cell adhesion molecules in growth cone guidance. Curr Opin Neurobiol 8(1):106–116
13. Sahasrabudhe A et al (2016) Formin 2 regulates the stabilization of filopodial tip adhesions in growth cones and affects neuronal outgrowth and pathfinding in vivo. Development 143(3): 449–460
14. Minegishi T et al (2018) Shootin1b mediates a mechanical clutch to produce force for neuronal migration. Cell Rep 25(3):624–639. e6
15. Denisin AK, Pruitt BL (2016) Tuning the range of polyacrylamide gel stiffness for mechanobiology applications. ACS Appl Mater Interfaces 8(34):21893–21902
16. Kandow CE et al (2007) Polyacrylamide hydrogels for cell mechanics: steps toward optimization and alternative uses. Methods Cell Biol 83:29–46
17. Tseng Q et al (2012) Spatial organization of the extracellular matrix regulates cell–cell junction positioning. Proc Natl Acad Sci 109(5): 1506–1511
18. Chan CE, Odde DJ (2008) Traction dynamics of filopodia on compliant substrates. Science 322(5908):1687–1691
19. Dent EW, Meiri KF (1992) GAP-43 phosphorylation is dynamically regulated in individual growth cones. J Neurobiol 23(8):1037–1053
20. Wang Y-L, Pelham RJ Jr (1998) [39] Preparation of a flexible, porous polyacrylamide substrate for mechanical studies of cultured cells. In: Methods in enzymology. Elsevier, pp 489–496
21. Betz T et al (2011) Growth cones as soft and weak force generators. Proc Natl Acad Sci 108(33):13420–13425
22. Tse JR, Engler AJ (2011) Stiffness gradients mimicking in vivo tissue variation regulate mesenchymal stem cell fate. PLoS One 6(1): e15978
23. Lee D et al (2015) Three-dimensional confocal microscopy indentation method for hydrogel elasticity measurement. Langmuir 31(35): 9684–9693
24. Gandin A et al (2021) Simple yet effective methods to probe hydrogel stiffness for mechanobiology. Sci Rep 11(1):22668
25. Butler JP et al (2002) Traction fields, moments, and strain energy that cells exert on their surroundings. Am J Phys Cell Phys 282(3):C595–C605
26. Sabass B et al (2008) High resolution traction force microscopy based on experimental and computational advances. Biophys J 94(1): 207–220

Chapter 18

Measuring Retrograde Actin Flow in Neuronal Growth Cones

Laura Pulido Cifuentes and Daniel M. Suter

Abstract

Actin flow refers to the motion of the F-actin cytoskeleton and has been observed in many different cell types, especially in motile cells including neuronal growth cones. The direction of the actin flow is generally retrograde from the periphery toward the center of the cell. Actin flow can be harnessed for forward movement of the cell through substrate-cytoskeletal coupling; thus, a key function of actin flow is in cell locomotion. In this chapter, we illustrate three different methods of quantifying retrograde F-actin flow in growth cones derived from cultured *Aplysia* bag cell neurons. These methods include tracking the movement of surface marker beads as well as kymograph analysis of time-lapse sequences acquired by differential interference contrast (DIC) imaging or fluorescent speckle microscopy (FSM). Due to their large size, *Aplysia* neuronal growth cones are uniquely suited for these methods; however, they can also be applied to any other growth cones with clear F-actin-rich peripheral domains.

Key words Retrograde F-actin flow, Actin cytoskeleton, Neuronal growth cone, Time-lapse imaging, Kymograph

Abbreviations

ASW	artificial seawater
Con A	Concanavalin A
DIC	differential interference contrast
EMCCD	Electron-multiplying gain charge-coupled device
FSM	fluorescent speckle microscopy
RBI	restrained bead interaction

1 Introduction

Neuronal growth cones are highly motile structures at the tip of growing axons and dendrites and are essential for the proper wiring of the nervous system. Growth cones sense a variety of information in the environment including molecular, mechanical, electrical, and topographical, and convert this information via signal transduction

Kazuhito Toyooka (ed.), *Neuronal Morphogenesis: Methods and Protocols*, Methods in Molecular Biology, vol. 2831, https://doi.org/10.1007/978-1-0716-3969-6_18,

cascades into changes of cytoskeletal and membrane dynamics [1–3]. Thus, the neuronal growth cone is a sophisticated cellular machine that can not only detect information but also transform it into dynamic and mechanic changes that enable directional movements. These essential growth cone properties have already been recognized by Ramon y Cajal, who described this structure for the first time over 100 years ago based on his observations of numerous fixed specimen [4]. Understanding how the neuronal growth cone works is not only critical for our knowledge of nervous system wiring but also for the development of treatments to enhance axonal growth and guidance in neurological disorders such as Lissencephaly and Parkinson's disease [5] or following traumatic injury [6].

Most of our knowledge about growth cones has been derived from studies of cultured neurons from different species including mice, rats, chickens, frogs, and *Aplysia*. Neuronal cell cultures provide a number of advantages over in vivo studies, such as quantitative high-resolution imaging and control of the environment. Whereas basic physical and chemical mechanisms are likely conserved between in vitro and in vivo systems, one has to keep in mind that certain growth cone properties may be different in vivo. The *Aplysia* bag cell neuronal growth cone has served as a valuable model system for many studies due to its large dimensions, which are 5–10× the size of other growth cones. The large dimensions and highly organized cytoskeletal structures make these growth cones suitable for high-resolution live cell imaging of protein dynamics [7, 8] as well as any biophysical manipulation and analysis [9, 10].

The neuronal growth cone can be divided into three different cytoplasmic regions: the peripheral (P) domain, the transition (T) zone, and the central (C) domain. The F-actin cytoskeleton is present in all three domains, but is most evident in the P domain and T zone, especially following chemical fixation and labeling with phalloidin [11]. Microtubules are prominent in the C domain but can also penetrate into the P domain and interact with the actin cytoskeleton [12, 13], which is critical for growth cone guidance [14]. The F-actin cytoskeleton in the growth cone P domain is organized into filopodial bundles and lamellipodial networks, whereas the main F-actin structures in the T zone are represented by arcs and intrapodia [15]. All F-actin structures in the P domain and T zone undergo constant retrograde motion, whereas intrapodia also exhibit assembly-driven motility in all directions [16]. The retrograde actin flow is maintained by a continuous assembly of actin filaments in filopodia and lamellipodia pushing against the membrane at the leading edge [17], retrograde pulling by non-muscle myosin II in the T zone [18], and recycling of actin filaments in the T zone via proteins such as cofilin [19]. The rate of retrograde actin flow is determined by the rate of these three processes as well as the level of substrate-coupling. The role of actin flow that is best understood is in adhesion-mediated growth cone advance. Adhesion molecules couple the retrograde actin flow to extracellular substrates via clutch proteins resulting in increased traction force and reduced retrograde flow [15]. Furthermore, actin flow maintains a low density of dynamic microtubules in the P domain [12]. However, there could be additional roles of retrograde actin flow that have not been characterized so far.

In this chapter, we first describe a protocol for culturing *Aplysia* bag cell neurons, which we have reported previously [20]. Then, we describe three different methods of quantifying retrograde F-actin flow in neuronal growth cones. These include tracking the movement of surface marker beads as well as kymograph analysis of time-lapse sequences acquired by differential interference contrast (DIC) imaging or fluorescent speckle microscopy (FSM). Although we have used *Aplysia* growth cones, these approaches can be applied to other growth cone preparations as well.

2 Materials

2.1 Aplysia *Bag Cell Neuronal Culture*

1. Leibovitz's L15 cell culture medium supplemented with artificial seawater (L15-ASW): L15 (powder, containing glutamine, Invitrogen #41300-039), pH 7.9, including 4 mM L-glutamine, 5 mM 4-(2-hydroxyethyl)-1-piperazineethanesulfonic acid (HEPES), 9 mM $CaCl_2$, 27 mM $MgSO_4$, 28 mM $MgCl_2$, 400 mM NaCl, and 50 μg/mL gentamicin. Osmolarity should be 950–1000 mmol/kg by using an osmometer. Filter through a 0.22 μm sterile filter. Store at 4 °C for up to 1 month (*see* **Note 1**).
2. Dispase II (Sigma # 04942078001).
3. *Aplysia californica*, adult, 100–200 g, from Marinus Scientific or the National Resource for *Aplysia* at the University of Miami.
4. 0.5 M $MgCl_2$. Store at 4 °C.
5. 6 N HCl for acid-cleaning of cover glass: Mix 1 volume of H_2O ultrapure with 1 volume of 36% HCl.
6. 60 mL syringe with Luer-lock tip; 18.5 gauge sterile hypodermic needle, 2 in. long.
7. Microdissection scissors: Vannas Spring Scissors, 8.5 cm long, 7 mm blade (WPI #500086); regular dissecting scissors; two pairs of forceps: Dumont Tweezers #5—11 cm long (WPI #500341).
8. Cover glass: #1.5, 22 × 22 mm, acid-cleaned and stored in 100% ethanol.
9. 35 mm Petri dishes (Sarstedt # 83.39).

10. Poly-L-lysine (10×): 200 μg/mL Poly-L-lysine (70–150 kDa; Sigma #P-6282) stock solution is prepared in sterile H_2O ultrapure. Store at −20 °C.
11. Laminar flow bench; stereo microscope; 14 °C incubator; water bath.

2.2 Preparing Con A-Coated Surface Marker Beads

1. 1 μm silica amino beads (Bangs laboratories, # SA04000).
2. For coupling of avidin, use 1 mL of 400 μg/mL Avidin D (Vector Laboratories, # A-2000-10) in 20 mM NaP_i pH 7.0 adjusted with NaOH.
3. 20 mM NaP_i pH 7.0 for protein coupling to amino beads.
4. 8% glutaraldehyde (EM grade, Electron Microscopy Sciences, #16019).
5. 10 mM Tris–HCl pH 8.0.
6. 1 mg/mL biotinylated Concanavalin A (Con A, Vector Laboratories, # B-1005-5) in 10 mM Tris–HCl pH 8.0.
7. 0.1 M NaOH.
8. 5 mg/mL bovine serum albumin (BSA).
9. 50 mM Tris–HCl pH 8.0.
10. Table-top refrigerated centrifuge Beckman Allegra X-22, water bath sonicator, tip sonicator, End-over-end rotator, e.g., Barnstead/Thermolyne Rotisserie Labquake.

2.3 DIC Time-Lapse Imaging with and Without Surface Bead Markers

1. L15-ASW containing 5 mg/mL BSA.
2. 10 mM Tris–HCl pH 8.0.
3. 1 μm Con A beads.
4. Chamber for live cell imaging.

2.4 Fluorescent Speckle Microscopy (FSM) of Actin Dynamics

This section was adapted from [20] with permission from Springer Nature.

1. Artificial seawater (ASW): 400 mM NaCl, 10 mM KCl, 55 mM $MgCl_2$, 10 mM $CaCl_2$, 15 mM HEPES, pH 7.8. Osmolarity should be 950–1000 mmol/kg by using an osmometer. Filter through a 0.45 μm sterile filter. Store at 4 °C.
2. Imaging medium: ASW supplemented with 2 mg/mL BSA, 1 mg/mL l-carnosine, and 0.25 mM vitamin E. Filter through a 0.22 μm sterile filter. Prepare freshly on the day of the experiment.
3. Injection buffer: 100 mM 1,4-Piperazinediethanesulfonic acid (PIPES) pH 7.0 adjusted with NaOH, 1 mM $MgCl_2$, 1 mM ethylene glycol tetraacetic acid (EGTA). Filter through a 0.22 μm sterile filter. Store at 4 °C.

4. G-buffer: 5 mM Tris–HCl pH 8.0, 0.2 mM $CaCl_2$, 0.2 mM dithiothreitol, 0.2 mM ATP, 10% (w/v) sucrose. Filter through a 0.22 μm sterile filter. Prepare freshly for dilution of G-actin.
5. Alexa Fluor 488-phalloidin (for microinjection): 20 mM Alexa Fluor 488-phalloidin (Invitrogen # A12379) freshly prepared in injection buffer. Therefore, transfer a small amount (10 μL) of phalloidin dissolved in methanol into a microcentrifuge tube, remove the methanol in a speed vacuum concentrator, and reconstitute the phalloidin in injection buffer, store at − 80 °C.
6. Alexa Fluor 568-G-actin (for microinjection): 1 mg/mL Alexa Fluor 568-G-actin (rabbit muscle actin; Invitrogen # A12374) freshly prepared in G-buffer; store stock aliquots of 7–8 mg/mL at −80 °C (*see* **Note 2**).
7. Imaging chamber: We used a custom-made imaging chamber consisting of a metallic holder and a plastic counterpart. Two coverslips separated by plastic spacers were positioned between the metallic and plastic holder (*see* **Note 3**).
8. Diamond pen; 22 × 2 mm plastic spacers (cut from 0.020 in. thick plastic shims); nylon grommets used as clips to hold the top and bottom part of the holder together; high vacuum grease (Dow Corning); Sparkle glass cleaner; Q-tips.
9. Microinjection needles: (Borosilicate) borosilicated glass capillaries (1B100F-4, World Precision Instruments). Microloader tips (Eppendorf #930001007).
10. Micropipette puller (Narishige PP830 or other pullers).
11. Micromanipulator: NP-2 patchman (Eppendorf); injection system: FemtoJet (Eppendorf).
12. Microscopy setup: Various combinations of imaging hardware and software are possible. The following setup has been used for most of our live cell imaging experiments: Eclipse TE2000E2 (Nikon) inverted microscope equipped with 10× objective for phase contrast imaging as well as 40× 1.3 NA oil, 60× 1.4 NA oil, and 100× 1.45 NA oil immersion objectives for DIC imaging. Flat-top motorized stage H117 ProScan™ (Prior) with linear encoders and joystick control. Electron-multiplying gain charge-coupled device (EMCCD) camera iXon Ultra 888 (Andor). X-cite 120 metal halide lamp (EXFO Photonic Solutions, Inc). Lambda 10-3 highspeed filter wheels and shutters (Sutter). Filter sets (Chroma) for rhodamine-tubulin: Tetramethylrhodamine isothiocyanate (TRITC) exciter (555/25nm) and emitter (HQ610/60 nm) of a FITC/TRITC set; for Alexa Fluor 488-phalloidin: Fluorescein isothiocyanate (FITC) exciter (485/20 nm) and emitter (517/30 nm) of a FITC/TRITC set; for Alexa Fluor

568-G-actin: TxRed exciter (575/15 nm) and emitter (610/40 nm) of a FITC/TxRed/Cy5 set. MetaMorph 7.8 software (Universal Imaging). PC with Core 2 Duo processor (3 GHz, Intel).

3 Methods

3.1 Aplysia *Bag Cell Neuronal Culture*

1. Prepare fresh Dispase-II solution by dissolving 10 mg of dispase II in a solution containing 900 mL of L15-ASW and 100 mL of sterile ultrapure water. Keep at room temperature (RT).
2. Fill a 60 mL syringe with 50 mL of 0.5 M $MgCl_2$ and attach an 18.5 gauge needle. Sterilize regular dissecting scissors, micro-dissection scissors, and two forceps with 70% ethanol and let air-dry.
3. Place an *Aplysia* over a Styrofoam board and inject all the $MgCl_2$ solution into the body cavity. Gently rub the animal to spread the solution throughout the body. Wait 1–3 min until the animal is completely anesthetized (*see* **Note 4**).
4. Pin the head and tail of the animal to the dissection board using two needles.
5. Use forceps to lift up the skin around the abdominal cavity and open the body wall with dissection scissors. Cut the body wall opening along the side toward the head and the tail.
6. Locate the abdominal ganglion around the mantle and gill (use Figs. 4 and 5 in [21]). Use forceps to lift up the connecting nerves rostrally to the ganglion and, cut the connecting nerves 1 cm above and below the ganglion with microdissection scissors. Store the residual body of the animal at −20 °C before incineration.
7. Transfer the ganglion into the tube with dispase II solution. In a temperature-controlled water bath, digest the tissue for 15–16 h at 22 °C (*see* **Note 5**).
8. In a laminar flow bench, prepare three 35 mm Petri dishes filled with 4 mL of L15-ASW each. Adjust the medium temperature to RT and use sterile forceps to transfer the ganglion to one of these dishes. From this step on, continue working in the laminar flow bench.
9. Preparing poly-L-lysine-coated coverslips:
    (a) Use acid-cleaned #1.5 square glass coverslips (22 × 22 mm) stored in 100% ethanol (EtOH) (*see* **Note 6**). Take a coverslip with sterile forceps and flame it with a Bunsen burner. Place the coverslip inside an empty 35 mm Petri dish (*see* **Note 6**). Depending on the desired number of

neurons per cover glass, 6–12 coverslips are usually enough to plate the cells derived from one animal.

(b) Coat each coverslip with 0.5 mL of 20 μg/mL poly-L-lysine and incubate for 20 min at RT.

(c) For each coverslip, conduct three wash cycles of water addition/vacuum aspiration using 0.5 mL of sterile ultrapure water per round. After the last wash, add 4 mL of L15-ASW into each dish.

10. Spray microdissection scissors and two forceps with 70% EtOH and let air-dry. While observing through a dissection microscope, cut the ganglion through the center along the rostral-caudal axis to separate the ganglion in two halves. Then, separate the cluster from the hemiganglion and the connecting nerves, making sure to not damage the cells with the tools and maintaining the cluster in the medium all the time.

11. Using a 20 μL pipette, transfer each cluster to one of the 35 mm Petri dishes filled with L15-ASW. From this step, continue working with one cluster at the time (*see* **Note 7**).

12. Use two forceps to remove the connective tissue sheath from the cluster. Hold the cluster with one forceps and use the other forceps to push against the connective tissue sheath. Find the opening where the cluster pops out from the connective tissue sheath and completely separate the cluster from the sheath. The cluster should appear as a "berry-like" structure.

13. Use forceps to bend a yellow P20 pipette tip at a 45° angle. Then, gently suck the cluster into the tip and push it out. Start with a low shear force and increase the force as needed. Healthy cells look bright in the center of the cell body, while dead cells appear transparent against a dark background.

14. Pick up a few healthy cells with medium and transfer them onto a poly-L-lysine-coated coverslip. Avoid transferring death cells or cellular debris (*see* **Note 8**).

15. Repeat **steps 13** and **14** until you get about 20–40 live neurons on each coverslip. Plate the cells well separated but close enough to the center of the coverslip, since it is difficult to image cells close to the edges with an oil-immersion objective. Then, remove dead cells or debris from the coverslips by picking them up with the pipette and, repeat **steps 12–15** for the second cluster.

16. After all healthy cells have been plated, keep the dishes 2 h at RT and avoid vibrations. Then, place the dishes into a 14 °C incubator until imaging is performed. Growth cones develop at neurite tips usually within 4–10 h after plating, but they get smaller with time in culture [22]. Thus, we typically use the cells within the first 48 h after plating.

3.2 Preparing Con A-Coated Surface Marker Beads

1. Resuspend 1% wt/vol 1 μm silica amino beads in H_2O ultrapure in a microcentrifuge tube. Use 10 mg bead powder in 1 mL H_2O ultrapure resuspended by vortexing and pipetting. If beads are purchased as a solution (10%), mix 100 μL beads with 900 μL H_2O ultrapure. Wash 1% beads at least once with H_2O ultrapure. Washing means: spinning the microcentrifuge tube at 21,000 RCF for 3 min, subsequently removing the supernatant, and resuspending the bead pellet in the new solution (in this case in H_2O ultrapure).
2. Tip sonication with Branson sonicator (micro tip): Output force 3, 50%, 10 pulses. This step is not necessary for 5 μm beads, since they do not cause clumping problems.
3. If you are using H_2O ultrapure, adjust pH to 6.5–7.0 by adding 0.1 M NaOH (the amount depends on pH; usually a few μL is sufficient), and check pH with stripes by quickly dipping. Be aware that you lose beads in this step (*see* **Note 9**).
4. Water bath sonication for 2 min.
5. Check beads under the scope for mono-dispersity.
6. If beads are dispersed, add 1 mL (equal volume as beads) of 8% glutaraldehyde (EM grade, Electron Microscopy Sciences, #16019; always use a freshly opened bottle). Vortex and incubate on the rotator at least for 6 h (or overnight) at RT (*see* **Note 10**).
7. Spin beads at 21,000 RCF for 3 min, and wash 3× with H_2O ultrapure (1 mL each).
8. 1× wash with 20 mM NaP_i pH 7.0.
9. Incubate beads with 1 mL of 400 μg/mL Avidin D (Vector Laboratories, # A-2000-10) in 20 mM NaP_i pH 7.0 for 4 h at RT or overnight at 4 °C.
10. Spin beads as above, remove supernatant, and incubate in 5 mg/mL BSA, 50 mM Tris–HCl pH 8.0, 0.02 % NaN_3 for 30 min at RT for blocking. Save supernatant for analysis of coupling efficiency, if you want.
11. After this blocking step, spin the beads and resuspend them again in the blocking agent to store the beads at 1% at 4 °C.
12. Take a reasonable volume of avidin beads into a new tube, e.g., 200–500 μL. Be aware that you do not usually need large volumes of beads for cell experiments and that protein-coated beads are not stable for a long time.
13. Wash avidin beads once with 1 mL of 10 mM Tris–HCl pH 8.0.
14. Pellet beads and resuspend them in biotinylated Con A solution: 1 mg/mL biotinylated lectin in 10 mM Tris–HCl pH 8.0. Use the same volume of the ligand solution as beads were in **step 1** to do binding at 1% beads. Incubate for at least 1 h at RT on a rotator or overnight at 4 °C.

15. Always store the beads in the ligand solution from now on, since this is a non-covalent interaction (ligand will come off the beads when stored in buffer only)!

3.3 DIC Time-Lapse Imaging with and Without Surface Bead Markers

1. Observe the cells under a microscope with a 10× objective 24 h after plating. Perform a half-volume medium change with L15-ASW adjusted previously to RT. If growth cones have not yet developed, it's possible to speed up the growth process by keeping the cultures at RT for 1–2 h.
2. To take high-resolution images, place the coverslip with the cells in either a custom-made or commercial imaging chamber, for example, the Series 20 platform from Warner Instruments. We use a custom-made chamber system originally designed and developed by Paul Forscher and Stephen Smith at Yale University [20] (*see* **Note 11**).
3. Mount the chamber holder onto the microscope stage of an inverted microscope. We use a Nikon Eclipse TE2000E2. Use the 10× phase objective to locate individual bag cells. Since the number of neurons is relatively low (~20–40) compared to other neuronal culture systems, it is helpful to use a motorized stage with memory function to mark the positions of individual cells.
4. Spin the stock of biotinylated ConA-coated beads in Allegra X-22 at 21,000 RCF (FX301.5 rotor), 4 °C for 5 min.
5. Remove the ligand solution (ConA) with a pipette and transfer it to a separate tube. Resuspend beads with 10 mM Tris–HCl pH 8.0 for washing (1 mL for 200 μL of 1% beads), spin again, and resuspend the beads in 10 mM Tris–HCl pH 8.0. Use the same volume as your ligand solution to get 1% beads.
6. Prepare 1 mL of washed Con A beads diluted at 1:1000 in 5 mg/mL BSA in L15-ASW. Keep this diluted bead solution at RT and use it on the day you prepared it. Add the bead dilution to the mounted imaging chamber.
7. Switch the objective to a high-resolution DIC oil-immersion objective. Set up the microscope for DIC imaging and perform Koehler illumination. Search appropriate growth cones for imaging.
8. The optimal camera settings depend on the type of CCD camera. We use either an Andor iXon 888 Ultra or a Roper Scientific CoolSnap HQ and an exposure time of 50 msec. Time-lapse sequences of growth cone dynamics are acquired in 5–10 s intervals for 10–30 min. Time-lapse sequences can be acquired with or without beads.
9. To make sure the chamber is always filled with medium, exchange with 1–3 mL of fresh medium every 30 min by

adding medium into one side of the chamber while removing medium on the other side with a vacuum line.

10. Analyze image stacks using appropriate spatial calibrations and image analysis software.

3.4 Retrograde Actin Flow Analysis Based on DIC Time-Lapse Sequences

There are two options for actin flow analysis based on DIC time-lapse sequences. The first option involves using kymographs of DIC time-lapse sequences. The second option involves tracking the centroid position of flow-coupled beads over time. In either case, ImageJ or Fiji is used as free image analysis software.

1. *Kymograph Analysis of DIC Time-Lapse Sequences*
   1. In ImageJ, import the time-lapse sequence by File>Import>Image Sequence.
   2. Apply the proper spatial calibration: Analyze>Set scale (enter the distance in pixels and actual distance in micrometers and click okay).
   3. Blur all images by selecting Process>Smooth.
   4. Apply a convolution filter to highlight the edges: Process>–Filters>Convolve and write the following kernel:

−1	−1	−1
−1	8	−1
−1	−1	−1

   5. Blur all images again as in **step 3**.
   6. From the toolbar, select the option "Straight Line."
   7. Draw a line along a filopodium in the P domain from the growth cone leading edge to the T zone. Select a line width of 1 pixel.
   8. Create a kymograph by one of the following two methods (Fig. 1):
      (a) Image>Stacks>Reslice. Select the output spacing as your spatial calibration to get a kymograph with the *x*-axis representing space and the *y*-axis representing time (both in μm). You get a kymograph that is spatially calibrated in μm or in pixels.
      (b) b) Analyze>Multi Kymograph>Multi Kymograph. Select a line width of 3 pixels. This kymograph will be in pixel by pixel.
   9. Draw a line along features that move at a constant speed. Measure the slope of this line. This could be done with the velocity measurement tool https://dev.mri.cnrs.fr/projects/imagej-macros/wiki/Velocity_Measurement_Tool.

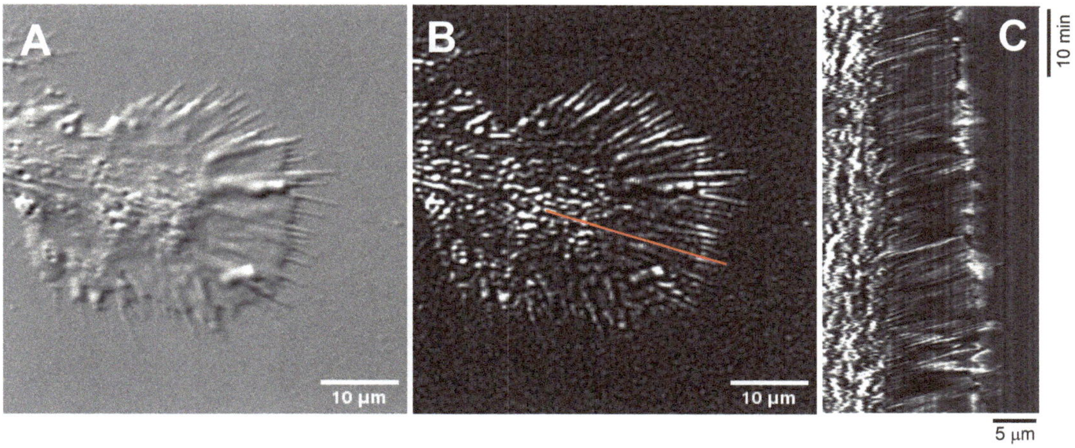

Fig. 1 Retrograde F-actin flow analysis based on DIC kymograph analysis. (**a**) DIC image of live *Aplysia* bag cell neuronal growth cone growing on coverslip coated with 50 µg/mL *Aplysia* cell adhesion molecule apCAM. (**b**) Processed DIC image as described in Subheading 3.4.1. The red line indicates where the kymograph was taken. (**c**) Kymograph of the intensity along the line indicated in (**b**). Time is going down. Scale bars as indicated

Therefore, copy the tool into the Macros/Toolsets folder and install it under Plugins>Macros>Install. It will show up as "Velocity." When you run the tool, the speed will be shown as the number of pixels along the *x*-axis (distance) divided by the number of pixels along the *y*-axis (time). This value needs to be multiplied with both the spatial (µm/pixel) and temporal calibration (pixel/min) values to get the velocity of actin flow in µm/min.

2. *Tracking Centroid Position of Flow-Coupled Beads over Time*
   1. Load the ImageJ plugin TrackMate by Plugins>Tracking>TrackMate.
   2. Import time-lapse sequence by File>Import>Image Sequence.
   3. Apply the proper spatial calibration: Analyze>Set scale (enter the distance in pixels and actual distance in micrometers and click okay).
   4. Select the frames you want to analyze in the crop settings and select the DoG detector.
   5. Set the estimated object diameter to 1 µm and the quality threshold to 100. Check the box sub-pixel localization. Click preview to verify that the desired bead has been detected by the software. If the bead of interest wasn't detected, change the quality threshold and/or object diameter. Click next and verify that the number of spots found is not smaller than the number of frames. Click "next" twice.
   6. Filter the spots based on the *x* and *y* positions. Click "next."

7. Select a simple LAP tracker, set gap closing max frame gap to 0, and let other parameters as default. Click "next" three times.
8. In the display options window, select spots and export to a . csv file. Use Excel or a script to analyze the bead position over time. An example data set taken from [23] is shown in Fig. 2.

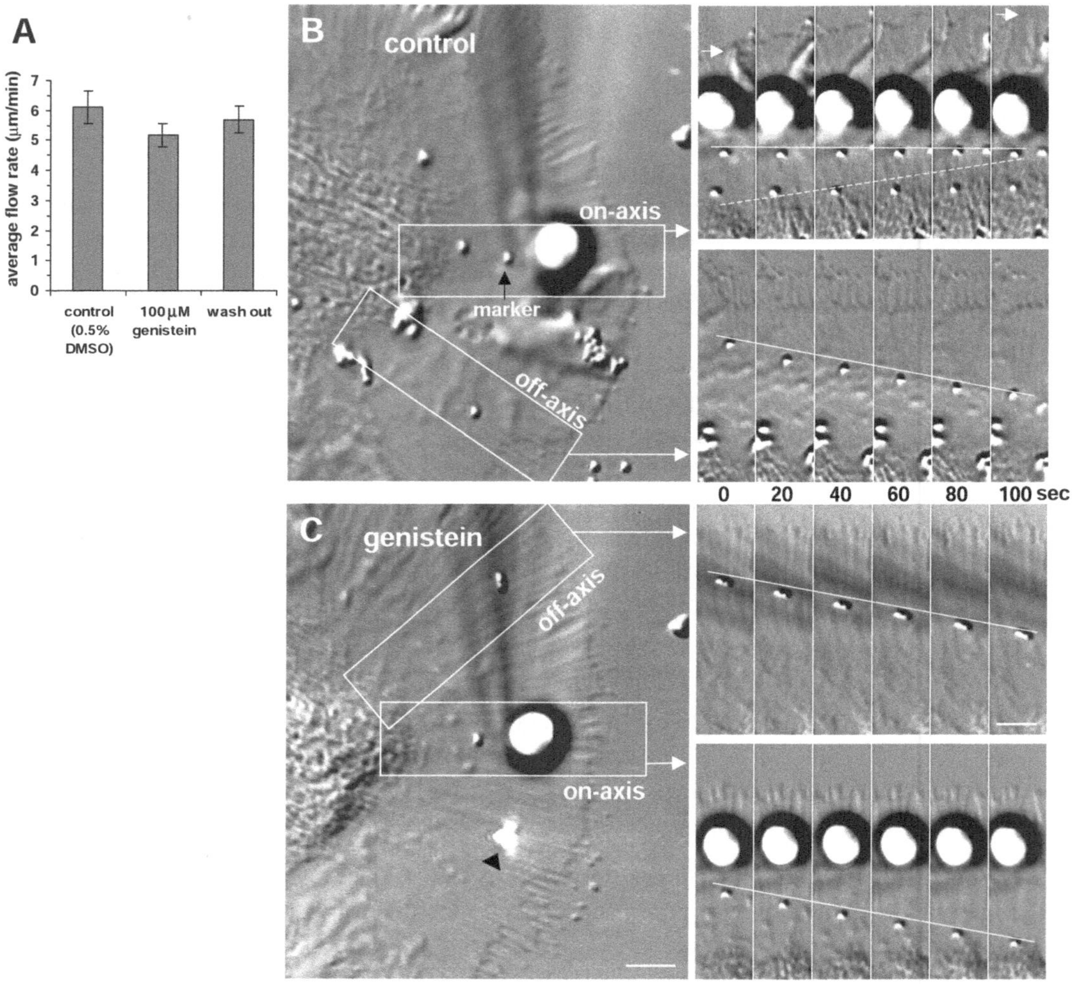

Fig. 2 Use of marker beads to determine F-actin flow in neuronal growth cones. (**a**) Quantification of retrograde F-actin flow effects by treatment with 100 μM genistein (tyrosine kinase inhibitor) for 25 min on *Aplysia* growth cones without restrained bead interactions (RBIs). Average values ± SEM. (**b** and **c**) Tyrosine kinase inhibition by genistein does not affect flow during uncoupled RBIs. All images refer to the same growth cone. Flow marker beads were placed with a laser tweezer both within the interaction corridor (on-axis) and on adjacent areas (off-axis) during an RBI under control conditions (**b**) and after pretreatment with 50 μM genistein for 20 min (**c**) Panels on the right show simultaneous DIC time sequences of on- and off-axis bead movements in areas of interests marked on the left. Flow rates for the displayed on- and off-axis beads are 0.31 versus 5.64 μm/min in control and 4.47 versus 5.14 μm/min in genistein. Central domain extension in control is indicated by dashed line and leading edge growth by a white arrow. Laser beam focus is marked with an arrowhead in **c**. Bars, 5 μm. (Reproduced from [23] with permission from Rockefeller University Press)

3.5 Fluorescent Speckle Microscopy (FSM) of Actin Dynamics

This section was adapted from [20] with permission from Springer Nature.

1. Pull needles for microinjection of cytoskeletal probes. A tip opening of 1 μm is suitable for injections into bag cell neuronal cell bodies. We use borosilicate glass capillaries and perform a double pull with all weights on the vertical puller PP830 using the heat setting "65" for the first pull and "53" for the second pull. Inspect the pulled needles on a microscope for desired tip shape before storing or using them.
2. Prepare cytoskeletal probes for microinjection as described in Subheading 2.4, **steps 5** and **6**.
3. Clarify cytoskeletal probes by centrifugation at 10,000 g for 30 min at 4 °C before microinjection. This reduces the clogging problem when performing microinjections (*see* **Note 12**).
4. Prepare your microinjection setup. We use the FemtoJet system for controlled pressure application. Start up the FemtoJet following the instructions of the manufacturer. Turn on the micromanipulator if you use a motorized one. We use an Eppendorf NP-2 patchman. Position the microneedle holder at a 40° angle to the stage.
5. Position the Petri dish containing a coverslip with cultured bag cell neurons onto the stage. Search for the position of individual neurons using a 10× phase objective and mark the cell positions.
6. Using microloader tips, backfill injection needles with a small amount of cytoskeletal probe (*see* **Note 13**).
7. Connect the microneedle to the needle holder. Keep compensation pressure at ~30–40 hPa.
8. Quickly lower the injection needle into the medium. Position the tip of the needle in the field of view without touching the surface. Switch to a 40× long working distance phase objective for microinjection and position the cell body of the neuron into the center of the field of view.
9. Check whether there is continuous flow coming out of the pipette either by using phase or fluorescence optics. A small "cloud" of fluorescent material should appear in front of the tip. If not the case, increase the compensation pressure or clear out the clog by pressing the "clean" function on the FemtoJet.
10. Position the needle tip adjacent to the cell body using the fine speed setting of the NP-2 micromanipulator. Inject the cell body by moving the needle tip quickly into an area of cytoplasm avoiding the nucleus. A fast movement increases the chance of actually penetrating the plasma membrane. An injection will result in a visible local expansion of cell volume. Leave needle in cell for ~1–3 s and then quickly move needle out since

a too large injection volume can kill the cell. Inject an estimated volume of no more than 10% of the cell body. Compensation pressure is usually sufficient for injections; if needed use an increased injection pressure.

11. To make sure the cell body was injected, switch to fluorescent mode using an appropriate fluorescent filter. A clearly injected cell will immediately exhibit fluorescent signals in the cell body that can be seen with the eyepiece.
12. After all cell bodies in one dish are injected, place the dish in a light-protected area for the cells to recover. Once all dishes are injected, carefully exchange half of the L15-ASW medium in each dish. Let the injected cells recover for at least 1 h in the dark before starting the imaging.
13. Assemble an imaging chamber as described under Subheading 3.3 and exchange the L15-ASW with imaging medium.
14. Mark cell positions with a 10× phase objective before switching to 60× or 100× oil immersion lenses to identify suitable growth cones for imaging.
15. Set up the appropriate illumination and acquisition parameters suitable for the injected fluorescent probes. We use an FITC exciter (485/20 nm) and emitter (517/30 nm) for Alexa Fluor 488-phalloidin. Reduce the fluorescence intensity with neutral density filters. The gain setting and exposure time will depend on the camera used. We typically use 500 ms exposure, gain 3, and additional EM gain on the iXon 888 ultra camera. Time intervals between images are typically 5–10 s. Healthy neurons can be imaged up to 1 h under such conditions. It is recommended to exchange fresh imaging medium every 30 min or perfuse with medium throughout the experiment. Make sure the sample remains in focus during time-lapse imaging and perform the necessary adjustments.

3.6 Retrograde Flow Analysis from Actin FSM Data

1. In ImageJ import the image sequence by File>Import>Image Sequence.
2. Apply the proper spatial calibration: Analyze>Set scale (enter the distance in pixels and actual distance in micrometers and click okay).
3. Adjust brightness/contrast by Image>Adjust>Brightness/Contrast to a level that you can see the actin signal against the background.
4. Remove noise by blurring: Process>Filters>Gaussian Blur (radius 1 on pixel basis).
5. Sharpen the images by using a sharpen filter: Process>Sharpen.
6. Select the line tool and draw a line along a filopodium in the P domain from the growth cone leading edge to the T zone. Select a line width of 1 pixel.

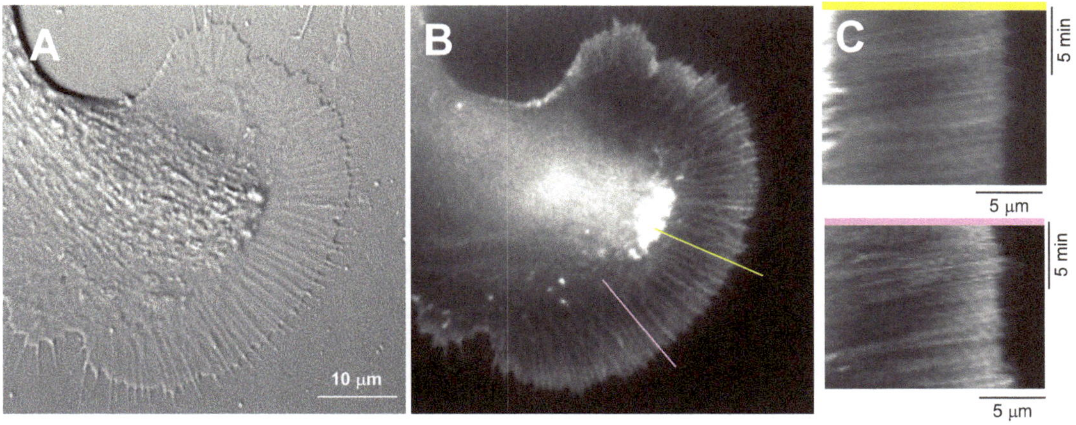

Fig. 3 Retrograde F-actin flow analysis based on actin FSM imaging. (**a**) DIC image of live *Aplysia* bag cell neuronal growth cone growing on a coverslip coated with poly-L-lysine. (**b**) Actin FSM image of the same growth cone. Cell was injected with Alexa 568 G-actin. Magenta and yellow lines indicate where kymographs were taken. (**c**) Kymographs of the intensity along the two lines indicated in (**b**). Time is going down (13 min total). Scale bars as indicated

7. Create a kymograph by one of two methods (Fig. 3):
   (a) Image>Stacks>Reslice. Select output spacing as your spatial calibration to get a kymograph with the *x*-axis being space and the *y*-axis being time (both in µm). You get a kymograph that is spatially calibrated in µm or pixels.
   (b) Analyze>Multi Kymograph>Multi Kymograph. Select a line width of 3 pixels. This kymograph will be in pixel by pixel.
8. Draw a line along features that move at constant speed. Measure the slope of this line. This could be done with the velocity measurement tool https://dev.mri.cnrs.fr/projects/imagej-macros/wiki/Velocity_Measurement_Tool Therefore, copy the tool in the Macros/toolsets folder and install under Plugins>Macros>Install. It will show up as "Velocity." When you run the tool, the speed will be shown as number of pixels along the *x*-axis (distance) divided by the number pixels along the *y*-axis (time). This value needs to be multiplied with both the spatial (µm/pixel) and temporal calibration (pixel/min) values to get the velocity of actin flow in µm/min.

4 Notes

1. All solutions are prepared with ultrapure water.
2. Besides using commercial cytoskeletal probes, tubulin, and actin can be purified in the lab [24]. Briefly, stock solutions are prepared in the buffers described in Subheading 2.4, and

aliquots between 1 and 5 μL are stored at −80 °C after freezing in liquid nitrogen.

3. There are a number of different options for imaging chambers, either custom-made or commercial products.
4. Avoid inking and muscle contractions by handling the animal gently. Make sure not to damage the internal organs with the needle and avoid getting ink into the body cavity.
5. The effectiveness of tissue digestion depends on the temperature, time of digestion, and amount of dispase. Thus, it is advisable to optimize these variables for each batch of the enzyme.
6. Acid cleaning of coverslips can be done by incubating the coverslips in 6 N HCl for at least 1 h, followed by extensive washing with H_2O ultrapure. Then, use sterile forceps to transfer individual coverslips to a container with 100% EtOH and store them at RT.
7. Try to avoid getting bubbles between the coverslip and the Petri dish bottom. This can be achieved by adding 10 μL of H_2O ultrapure to the center of an empty Petri dish. Then, place the coverslip over the drop. In this way, you can stick the coverslip to the bottom of the Petri dish.
8. Keep the cells and the cluster in medium all the time because contact with the air interface can damage cells.
9. Glutaraldehyde activation works best under slightly acidic conditions (pH 6–7).
10. If beads are not dispersed well, do not proceed, otherwise, you will get too much clumping. Try to make the solution disperse by additional sonication. If this does not help, start with activation, and remove large, crosslinked clumps later by a low-spin centrifugation step.
11. It is essential that the two coverslips are well aligned with each other as well as with the groove of the aluminum bottom part before the top part is mounted. Uneven positioning will generate local pressure after applying the clips and result in breaking the cover glass.
12. In case of clogging, try the following: (a) spin sample at higher g forces; (b) coat needles with hexamethyldisilazane (HMDS, Pierce Chemical) by placing a few drops of HMDS on the bottom of a needle-holding box, and allow the vapor to coat the needles overnight in a fume hood; (c) clear out the clog by increasing the pressure of the microinjection system.
13. Filling about 1 cm of the narrow microloader tip provides a sufficient amount for injecting several cells. Make sure not to touch the bottom of the tube where protein aggregates could

have accumulated. Gently tap needle to remove any potential air bubbles from the solution in the microneedle. Alternative to pulling injection needles: commercial injection capillaries such as Femtotips (Eppendorf). Alternative to microloader tips: long and thin pulled glass capillaries.

Acknowledgments

This project was supported, in part, by the Indiana Spinal Cord & Brain Injury Research Fund from the Indiana State Department of Health (Contract 26351) and by a grant from NIH (5R01NS117701). We are thankful to Dr. Soumyajit Dutta for proofreading the manuscript.

References

1. Dent EW, Gupton SL, Gertler FB (2011) The growth cone cytoskeleton in axon outgrowth and guidance. Cold Spring Harb Perspect Biol 3(3). https://doi.org/10.1101/cshperspect.a001800
2. Tojima T, Hines JH, Henley JR, Kamiguchi H (2011) Second messengers and membrane trafficking direct and organize growth cone steering. Nat Rev Neurosci 12(4):191–203. https://doi.org/10.1038/nrn2996
3. Zang Y, Chaudhari K, Bashaw GJ (2021) New insights into the molecular mechanisms of axon guidance receptor regulation and signaling. Curr Top Dev Biol 142:147–196. https://doi.org/10.1016/bs.ctdb.2020.11.008
4. Cajal R (1890) A quelle époque apparaissent les expansions des cellules nerveuses de la moelle épinière du poulet. Anatomischer Anzeiger 5:609–613
5. Munoz-Lasso DC, Roma-Mateo C, Pallardo FV, Gonzalez-Cabo P (2020) Much more than a scaffold: cytoskeletal proteins in neurological disorders. Cells 9(2). https://doi.org/10.3390/cells9020358
6. Blanquie O, Bradke F (2018) Cytoskeleton dynamics in axon regeneration. Curr Opin Neurobiol 51:60–69. https://doi.org/10.1016/j.conb.2018.02.024
7. He Y, Ren Y, Wu B, Decourt B, Lee AC, Taylor A, Suter DM (2015) Src and cortactin promote lamellipodia protrusion and filopodia formation and stability in growth cones. Mol Biol Cell 26(18):3229–3244. https://doi.org/10.1091/mbc.E15-03-0142
8. Lee AC, Suter DM (2008) Quantitative analysis of microtubule dynamics during adhesion-mediated growth cone guidance. Dev Neurobiol 68(12):1363–1377. https://doi.org/10.1002/dneu.20662
9. Athamneh AI, Cartagena-Rivera AX, Raman A, Suter DM (2015) Substrate deformation predicts neuronal growth cone advance. Biophys J 109(7):1358–1371. https://doi.org/10.1016/j.bpj.2015.08.013
10. Suter DM, Errante LD, Belotserkovsky V, Forscher P (1998) The Ig superfamily cell adhesion molecule, apCAM, mediates growth cone steering by substrate-cytoskeletal coupling. J Cell Biol 141(1):227–240. https://doi.org/10.1083/jcb.141.1.227
11. Forscher P, Smith SJ (1988) Actions of cytochalasins on the organization of actin filaments and microtubules in a neuronal growth cone. J Cell Biol 107(4):1505–1516. https://doi.org/10.1083/jcb.107.4.1505
12. Schaefer AW, Kabir N, Forscher P (2002) Filopodia and actin arcs guide the assembly and transport of two populations of microtubules with unique dynamic parameters in neuronal growth cones. J Cell Biol 158(1):139–152. https://doi.org/10.1083/jcb.200203038
13. Tanaka EM, Kirschner MW (1991) Microtubule behavior in the growth cones of living neurons during axon elongation. J Cell Biol 115(2):345–363. https://doi.org/10.1083/jcb.115.2.345
14. Tanaka E, Sabry J (1995) Making the connection: cytoskeletal rearrangements during growth cone guidance. Cell 83(2):171–176. https://doi.org/10.1016/0092-8674(95)90158-2
15. Miller KE, Suter DM (2018) An integrated cytoskeletal model of neurite outgrowth.

Front Cell Neurosci 12:447. https://doi.org/10.3389/fncel.2018.00447

16. Rochlin MW, Dailey ME, Bridgman PC (1999) Polymerizing microtubules activate site-directed F-actin assembly in nerve growth cones. Mol Biol Cell 10(7):2309–2327. https://doi.org/10.1091/mbc.10.7.2309
17. Mallavarapu A, Mitchison T (1999) Regulated actin cytoskeleton assembly at filopodium tips controls their extension and retraction. J Cell Biol 146(5):1097–1106. https://doi.org/10.1083/jcb.146.5.1097
18. Medeiros NA, Burnette DT, Forscher P (2006) Myosin II functions in actin-bundle turnover in neuronal growth cones. Nat Cell Biol 8(3):215–226. https://doi.org/10.1038/ncb1367
19. Marsick BM, Flynn KC, Santiago-Medina M, Bamburg JR, Letourneau PC (2010) Activation of ADF/cofilin mediates attractive growth cone turning toward nerve growth factor and netrin-1. Dev Neurobiol 70(8):565–588. https://doi.org/10.1002/dneu.20800
20. Suter DM (2011) Live cell imaging of neuronal growth cone motility and guidance in vitro. Methods Mol Biol 769:65–86. https://doi.org/10.1007/978-1-61779-207-6_6
21. Kandel ER (1976) Cellular basis of behavior. 727 pages.
22. Ren Y, Suter DM (2016) Increase in growth cone size correlates with decrease in neurite growth rate. Neural Plast 2016:3497901. https://doi.org/10.1155/2016/3497901
23. Suter DM, Forscher P (2001) Transmission of growth cone traction force through apCAM-cytoskeletal linkages is regulated by Src family tyrosine kinase activity. J Cell Biol 155(3):427–438. https://doi.org/10.1083/jcb.200107063
24. Waterman-Storer C (2002) Fluorescent speckle microscopy (FSM) of microtubules and actin in living cells. Curr Protoc Cell Biol Chapter 4:Unit 4 10. https://doi.org/10.1002/0471143030.cb0410s13

Chapter 19

Morphological Analysis of Neurons and Glia Using Mosaic Analysis with Double Markers

Osvaldo A. Miranda, Giselle Cheung, and Simon Hippenmeyer

Abstract

Mosaic Analysis with Double Markers (MADM) is a powerful genetic method typically used for lineage tracing and to disentangle cell autonomous and tissue-wide roles of candidate genes with single cell resolution. Given the relatively sparse labeling, depending on which of the 19 MADM chromosomes one chooses, the MADM approach represents the perfect opportunity for cell morphology analysis. Various MADM studies include reports of morphological anomalies and phenotypes in the central nervous system (CNS). MADM for any candidate gene can easily incorporate morphological analysis within the experimental workflow. Here, we describe the methods of morphological cell analysis which we developed in the course of diverse recent MADM studies. This chapter will specifically focus on methods to quantify aspects of the morphology of neurons and astrocytes within the CNS, but these methods can broadly be applied to any MADM-labeled cells throughout the entire organism. We will cover two analyses—soma volume and dendrite characterization—of physical characteristics of pyramidal neurons in the somatosensory cortex, and two analyses—volume and Sholl analysis—of astrocyte morphology.

Key words Mosaic Analysis with Double Markers (MADM), Cellular morphology, Neuron, Glia, Sholl analysis, Single cell, Development, Neurodevelopmental disease

1 Introduction

Neurodevelopmental disorders like autism and fragile X syndrome have been associated with altered neuronal and dendritic spine morphology [1–3]. Changes in neuronal cell volume affects many intracellular processes like cell signaling and ion diffusion [4, 5] whereas dendritic spines are directly involved in the establishment of synaptic contacts and neuronal communications [6]. Besides neuronal dysfunctions, glial cell-types like astrocytes are often also affected in pathologies associated with brain development [7, 8]. The extent of the elaborate organization of astroglial fine

Authors Osvaldo A. Miranda and Giselle Cheung have equally contributed to this chapter.

Kazuhito Toyooka (ed.), *Neuronal Morphogenesis: Methods and Protocols*, Methods in Molecular Biology, vol. 2831,
https://doi.org/10.1007/978-1-0716-3969-6_19,

processes largely reflects their maturation state and, in turn, their functions within astrocytic networks or with neighboring neurons [9, 10]. Therefore, in order to understand pathological alterations in neurons and astrocytes, their morphological parameters down to the finest structures are commonly assessed and compared.

Mosaic Analysis with Double Markers (MADM) is broadly used to concomitantly label and genetically manipulate genetically defined single cells within a mouse tissue of interest [11]. MADM relies on Cre-mediated interchromosomal recombination of two reciprocal chimeric marker genes inserted at identical genomic loci on homologous chromosomes. A tissue of interest can be targeted by placing *Cre* under the control of a tissue-specific promoter. Recently, we generated a major resource [12] detailing the creation of the "MADM library," a repository of 19 MADM reporter mouse lines, which enable MADM analysis of over 96% of all the genes in the mouse genome. MADM mice have been used to study genes associated with microcephaly [13], epigenetic molecular machinery regulating gene silencing [14], radial neuronal migration in the neocortex [15], among other studies. MADM provides a significant advantage in studying candidate genes of interest in that only a very small subset of cells are labeled and genetically manipulated, bypassing any lethality issues present in more "traditional" knockout (KO) or conditional knockout (cKO) models. The fluorophores expressed by the reconstituted MADM cassettes, tandem dimer Tomato (tdTomato) and green fluorescent protein (GFP), very efficiently fill even the finest cellular structures, such as dendritic spines [11]. Additional staining with primary antibodies against GFP and tdTomato further enhances the results of confocal imaging, and is necessary in the case of embryonic tissue. This technology provides the perfect platform for a multifaceted study of the role of a gene of interest, not only in the brain but in any organ or tissue in the mouse.

2 Materials

The following is a list of materials and equipment that are necessary for the processing, imaging, and analysis of MADM samples described in this protocol.

2.1 Reagents and Equipment

1. Phosphate buffered saline (PBS, Sigma-Aldrich).
2. Triton X-100 (Sigma-Aldrich).
3. Donkey Serum (Thermo Fisher).
4. Chicken anti-GFP (AVES, dilution 1:500).
5. Goat anti-tdTomato (SICGEN, dilution 1:500).

6. Donkey anti-chicken Alexa Fluor 488 secondary antibody (Jackson Immuno., dilution 1:500).
7. Donkey anti-goat Alexa Fluor 568 secondary antibody (Molecular Probes, dilution 1:1000).
8. DAPI (4′,6-diamidino-2-phenylindole) (Thermo Fisher Scientific).
9. Glass cover slides (Thermo Fisher Scientific).

2.2 Microscopes and Software

1. Confocal inverted microscope with camera and 40× or 63× objective: Zeiss LSM-800.
2. Confocal acquisition software: ZEN 2.6 (Zeiss).
3. Image analysis and 3D reconstruction software: Imaris v9.9.
4. Statistical analysis and data visualization program: GraphPad Prism v8.

3 Methods

The following protocol lists the steps required for tissue preparation, image acquisition, and final data visualization with minimal coding experience. Naturally, the coding-inclined can perform the statistical analysis and data visualization using their platform of choice (such as RStudio). Here, we focus on quantitative analysis of neuronal soma volume dendritic spine morphology, and astrocyte morphology.

3.1 Preparation of Experimental Tissue Containing MADM-Labeled Cells

3.1.1 Generation and Cryosectioning of MADM Brains

For the generation of MADM-labeled mice and the preparation of cryosections of MADM brains, follow previously published protocols [16, 17] (*see* **Notes 1** and **2**).

3.1.2 Immunostaining to Enhance Endogenous MADM Signals

The amplification of endogenous MADM signals is recommended for dendritic spine morphology analysis but not required for neuronal soma volume and astrocyte morphology analysis.

1. Rehydrate and wash mounted cryosections on glass coverslips with PBS.
2. Incubate sections in blocking solution (0.5% Triton-X and 5% donkey serum in PBS) for 2 h at room temperature (RT). Protect from light.
3. Incubate sections in blocking solution containing primary antibodies overnight at 4 °C. Protect from light. The following

primary antibodies are recommended: Chicken anti-GFP (AVES, 1:500) and Goat anti-tdTomato (SICGEN, 1:500).

4. Wash in PBS and incubate sections in PBS containing 0.5% Triton-X and secondary antibodies for 2 h at RT. Protect from light. The following secondary antibodies are recommended: Donkey anti-chicken Alexa Fluor 488 (Jackson Immuno. 1: 500) and Donkey anti-goat Alexa Fluor 568 (Molecular Probes, 1:500).
5. Wash in PBS and incubate sections in PBS with 2.5% DAPI (4′,6-diamidino-2-phenylindole) (Thermo Fisher Scientific) for 10 min at RT. Protect from light.
6. Wash in PBS and mount with Mowiol 4-88 with DABCO (Carl Roth) and cover with glass coverslip (24 × 60 mm, Thermo Scientific).
7. Allow slides to air-dry at 4 °C protected from light until used.

3.2 Neuron Soma and Astrocyte Volume Measurements

3.2.1 Image Acquisition

The following steps are described for the use of an LSM 800 inverted confocal microscope (Zeiss) operated using ZEN Blue 2.6 software (Zeiss). Adapt accordingly to other confocal microscopes and software.

1. Initiate confocal microscope according to standard operating procedures. Be sure to not overexpose the soma, as it can lead to issues with reconstruction later on.
2. Set up the system to acquire Z-stacks so that the entire cell is within the first and last Z-planes. For the subsequent analysis, avoid having multiple soma in the field of view (*see* **Notes 1** and **2**).

3.2.2 Image Processing for Neuron and Astrocyte Volume Measurements

The following image processing steps are described for the use of Imaris 9.9.1 (Oxford Instruments). Adapt accordingly to other image analysis software like ImageJ (National Institutes of Health).

1. Use the Imaris File Converter tool to convert source files (.czi in this case) to .ims files.
2. Open Imaris 9.9.1 and drag and drop the .ims file into the "Arena." Click on the file to be analyzed. Then navigate to the "Surpass" tab.
3. On the left side of the screen, select "Add new Surfaces" (Fig. 1a orange highlight).
4. Next the boundaries of the cell will be defined under the "Create" tab (Fig. 1b blue highlight).
5. If the image is dense with cells, segmentation of only a single cell is recommended. To do this, select the "Segment only a region of interest" box. Otherwise skip to 6.

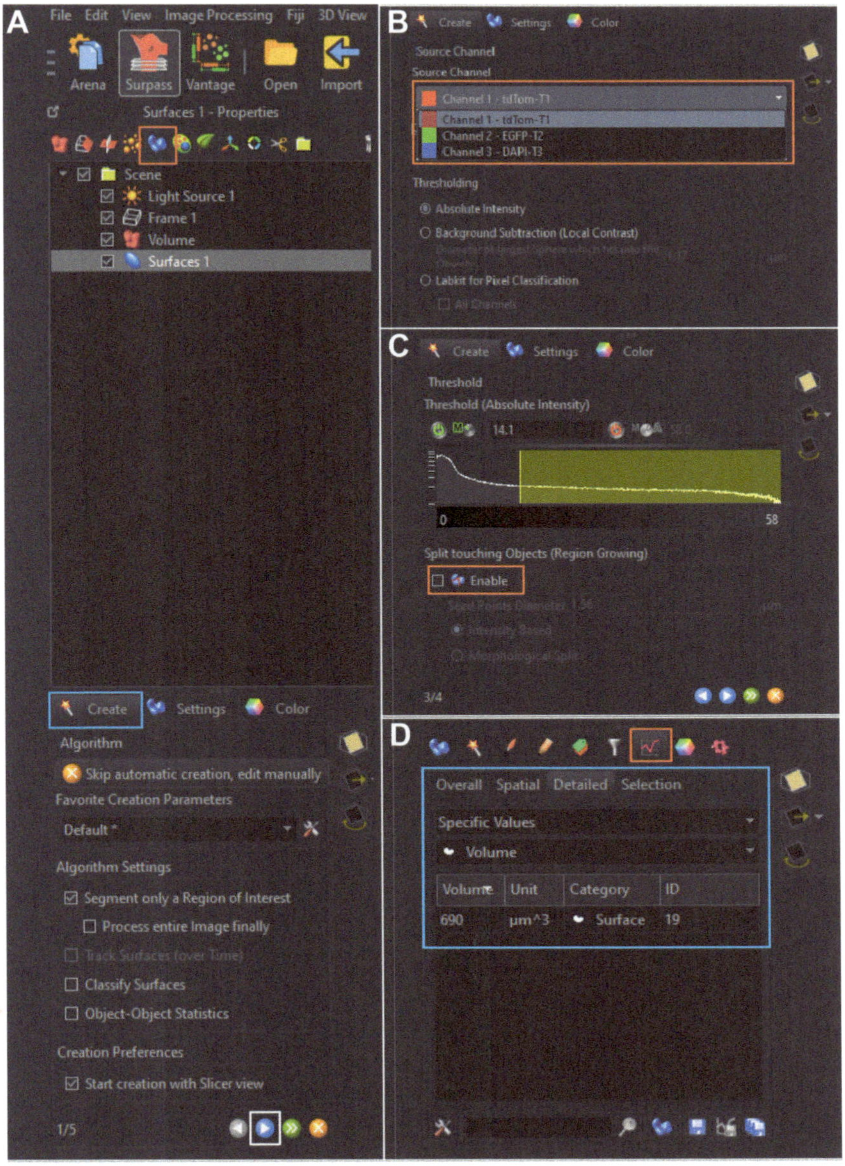

Fig. 1 Relevant steps to measure the volume of a neuron cell body

(**a**) Upon importing an .ims file, all processing takes place in the "surpass" tab. The orange box highlights the "create a new surface" function; all subsequent steps will take place under the "create" tab (blue box)

(**b**) A source channel must be selected for further analysis; thresholding is always set to "Absolute intensity"

(**c**) Imaris is able to automatically threshold the image, this can be manually adjusted. Additionally, if multiple objects are in close proximity, the "split touching objects" function (orange box) can separate these

(**d**) The resulting data are extensive, even for a straightforward analysis such as this, click on the graph (orange box) and select the "detailed" tab>"specific values">"volume." The volume will now be displayed along with the corresponding unit of measurement

6. To segment the image, click on the blue arrow to continue. In the next step left-click on the arrows to drag the box so that it only contains the soma of the neuron of interest.
7. Select the "Start creation with Slicer View" box. Then click on the blue arrow on the bottom of this area to continue.
8. On the next page select the source channel (Fig. 1b orange highlight), in the present example we use the red tdTomato channel. Next, click the blue arrow to continue.
9. Set the threshold for detection of the neuron. Use the same setting for all samples in the experiment. In the present example an automatically detected value of 14.1 is used (*see* **Note 3**). Should there be multiple cells in the image, choose to "Split touching Objects" by selecting the "Enable" box (Fig. 1c orange highlight). Click on the blue arrow to continue.
10. On this page filter out any objects smaller than a certain value (number of voxels), such as artifacts. Otherwise just continue.
11. Skip machine-learning object classification, click the green double arrow to finish this process, and render the surface of the neuron (*see* **Note 4**).
12. To get the target data, go to the tab with the graph icon (Fig. 1d orange highlight) and then choose the "Detailed" tab, then in the first drop-down menu choose "Specific Values," and in the second drop-down menu select "Volume" (Fig. 1d blue highlight).
13. The data is now displayed. Here the example neuron has a volume of 690 cubic microns. Other metrics from this surface can be viewed by navigating the drop-down menus, such as area and sphericity.
14. The measurement will be recorded and later plotted with Graphpad Prism software (Fig. 2).

3.3 Analysis of Dendritic Spines in MADM-Labeled Neurons

The following steps are described for the use of an LSM 800 inverted confocal microscope operated using ZEN Blue 2.6 software. Adapt accordingly to other confocal microscopes and software.

3.3.1 Image Acquisition

1. Configure confocal software for the following fluorophores: Alexa Fluor 488 (excitation laser at 488 nm), Alexa Fluor 568 (excitation laser at 561 nm) and DAPI (excitation laser at 405 nm). (*See* **Note 5** for an optional step.)
2. To image dendritic spines, use objectives and zoom settings to obtain imaging resolution of 0.08 mm/pixel. For example: a 63× oil immersion objective (NA 1.4) with 2.5× zoom, or a 40× oil immersion objective (NA 1.4) with 4× zoom.

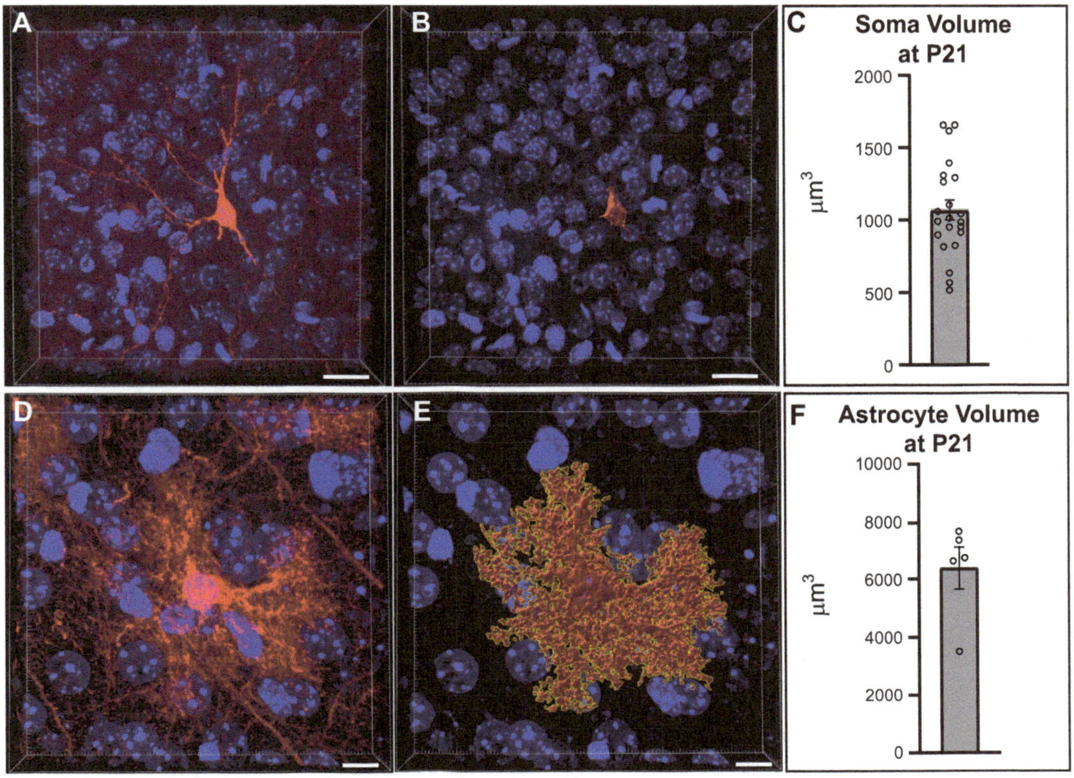

Fig. 2 Volume measurements of neuron soma and astrocyte
(**a**) Source data displayed upon importing of the corresponding .ims file. The red neuron is labeled using MADM, hence the lack of neighboring cells
(**b**) Once all processing has concluded, Imaris will render the surface; here we have cropped the neurons pre-processing to measure only the volume of the soma
(**c**) The resulting soma volume data is plotted, in this case using Graphpad Prism
Plots are shown as mean ± SEM ($n = 22$ neurons, $n = 5$ astrocytes)
Scale bars = 20 μm (A–B); 10 μm (D–E)

3. Set up image acquisition parameters for the relevant channel for dendritic spine labeling with pinhole of 1 Airy Unit and optimal laser intensity and master gain (500–600 V). To reduce background noise, averaging of acquisition by 2–4 times can be used (*see* **Note 6**).
4. Set up z-stack and tiles to cover area of interest with 0.20 μm z-interval (*see* **Note 7**).
5. Start image acquisition.
6. After image acquisition, stitch image stacks using “Stitching” function under processing tab.

3.3.2 Image Deconvolution

The following image processing steps are described for the use of Imaris 9.9.1. Adapt accordingly to other image analysis software like ImageJ or Huygens (Scientific Volume Imaging).

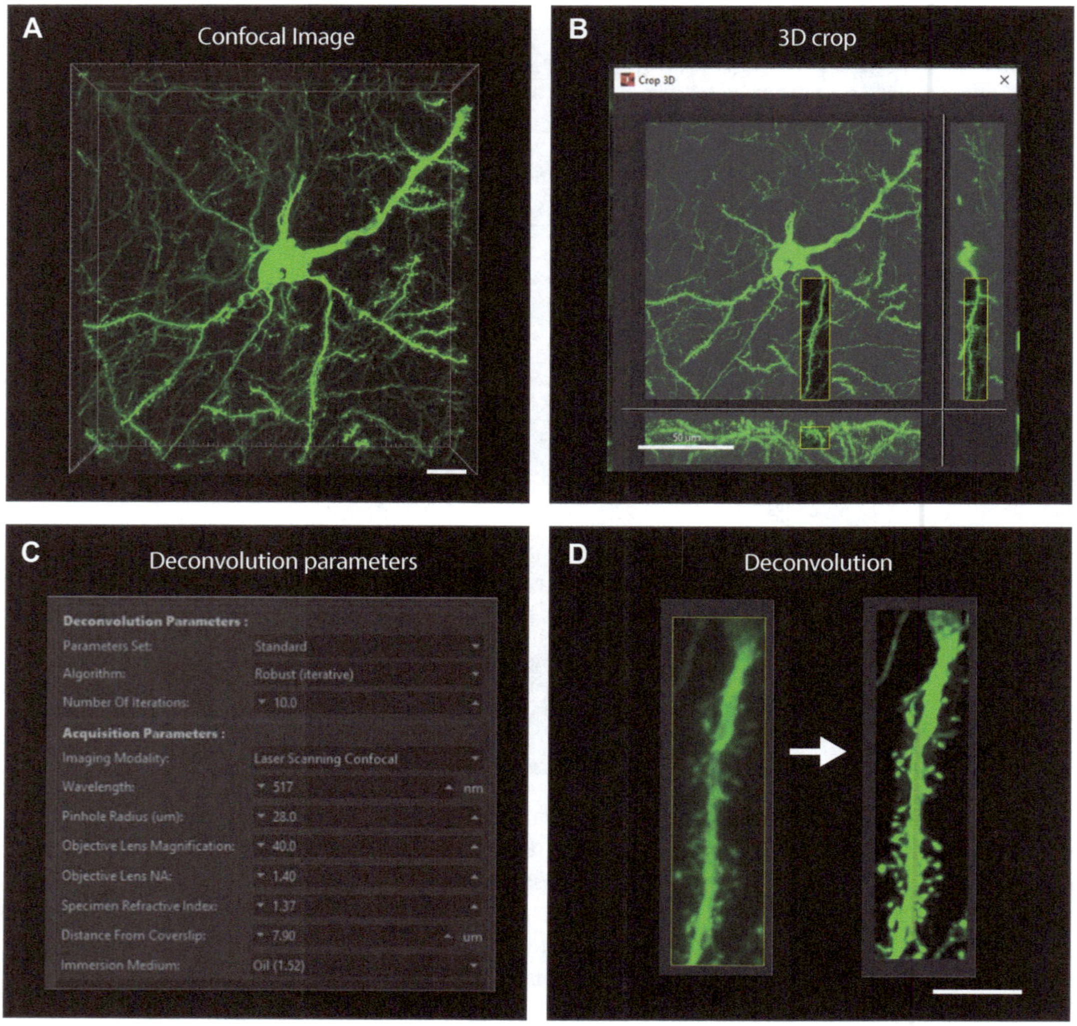

Fig. 3 3D Cropping and deconvolution of dendritic segments
(**a**) Confocal image of a cortical neuron labeled in green GFP fluorescence
(**b**) 3D Crop window in 3D slice mode with selected area (yellow box) highlighted to be cropped
(**c**) Deconvolution parameter windows with settings specific to confocal acquisition and processed image
(**d**) Magnification of region in yellow box shown in (**b**) before (left) and after (right) deconvolution. Scale bars: 15 μm (**a**); 50 μm (**b**); 5 μm (**d**)

1. Open .czi image in Imaris and convert to .ims file (Fig. 3a).
2. Choose Crop 3D under Edit and crop into smaller segments of dendritic spines (Fig. 4a). Save as separate .ims files (*see* **Note 8**).
3. Open individual .ims files of cropped fragments and choose "Image Processing" under "View."
4. For each channel, choose "Deconvolution" in the drop-down menu.

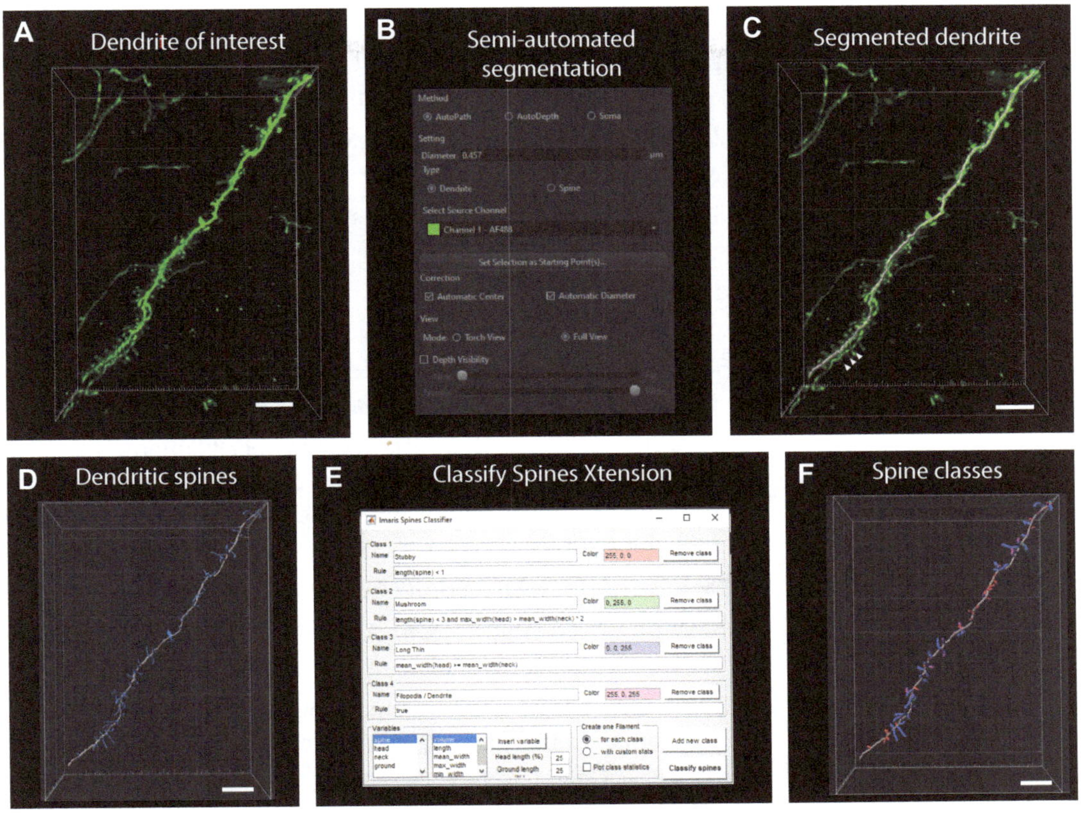

Fig. 4 Segmentation and reconstruction of dendritic spines
(**a**) Deconvolved image of a cropped dendrite of interest
(**b**) Settings for semi-automated (Auto-Path) method to segment dendrite. Note that "Automatic Center" and "Automatic Diameter" options are selected
(**c**) Segmented dendrite is shown in thick line. Arrowheads indicate a branched dendritic segment behind the dendrite segment of interest which was not segmented
(**d**) Filament view of segmented dendrite (gray) with its dendritic spines detected in automatic mode (blue)
(**e**) MATLAB window extension allowing user-defined parameters to classify spines
(**f**) Classified spines are highlighted in filament view with color codes for different spine class
Scale bars: 5μm

5. Specify deconvolution parameters according to image acquisition parameters (e.g., Fig. 3c). Image acquisition parameters can be found associated with each .czi file listed under information tab in ZEN software.
6. Check preview window for quality (Fig. 3d) and press "OK" to perform deconvolution if parameters are correct. Save as new .ims file.

3.3.3 Reconstruction of Dendrites and Dendritic Spines

The following image processing steps are described for the use of Imaris 9.9.1. Adapt accordingly to other image analysis software like Neuronstudio (BioImage Informatics Index).

(a) Segmentation of Dendrites (Semi-Automated Mode)

1. In Imaris, open deconvolved .ims file from "Arena." In 3D View, press "Add New Filament."
2. Press "Skip automatic creation, edit manually" (*see* **Note 9**).
3. Select "AutoPath" method, automatic center, and diameter correction (Fig. 4b).
4. Set the start of dendrite segment (hold Shift + right mouse click).
5. Move cursor slowly along dendrite segment and mark the end point (hold Shift + left mouse click).
6. Reconstruct one dendritic segment per filament (Fig. 4c). As shown in this image, another dendritic branch denoted by white arrows are not segmented and therefore removed in the analysis.

(b) Dendritic Spines Detection (Automated Mode)

1. Within "Creation" tab, under "Recompute," choose "Dendrite Diameters."
2. For the following automated mode of detection, use automated thresholding or set to specific values (*see* **Note 10**).
3. Detect spines by setting thinnest diameter and maximum length (*see* **Note 11**). In this example, default values = 0.156 and 3.9 mm are used, respectively.
4. Use automated thresholding for seed points and spine diameter of set to specific values (*see* **Notes 10** and **11**). Proceed and complete the reconstruction wizard following on-screen instructions.
5. An example of a reconstructed dendritic segments with spines is illustrated (Fig. 4d).

3.3.4 Classification of Dendritic Spines

1. Under filament tab, select "Classify Spines" tool, a MATLAB window will open.
2. Use default parameters or change accordingly to define different categories of dendritic spines (*see* **Note 11**). New classes of spines can also be created. In the present example, four types of spine morphology are considered: filopodia/dendrite, long thin, mushroom, or stubby.
3. When the correct parameters are set, choose "Classify spines." An example using default settings are shown (Fig. 4e).
4. All spines are automatically classified based on defined parameters and color coded in viewing panel (Fig. 4f).
5. Save all reconstruction and analysis performed on the dendritic segment by saving the .ims file.
6. An example of a reconstructed dendritic segment is shown in Fig. 5a in stepwise manner.

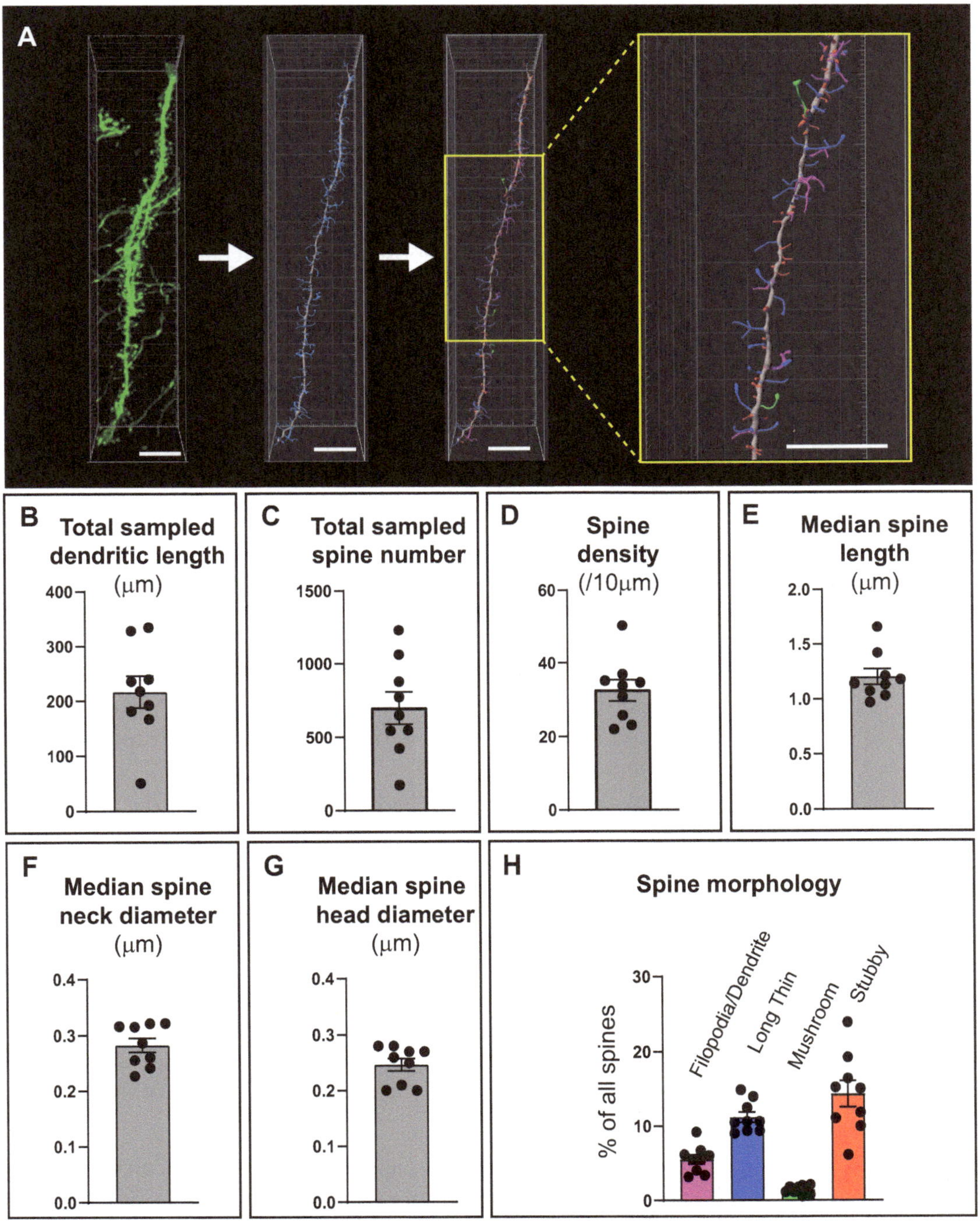

Fig. 5 Reconstruction and analysis of dendritic spines
(**a**) Images showing progressing analysis of a dendritic segment to detect and subsequently classify spines. Magnified view of classified spines in filament view mode is shown in yellow box
(**b**–**h**) After extraction from exported parameters total dendritic length per cell (**b**), total spine number per cell (**c**), spine density per cell (**d**), median spine length (**e**), median spine neck diameter (**f**), median spine head diameter (**g**), and % of all spines for each spine class (**h**) are shown as mean ± SEM (*n* = 9 neurons, 1–9 dendritic segments analyzed per cell)
Scale bars: 7μm

3.3.5 Data Extraction and Statistical Analysis

1. In the "Arena" of Imaris, create a new folder containing all dendritic fragments files to be analyzed.
2. Select folder and choose "New Plot" option.
3. Under plot type, choose desired parameters to be exported. Some basic parameters include: Dendrite length, Dendrite no. spines, Dendrite Spine Density, Spine Length, Spine Neck Max Diameter, and Spine Terminal Pt Diameter.
4. For each parameter, choose to display detailed plot and export data as individual .xls files.
5. For each neuron, analyze multiple dendritic segments depending on imaged area and combine dendritic length and spine number per cell. Example quantifications obtained from nine neurons each with one to nine dendritic segments analyzed are shown in Fig. 5b–h. Individual values and mean ± SEM are shown in bar plots.
6. For example, from the total dendritic length (Fig. 5b) and spine number (Fig. 5c) sampled, spine density can be determined as the total spine number per 10 μm of total dendritic length per cell (Fig. 5d).
7. For the analysis of spine parameters, determine the median value of spine length (Fig. 5e), neck diameter (Fig. 5f), and head diameter (Fig. 5g) per cell. Alternatively, distribution of these parameters can also be analyzed as histograms and compared.
8. Spine morphology can be compared based on the number of spines of each category per cell (Fig. 5h). The relative proportion of spines belonging to different classes can be determined.
9. Repeat analysis for neurons of different experimental conditions and perform appropriate statistical analysis to compare different parameters.

3.4 Astrocyte Sholl Analysis

3.4.1 Image Acquisition

The following steps are described for the use of an LSM 800 inverted confocal microscope (Zeiss) operated using ZEN Blue 2.6 software (Zeiss). Adapt accordingly to other confocal microscopes and software.

1. Initiate confocal microscope according to standard operating procedures. Be sure to not overexpose the soma, as it can lead to issues with reconstruction later on.
2. Set up the system to acquire Z-stacks so that the entire cell is within the first and last Z-planes. In this analysis, avoid having multiple astrocytes in the field of view (*see* **Notes 1** and **2**).

3.4.2 Image Deconvolution

The following image processing steps are described for the use of Imaris 9.9.1. Adapt accordingly to other image analysis software like ImageJ or Huygens (Scientific Volume Imaging).

1. Open .czi image in Imaris and convert to .ims file (Fig. 3a).
2. Open individual .ims files of cropped fragments and choose "Image Processing" under "View."
3. For each channel, choose "Deconvolution" in the drop-down menu.
4. Specify deconvolution parameters according to image acquisition parameters (e.g., Fig. 3c). Image acquisition parameters can be found associated with each .czi file listed under information tab in ZEN software.
5. Check preview window for quality (Fig. 3d) and press "OK" to perform deconvolution if parameters are correct. Save as new . ims file.

3.4.3 Render Astrocyte Filaments

The following image processing steps are described for the use of Imaris 9.9.1. Adapt accordingly to other image analysis software like ImageJ or Huygens (Scientific Volume Imaging).

1. In Imaris, open deconvolved .ims file from "Arena." In 3D View, press "Add New Filament."
2. Under "Creation Preferences" turn off "Start creation with Slicer view" and press next (Fig. 6a blue highlight).
3. Select source channel and continue. Indicate 8.0 μm as the largest diameter for "Starting Points," and 0.7 μm for "Seed Points." These values have generated the best results in our analyses.
4. Adjust the "Starting Points Threshold" so that only a single starting point is present in the body of the astrocyte. Additionally, adjust the "Diameter of Sphere Region(s)" to clear seed points within the astrocyte cell body.
5. Adjust "Seed Points Threshold" as necessary, so that, if possible, seed points remain only within the cell of interest.
6. Skip dendrite diameter thresholding, the software's automatic settings work well. Adjust if necessary.
7. Skip spine detection. Click green double arrow to finish.
8. The filaments should rather faithfully reflect the morphology of the astrocyte. Trimming of excess filaments is common. To do this, click on the "edit" tab (pencil).
9. Select "Branch" under the "Mouse selects," this command instructs the software to highlight the branch distal to its nearest branch point.
10. Delete selected branch with the "delete" key.

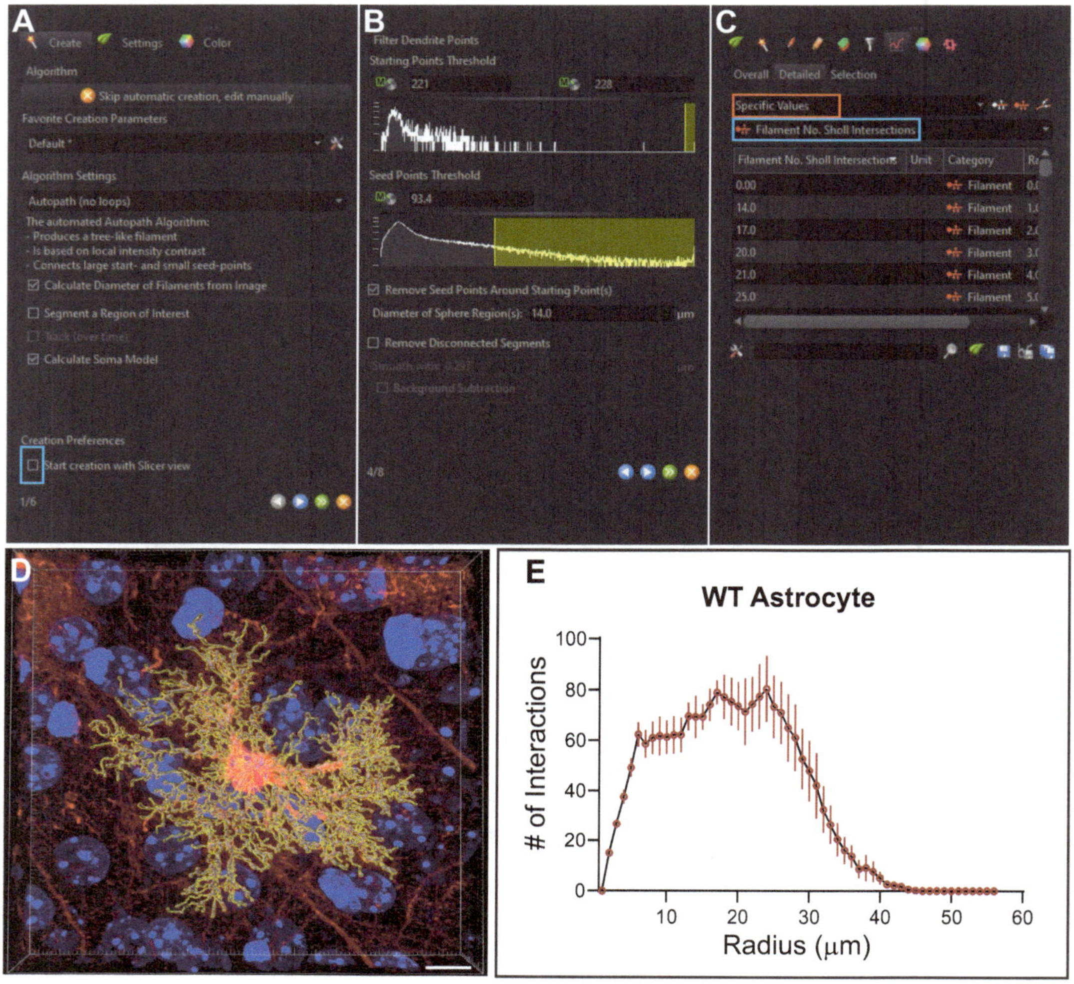

Fig. 6 Reconstruction and analysis of astrocyte filaments
(**a**, **b**) Images depicting the stages of processing of images for filament detection
(**c**) The resulting data from this analysis displaying the number of filament intersections and the radius of each sholl boundary
(**d**) Visual representation of the filaments reconstructed by the Imaris software
(**e**) Plot of the sholl intersections data. Plots are shown as mean ± SEM ($n = 5$ astrocytes)
Scale bar = 10 μm

11. Once the render is accurate, select the "statistics" tab, depicted by the graph icon.
12. Select the "detailed" tab, under the drop-down menu select "specific values" (Fig. 6c orange highlight). A second drop-down menu will appear, select "filament no. sholl intersections" (Fig. 6c blue highlight).
13. The selected data will be displayed, it can then be exported and plotted (Fig. 6d, e).

4 Notes

1. It is important to consider sparseness of labeled neurons and astrocytes in order to obtain individual isolated labeled cells for accurate morphological measurements. When neurites or glial processes overlap with another cell, a semi-automated reconstruction to segment out undesired background is required.
2. Thickness of cryosections should be optimized based on the size of neurons or astrocytes to be analyzed and maximal penetration for immunostaining and confocal microscopy. Typically, sections between 30 and 50 μm are used.
3. The same, or very similar, settings should be used to analyze the dataset of a single experiment. This helps ensure consistency in the analysis, BUT manual intervention is sometimes necessary when the software does not get the threshold correct. In this case, increase/decrease the threshold as required for obtaining the proper rendering of the cell.
4. For batch analysis of cells, save the parameters after the first cell is completely processed and go to the magic wand tab and click "Store Parameters for Batch." This can then be loaded into the program to automatically set the parameters for a batch of .ims files.
5. (Optional step) It is recommended that a low resolution image (with 10–20× objectives) of the entire cell should be taken in three different channels in order to capture the neuronal morphology and document MADM-labeling. Subsequent image acquisition of spines should be taken of the relevant channel only to save time.
6. In order to capture fine dendritic spine structures, neuronal cell bodies are often overexposed.
7. During confocal sampling, it is crucial for subsequent deconvolution steps that sampling interval is less than the optimal interval determined automatically by confocal software.
8. Processing the sample in smaller fragments can increase the efficiency of deconvolution as well as accuracy of segmentation of dendrites of interest.
9. For dendrite segmentation, automatic creation can also be performed but only recommended for simple images with clearly defined neurites without background.
10. It is important to use the same threshold for the same image and preferably also across different experiments. Therefore, it is important to determine whether automated thresholding or specific values should be used.

11. Dendritic spines of different neuronal types across different brain regions may differ by length and thickness. It is important to adapt to the specific type of neuron of interest. When unsure, it is recommended that the dimension of a number of spines is measured from the same image under "Slice" view.

Acknowledgments

We thank all Hippenmeyer lab members for support and discussions. This work was supported by the Scientific Service Units (SSU) at ISTA through resources provided by the Imaging & Optics Facility (IOF). O.A.M was a recipient of a DOC Fellowship (26253) of the Austrian Academy of Sciences. This work was supported by ISTA institutional funds, and The Austrian Science Fund Special Research Programmes (FWF SFB F78 Neuro Stem Modulation) to S.H.

Competing Interests The authors declare no competing interests.

Author Contributions Project Management: S.H.; Conceptualization and Design: S.H., O.A.M., and G.C.; Experimental Data Collection and Image Analysis: O.A.M. and G.C.; Data Interpretation: O.A.M. and G.C.; Data Visualization: O.A.M. and G.C.; Manuscript Drafting: O.A.M. and G.C.; Critical Revision of the Manuscript: All authors; Final version was read by all authors and approved to be published.

References

1. Phillips M, Pozzo-Miller L (2015) Dendritic spine dysgenesis in autism related disorders. Neurosci Lett 601:30–40. https://doi.org/10.1016/j.neulet.2015.01.011
2. Martinez-Cerdeno V (2017) Dendrite and spine modifications in autism and related neurodevelopmental disorders in patients and animal models. Dev Neurobiol 77(4):393–404. https://doi.org/10.1002/dneu.22417
3. Rademacher S, Eickholt BJ (2019) PTEN in autism and neurodevelopmental disorders. Cold Spring Harb Perspect Med 9(11). https://doi.org/10.1101/cshperspect.a036780
4. Lang F, Busch GL, Ritter M, Volkl H, Waldegger S, Gulbins E, Haussinger D (1998) Functional significance of cell volume regulatory mechanisms. Physiol Rev 78(1):247–306. https://doi.org/10.1152/physrev.1998.78.1.247
5. Hoffmann EK, Lambert IH, Pedersen SF (2009) Physiology of cell volume regulation in vertebrates. Physiol Rev 89(1):193–277. https://doi.org/10.1152/physrev.00037.2007
6. Yuste R, Denk W (1995) Dendritic spines as basic functional units of neuronal integration. Nature 375(6533):682–684. https://doi.org/10.1038/375682a0
7. Molofsky AV, Krencik R, Ullian EM, Tsai HH, Deneen B, Richardson WD, Barres BA, Rowitch DH (2012) Astrocytes and disease: a neurodevelopmental perspective. Genes Dev 26(9):891–907. https://doi.org/10.1101/gad.188326.112
8. Sloan SA, Barres BA (2014) Mechanisms of astrocyte development and their contributions to neurodevelopmental disorders. Curr Opin Neurobiol 27:75–81. https://doi.org/10.1016/j.conb.2014.03.005

9. Torres-Ceja B, Olsen ML (2022) A closer look at astrocyte morphology: development, heterogeneity, and plasticity at astrocyte leaflets. Curr Opin Neurobiol 74:102550. https://doi.org/10.1016/j.conb.2022.102550
10. Santello M, Toni N, Volterra A (2019) Astrocyte function from information processing to cognition and cognitive impairment. Nat Neurosci 22(2):154–166. https://doi.org/10.1038/s41593-018-0325-8
11. Zong H, Espinosa JS, Su HH, Muzumdar MD, Luo L (2005) Mosaic analysis with double markers in mice. Cell 121(3):479–492. https://doi.org/10.1016/j.cell.2005.02.012
12. Contreras X, Amberg N, Davaatseren A, Hansen AH, Sonntag J, Andersen L, Bernthaler T, Streicher C, Heger A, Johnson RL, Schwarz LA, Luo L, Rulicke T, Hippenmeyer S (2021) A genome-wide library of MADM mice for single-cell genetic mosaic analysis. Cell Rep 35(12):109274. https://doi.org/10.1016/j.celrep.2021.109274
13. Beattie R, Postiglione MP, Burnett LE, Laukoter S, Streicher C, Pauler FM, Xiao G, Klezovitch O, Vasioukhin V, Ghashghaei TH, Hippenmeyer S (2017) Mosaic analysis with double markers reveals distinct sequential functions of Lgl1 in neural stem cells. Neuron 94(3):517–533 e513. https://doi.org/10.1016/j.neuron.2017.04.012
14. Amberg N, Pauler FM, Streicher C, Hippenmeyer S (2022) Tissue-wide genetic and cellular landscape shapes the execution of sequential PRC2 functions in neural stem cell lineage progression. Sci Adv 8(44):eabq1263. https://doi.org/10.1126/sciadv.abq1263
15. Hansen AH, Pauler FM, Riedl M, Streicher C, Heger A, Laukoter S, Sommer C, Nicolas A, Hof B, Tsai LH, Rülicke T, Hippenmeyer S (2022) Tissue-wide effects override cell-intrinsic gene function in radial neuron migration. Oxford Open Neurosci 1. https://doi.org/10.1093/oons/kvac009
16. Beattie R, Streicher C, Amberg N, Cheung G, Contreras X, Hansen AH, Hippenmeyer S (2020) Lineage tracing and clonal analysis in developing cerebral cortex using mosaic analysis with double markers (MADM). J Vis Exp 159. https://doi.org/10.3791/61147
17. Amberg N, Hippenmeyer S (2021) Genetic mosaic dissection of candidate genes in mice using mosaic analysis with double markers. STAR Protoc 2(4):100939. https://doi.org/10.1016/j.xpro.2021.100939

Chapter 20

Isolation and Culture of Adult Rat Dorsal Root Ganglion Neurons for the In Vitro Analysis of Peripheral Nerve Degeneration and Regeneration

Kazunori Sango, Shizuka Takaku, Naoko Niimi, and Hideji Yako

Abstract

Isolation and culture of dorsal root ganglion (DRG) neurons from adult animals is a useful experimental system for evaluating neural plasticity after axonal injury, as well as the neurological dysfunction resulting from aging and various types of disease. In this chapter, we will introduce a detailed method for the culture of mature rat DRG neurons. About 30–40 ganglia are dissected from a rat and mechanically and enzymatically digested. Subsequently, density gradient centrifugation of the digested tissue using 30% Percoll efficiently eliminates myelin debris and non-neuronal cells, to afford neuronal cells with a high yield and purity.

Key words DRG neurons, Adult rat, Cell culture, Percoll, Cell survival, Neurite outgrowth, Immunofluorescence, Transmission electron microscopy, Western blotting

1 Introduction

Dorsal root ganglion (DRG) neurons are constituents of the sensory system, and methods for their primary culture from embryonic, neonatal, young-adult, and aged rodents have been established [1, 2]. These neurons have been shown to express types of membrane channels and receptors that are similar to those of cutaneous receptors in vivo. Because some of the biological properties of DRG neurons change with maturation and aging, it would be fair to state that the culture system of adult DRG neurons mimics peripheral nerve degeneration and regeneration better than does the culture system of immature neurons [3]. In addition to normal adult rodents, the methods of DRG neuron culture have been applied to the animal models of a variety of disorders (e.g., peripheral neuropathies [4], axonal injuries [5], and lysosomal storage diseases [6]).

Kazuhito Toyooka (ed.), *Neuronal Morphogenesis: Methods and Protocols*, Methods in Molecular Biology, vol. 2831,
https://doi.org/10.1007/978-1-0716-3969-6_20,

Here, we propose detailed methods for the primary culture of adult rat DRG neurons and for the functional, morphological, molecular, and biochemical analyses of these cells. The culture protocol is easy and straightforward; therefore, it can be mastered with 1 month of practice. Primary culture of adult mouse DRG neurons can be carried out in a similar manner. However, it is needless to say that the findings of in vitro analyses do not always apply to the biological phenomena in vivo. For instance, in vitro neurite outgrowth lacks the processes of Wallerian degeneration and functional reinnervation to the target tissues, which are essential for axonal regeneration in vivo [3]. In addition, the mechanical and enzymatic treatments of DRG tissues that are necessary for the isolation of neurons disrupt the contact of neurons with the surrounding satellite glial cells and other non-neuronal cells, which may affect the neuronal phenotype [7].

2 Materials

2.1 Equipment

1. Laminar flow cabinet.
2. CO_2 incubator.
3. Binocular dissecting microscope.
4. Phase-contrast microscope.
5. Constant-temperature shaking bath.
6. Benchtop centrifuge.
7. Dissection instruments (kitchen shears, straight surgical scissors, ophthalmic scissors, tweezers, precision tweezers (INOX No. 5), blade holders, and Feather Hi-Stainless Double Edge razor blades) (Fig. 1). The razor blades are cut into strips with a width of 3–5 mm and attached to the blade holders (Fig. 2).
8. Centrifuge tube (15 mL), Pasteur pipette (sterilized), and disposable plastic transfer pipette.
9. Cell culture dishes (35, 60, and 100 mm in diameter), 4-well and 12-well culture plates, and 8-well chamber slides. In addition to these manufactured products, Aclar fluorocarbon film is punched into a circle (0.9 cm in diameter) and attached to a well of a four-well-plate coated with poly-L-lysine (*see* Subheading 2.2), for use in immunocytochemistry and electron microscopy experiments [6] (Fig. 3).

2.2 Culture Medium, Reagents, Etc.

1. Dulbecco's Modified Eagle's Medium/Ham's F-12 (DMEM/F12) supplemented with antibiotics.
2. Fetal bovine serum (FBS): complement inactivated (56 °C, 30 min: required according to the product), dispensed into appropriate amounts, and stored frozen.

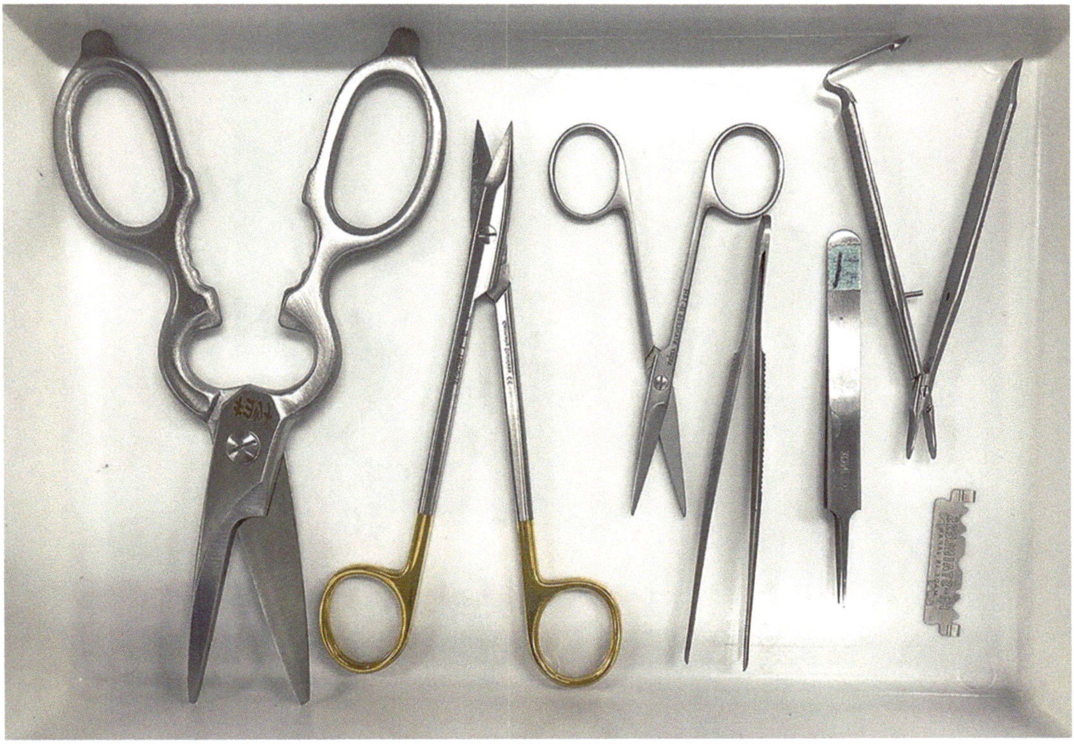

Fig. 1 Dissection instruments (from left to right): kitchen shears, straight surgical scissors, ophthalmic scissors, tweezers, precision tweezers (INOX No.5), blade holder, and Feather Hi-Stainless Double Edge razor blades

3. Serum-free reagent (B27 supplement).
4. Hanks' balanced salt solution (HBSS) without Ca^{2+} and Mg^{2+}.
5. Poly-L-lysine (100× stock): adjusted to 1 mg/mL with sterilized water, filtered, dispensed into appropriate amounts, and stored frozen (*see* **Note 1**).
6. Collagenase type III (5× stock): adjusted to 10 mg/mL with DMEM/F12, filtered, dispensed into appropriate amounts, and stored frozen.
7. Trypsin type I (4× stock): adjusted to 10 mg/mL with HBSS, filtered, dispensed into appropriate amounts, and stored frozen.
8. Trypsin inhibitor: adjusted to 50 μg/mL with HBSS, filtered, dispensed into appropriate amounts, and stored in a refrigerator.
9. Percoll (which is a tool used for density gradient centrifugation): diluted to 30% with HBSS (e.g., 30 mL of Percoll + 3.3 mL of 10× HBSS + 67 mL of 1× HBSS) and stored in a refrigerator.

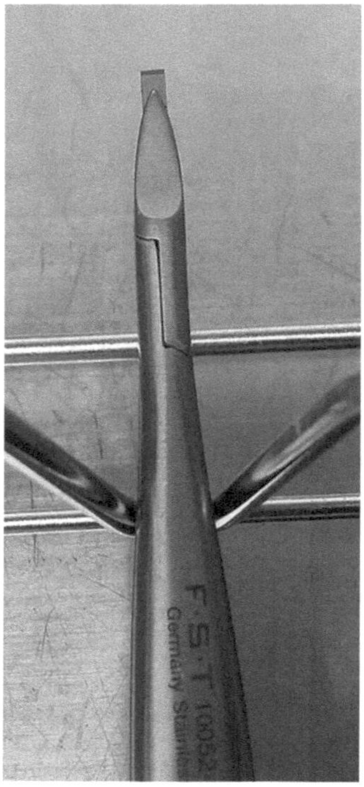

Fig. 2 A razor blade strip attached to the blade holder is used for the transection of spinal nerve fiber bundles from the ganglia

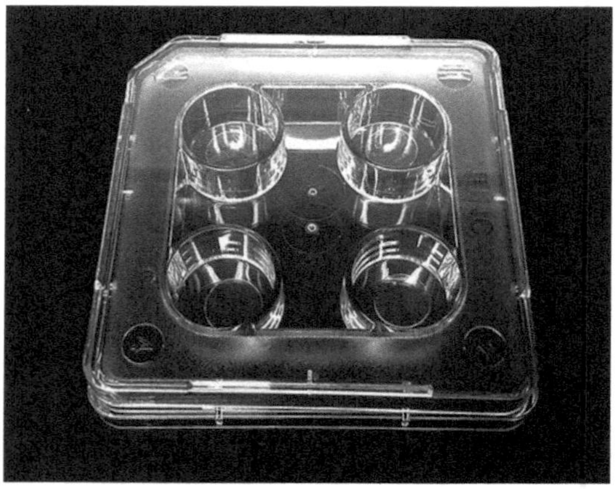

Fig. 3 Aclar fluorocarbon cover slips (0.9 cm in diameter) attached to the wells of a four-well-plate are used for immunocytochemistry and transmission electron microscopy

3 Methods

All experiments using rats should be conducted in accordance with the animal research: reporting of in vivo experiments (ARRIVE) [8] and institutional guidelines.

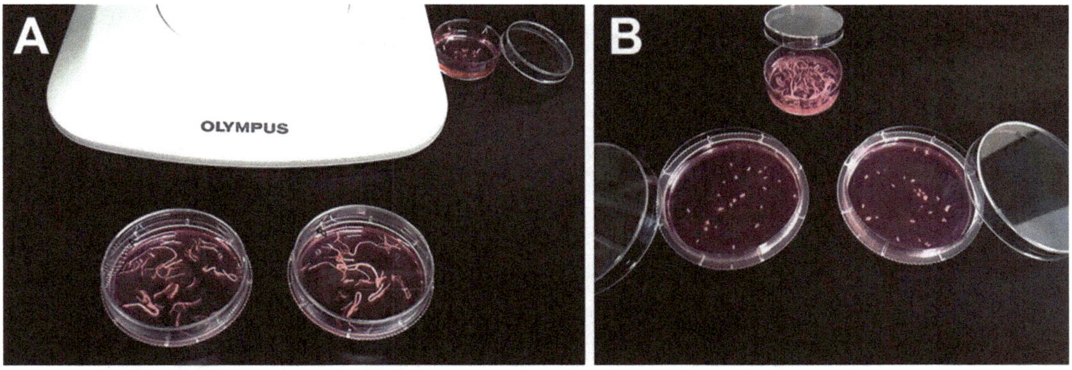

Fig. 4 Transection of spinal nerve fiber bundles from the ganglia under a dissecting microscope. (**a**) Dissected DRGs are placed in two dishes filled with DMEM/F12. (**b**) Transected spinal nerve fiber bundles (upper dish) are discarded, and isolated ganglia (lower dishes) are subjected to enzymatic and mechanical digestion

3.1 Dissection

1. A 3-month-old rat is deeply anesthetized for euthanasia using 3% isoflurane.
2. The skin of the back is sterilized with 80% ethanol and cut longitudinally using surgical scissors.
3. The spine (from the thoracic to the lumbar level) is dissected using kitchen shears. The surrounding muscle, connective tissues, and fat are fully eliminated, to clean the spine sample.
4. With the dorsal side facing up, the bottom blade of ophthalmic scissors is inserted into the rostral side of spinal canal. The center zone of the canal is removed from the rostral to the caudal side using the scissors, and the spinal cord is removed using forceps.
5. The spine is symmetrically divided into left and right parts by cutting the center of the ventral side with scissors. The location of each DRG is easily identified under a dissecting microscope.
6. Approximately 25–30 DRGs are dissected from the thoracic to lumbar level using precision tweezers (*see* **Note 2**) and placed in a dish (60 mm in diameter) filled with DMEM/F12 (Fig. 4a).
7. Under a dissecting microscope, spinal nerve fiber bundles are transected at sites close to the ganglia using strips of razor blade attached to the blade holders (Figs. 2 and 4b) (*see* **Note 3**).

3.2 Isolation of DRG Neurons

1. The ganglia collected in dishes are transferred to 15 mL centrifuge tubes (approximately 15 DRGs/tube) using a sterile glass pipette or a disposable plastic transfer pipette.
2. The ganglia are rinsed twice with DMEM/F12 and incubated in DMEM/F12 containing 2 mg/mL collagenase for 120 min at 37 °C. Efficient enzymatic treatment can be achieved using a constant-temperature shaking bath (Fig. 5) (*see* **Note 4**).

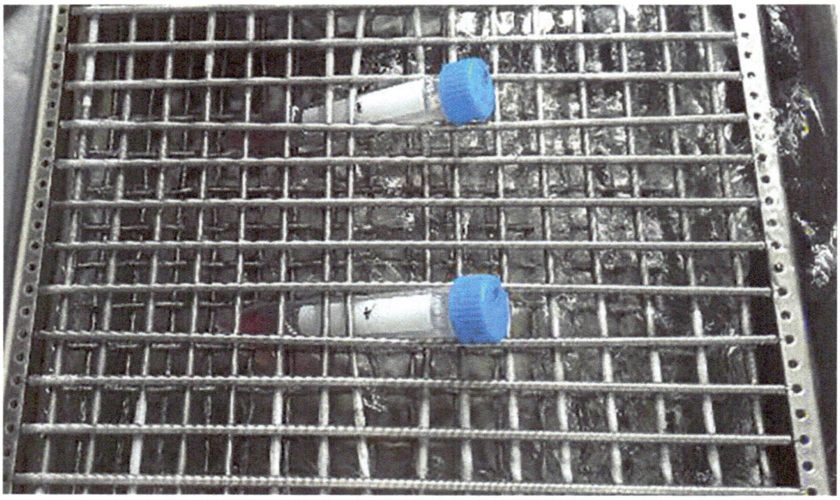

Fig. 5 The ganglia (Fig. 4b) are subjected to collagenase treatment at 37 °C in a constant-temperature shaking bath

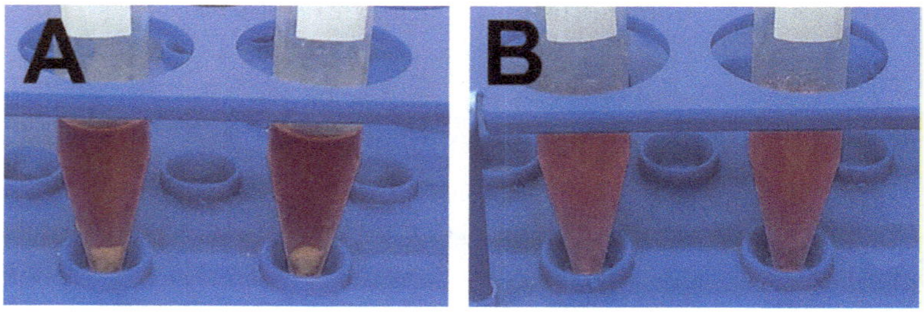

Fig. 6 Ganglia treated with collagenase and trypsin are observed at the bottom of 15 mL tubes (**a**) but become invisible after repeated gentle pipetting using a thin flame-polished glass pipette in HBSS containing a trypsin inhibitor (50 μg/mL) (**b**)

3. Collagenase-treated ganglia are centrifuged (180 × *g*, 3 min) and the supernatant is discarded. After rinsing twice with Ca^{2+}-Mg^{2+}-free HBSS (180× *g*, 3 min), the ganglia are incubated in HBSS containing 2.5 mg/mL trypsin in a constant-temperature shaking bath at 37 °C for 15 min.
4. Trypsin-treated ganglia are triturated by gentle pipetting using a thin flame-polished glass pipette in HBSS containing a trypsin inhibitor (50 μg/mL). If the collagenase and trypsin treatments are sufficient, almost all tissue clumps will become invisible after this pipetting operation (Fig. 6).
5. The digested tissue is gently layered onto 6 mL of 30% Percoll solution in a 15 mL centrifuge tube. Subsequent density gradient centrifugation (180× *g*, 5 min) results in the separation of almost all neurons (in the pellet) from myelin debris and non-neuronal cells (which remain in the supernatant) [9] (*see* **Note 5**).

3.3 Plating of Neurons

1. After the supernatant is discarded, the harvested neurons are rinsed twice in DMEM/F12 containing 10% FBS (180× *g*, 3 min), and suspended in 2 mL of serum-containing medium. The neurons are then seeded onto poly-L-lysine-coated culture dishes or chamber slides (*see* **Note 6**). Because an extremely high or low cell density will prevent the accurate evaluation of neuronal cell survival and neurite outgrowth, care must be taken to adjust the cell density to approximately 1–2 × 10^3 cells/cm^2) (*see* **Note 7**).
2. The cells seeded onto dishes are maintained overnight at 37 °C in a CO_2 incubator, then fed with serum-containing (DMEM/F12/10% FBS) or serum-free (DMEM/F12/B27) medium, depending on the experiment. The medium is changed every 2–3 days (*see* **Note 8**).

3.4 Functional Analyses

3.4.1 Cell Viability Assay

1. Prior to the seeding of neurons, circles with a diameter of 0.9 cm are delineated on the bottom of each well of 12-well plates using black thin lines.
2. The neuronal cell density is adjusted to approximately 500–700 cells within a circle (*see* **Note 9**). After an overnight incubation in DMEM/F12/10% FBS, the cells are maintained in DMEM/F12/B27 (*see* **Note 10**).
3. The relative cell viability at specific time points (e.g., 1, 3, 5, and 7 days in culture) is determined by counting the number of neurons. At each stage of counting, neurons that have lost phase-contrast brightness or clear cell membrane shapes are considered dead cells. At the final stage of counting, dead neurons are confirmed by positive trypan blue staining (Fig. 7).
4. The number of viable neurons at each stage is expressed as a relative value to the original number measured 1–12 h after seeding, which is set at 100 [10].

3.4.2 Neurite Outgrowth Assay

1. DRG neurons are seeded on the poly-L-lysine-coated wells of 8-well chamber slides at approximately 2 × 10^3 cells/well (*see* **Note 11**).
2. After an overnight incubation in DMEM/F12/10% FBS, the cells are maintained in DMEM/F12/B27 in the presence or absence of reagents, for evaluation.
3. Neurite outgrowth at 2 days in culture is evaluated by immunocytochemistry (*see* **Note 12**). The cells are fixed with 100% methanol or acid–alcohol solution (95% ethanol/5% acetic acid) at 4 °C for 10 min, then incubated overnight at 4 °C with an antibody against βIII tubulin, which is a specific neuronal marker. The cells are then incubated for 1 h at 37 °C with a peroxidase-conjugated secondary antibody. Immunoreactions are visualized under a light microscope with 0.01%

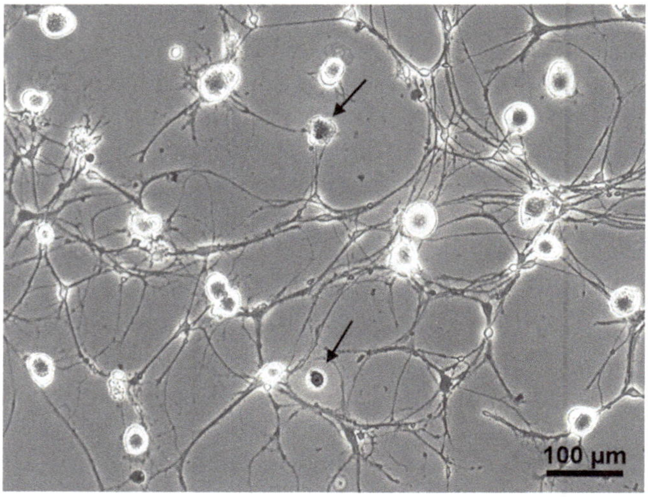

Fig. 7 Phase-contrast micrograph of DRG neurons after 7 days in culture. The arrows indicate trypan-blue-positive dead neurons

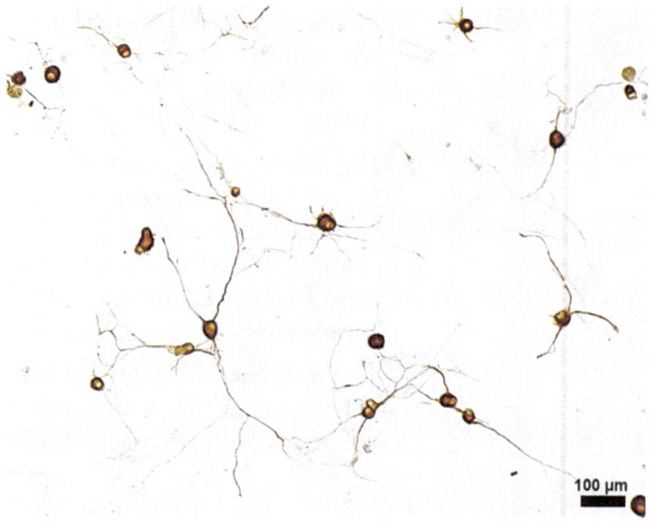

Fig. 8 Light photomicrograph of cultured DRG neurons stained with an anti-βIII tubulin antibody. Neurite length is measured from digital images of stained neurites using an image analysis system

diaminobenzidine tetrahydrochloride and 0.01% hydrogen peroxide in 50 mM Tris–HCl buffer (pH 7.4) at 37 °C for 10 min (Fig. 8).

4. The average neurite length (in μm) is calculated from the measurements of 300–500 neurites from at least three different cultures of each experimental group using digital images, followed by normalization relative to the control group (without reagents).

3.4.3 Patch-Clamp Recording, Calcium Flux Imaging, Etc.

Because of their large cell bodies and the presence of abundant ion channels on their cell membranes, cultured DRG neurons are suitable for electrophysiological analyses, such as patch-clamp recording and calcium flux imaging. For details, refer to the individual publications [11, 12].

3.5 Morphological Analyses

3.5.1 Immunofluorescence

1. DRG neurons are fixed with 4% paraformaldehyde for 10 min at 4 °C, then treated with 0.1% Triton X-100 in phosphate-buffered saline (PBS) for 5 min at room temperature.
2. The cells are incubated overnight at 4 °C with a primary antibody (*see* **Note 13**) diluted with 20 mM PBS.
3. After rinsing with PBS, the cells are incubated for 1 h at 37 °C with an Alexa Fluor®-conjugated secondary antibody (*see* **Note 14**).
4. After processing for immunofluorescence, the samples are observed under a light microscope. A representative immunofluorescence image of DRG neurons obtained using an anti-neural cell adhesion molecule (NCAM) antibody [13] is shown in Fig. 9a.
5. To examine the co-distribution of two different antigens in the same sample, a double immunofluorescence procedure can be conducted using a mixture of the two primary antibodies raised in different species (usually rabbit and mouse). A confocal microscopic system is useful for the observation and recording of the samples processed for double immunofluorescence. A representative double immunofluorescence image of DRG neurons and non-neuronal cells (mainly Schwann cells) obtained using anti-neurofilament and anti-vimentin antibodies is shown in Fig. 9b.

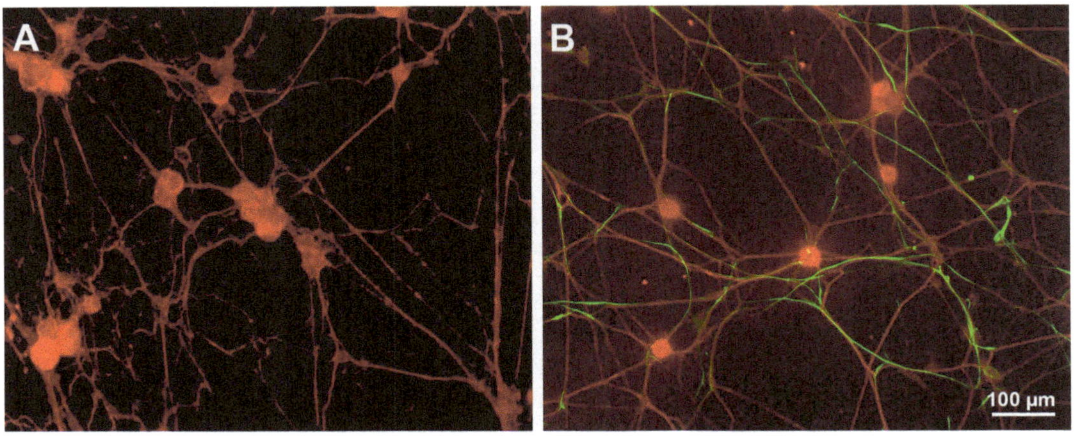

Fig. 9 Immunofluorescence micrographs of cultured DRG cells stained with (**a**) an anti-NCAM antibody and (**b**) anti-neurofilament (red) and anti-vimentin (green) antibodies. Both neuronal cell bodies and neurites are immunoreactive for NCAM and neurofilament, whereas non-neuronal cells (mainly Schwann cells) are immunoreactive for vimentin

3.5.2 Transmission Electron Microscopy

1. DRG neurons in Aclar cover slips (*see* **Note 15**) are fixed in 3% glutaraldehyde in 0.1 M phosphate buffer (PB) at room temperature for 20 min.
2. After washing in 0.1 M PB, the cells are postfixed in 1% buffered osmium tetroxide at room temperature for 30 min, rinsed, dehydrated through an ethanol gradient, and embedded in epoxy resin.
3. Sections (1 μm thick) are examined using a light microscope, and areas appropriate for further ultrathin sectioning are selected.
4. Ultrathin sections are double stained with uranyl acetate and lead citrate and examined using an electron microscope [6].

3.6 Molecular and Biochemical Analyses

3.6.1 Gene Expression Analyses

Real-time RT-PCR analysis is applicable to cultured adult rat DRG neurons [14]; however, care must be taken regarding the interpretation of the data, because a small number of Schwann cells and other non-neuronal cells are present in the culture. The number of DRG neurons obtained by primary culture is insufficient for analyses that require an adequate amount of DNA and RNA, such as microarray analyses and other omics approaches. For those analyses, established DRG neuron cell lines (e.g., 50B11 and ND7/23) [15] can serve as a substitute.

3.6.2 Western Blotting

1. Western blot analysis is applicable to samples containing a large number of DRG neurons ($>5 \times 10^3/\text{cm}^2$) isolated from 80–90 ganglia from 2 to 3 rats (Fig. 10).
2. Protein is extracted from the cells using 1× sodium dodecyl sulfate (SDS) sample buffer. The extracts are resolved using SDS polyacrylamide gel electrophoresis (SDS–PAGE) using 5–20% SDS–PAGE gels and are transferred onto polyvinylidene fluoride membranes using an electroblotting apparatus.

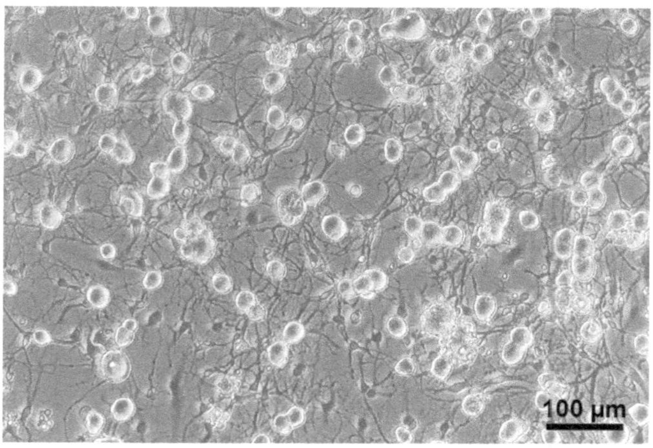

Fig. 10 Phase-contrast micrograph of a neuron-enriched culture used for Western blot analysis

3. The membrane is incubated in PBS with 0.1% Tween 20 (including 5% skimmed milk or 3% bovine serum albumin) for 1 h at room temperature, and then overnight at 4 °C with a primary antibody.
4. After rinsing in PBS, the membrane is incubated with a horseradish-peroxidase-conjugated secondary antibody for 1 h.
5. After rinsing in PBS, immunocomplexes on the membrane are visualized using enhanced chemiluminescence. The signal intensity is quantified using a chemiluminescence imaging system, and the relative intensity of each protein is expressed as its intensity divided by the intensity of β-actin [16].

3.6.3 Gene Manipulation Techniques

Although the knockdown experiments on adult DRG neurons using gene transfer or RNA interference are more difficult than those on immature neurons and lined neurons, successful cases have been reported. For details, refer to the individual publications [17, 18].

4 Notes

1. Poly-L-lysine (10 μg/mL) is used to coat the central area of the dishes onto which the cells are seeded. The dishes are kept on a clean bench for 30 min, then poly-L-lysine is removed from the dishes using a Pasteur pipette. The dishes are dried on the clean bench for at least 30 min prior to use (no need to wash with sterile water etc.).
2. To minimize the damage to neurons, do not directly hold the ganglia with tweezers; rather, lightly pinch the nerve fiber bundles and lift them up.
3. This treatment may be omitted but promotes a good yield and high purity of neurons.
4. In the culture of adult mouse DRG neurons, a shorter incubation time (60–90 min) is recommended.
5. The whole isolation procedure results in a yield of more than 5×10^4 neurons per rat, accompanied by a smaller number of non-neuronal cells.
6. Under a phase-contrast microscope, an isolated DRG neuron is identified as having a round and bright cell body with a diameter of 20–40 μm, which is easily distinguishable from Schwann cells and other non-neuronal cells.
7. Because the number of DRG neurons obtained from the primary culture is much smaller than that obtained from established cell lines, it is difficult to estimate the cell density using a

hemocytometer. In our laboratory, 100 μL of cell suspension is dropped onto an Aclar cover slip to check and adjust the cell density.

8. Nerve growth factor promotes neurite outgrowth but is not essential for the survival of mature DRG neurons.
9. The number of cells within this range is suitable for cell counting.
10. The proliferation of non-neuronal cells (particularly fibroblasts) can be suppressed under a serum-free culture environment.
11. The use of a removable silicon chamber with eight wells on a standard glass or plastic slide is a time-saving step for immunostaining and observation under a light microscope.
12. To evaluate the neurite-outgrowth-promoting activity of reagents, the assay should be conducted at about 48 h after the seeding of the neurons. Longer incubation may cause excess neurite outgrowth in a standard culture and make it difficult to conduct the study.
13. It is desirable to use antibodies for which specificity has been confirmed by knockout validation.
14. For nuclear staining, the cells are incubated with 4′,6-diamidino-2-phenylindole (DAPI; 300 nM) for 5 min at room temperature.
15. Aclar cover slips are a time-saving tool for the whole process of electron microscopy.

Acknowledgments

We thank Dr. Hidenori Horie for helpful advice on culture techniques as my mentor, and the late Kyoko Ajiki for technical support with light and electron microscopic analyses.

References

1. Horie H, Ikuta S, Takenaka T (1990) Membrane elasticity of mouse dorsal root ganglion neurons decreases with aging. FEBS Lett 269: 23–25
2. Dedov VN, Armati PJ, Roufogalis BD (2000) Three-dimensional organisation of mitochondrial clusters in regenerating dorsal root ganglion (DRG) neurons from neonatal rats: evidence for mobile mitochondrial pools. J Peripher Nerv Syst 5:3–10
3. Sango K, Mizukami H, Horie H et al (2017) Impaired axonal regeneration in diabetes. Perspective on the underlying mechanism from in vivo and in vitro experimental studies. Front Endocrinol (Lausanne) 8:12
4. Melli G, Höke A (2009) Dorsal root ganglia sensory neuronal cultures: a tool for drug discovery for peripheral neuropathies. Expert Opin Drug Discov 4:1035–1045
5. Varier P, Raju G, Madhusudanan P et al (2022) A brief review of in vitro models for injury and regeneration in the peripheral nervous system. Int J Mol Sci 23:816

6. Sango K, Yamanaka S, Ajiki K et al (2002) Lysosomal storage results in impaired survival but normal neurite outgrowth in dorsal root ganglion neurones from a mouse model of Sandhoff disease. Neuropathol Appl Neurobiol 28:23–34
7. Christie K, Koshy D, Cheng C et al (2015) Intraganglionic interactions between satellite cells and adult sensory neurons. Mol Cell Neurosci 67:1–12
8. Kilkenny C, Browne WJ, Cuthill IC et al (2010) Improving bioscience research reporting: the ARRIVE guidelines for reporting animal research. PLoS Biol 8:e1000412
9. Goldenberg SS, De Boni U (1983) Pure population of viable neurons from rabbit dorsal root ganglia, using gradients of Percoll. J Neurobiol 14:195–206
10. Sango K, Yanagisawa H, Komuta Y et al (2008) Neuroprotective properties of ciliary neurotrophic factor for cultured adult rat dorsal root ganglion neurons. Histochem Cell Biol 130:669–679
11. Hong S, Morrow TJ, Paulson P et al (2004) Early painful diabetic neuropathy is associated with differential changes in tetrodotoxin-sensitive and -resistant sodium channels in dorsal root ganglion neurons in the rat. J Biol Chem 279:29341–29350
12. Ibi M, Matsuno K, Shiba D et al (2008) Reactive oxygen species derived from NOX1/NADPH oxidase enhance inflammatory pain. J Neurosci 28:9486–9494
13. Ono K, Asou H, Yamada M et al (1992) Gradient expression of neural cell adhesion molecule (NCAM) in the pontine migratory stream of fetal rats. Neurosci Res 15:221–223
14. Dodge ME, Wang J, Guy C et al (2006) Stress-induced heat shock protein 27 expression and its role in dorsal root ganglion neuronal survival. Brain Res 1068:34–48
15. Haberberger RV, Barry C, Matusica D (2020) Immortalized dorsal root ganglion neuron cell lines. Front Cell Neurosci 14:184
16. Takaku S, Yanagisawa H, Watabe K et al (2013) GDNF promotes neurite outgrowth and upregulates galectin-1 through the RET/PI3K signaling in cultured adult rat dorsal root ganglion neurons. Neurochem Int 62:330–339
17. Miao T, Wu D, Zhang Y et al (2006) Suppressor of cytokine signaling-3 suppresses the ability of activated signal transducer and activator of transcription-3 to stimulate neurite growth in rat primary sensory neurons. J Neurosci 26:9512–9519
18. Sabbir MG, Fernyhough P (2018) Muscarinic receptor antagonists activate ERK-CREB signaling to augment neurite outgrowth of adult sensory neurons. Neuropharmacology 143:268–281

Chapter 21

A Live Imaging Assay for Regenerating Peripheral Neurons

Alice E. Mortimer, Raman M. Das, and Adam J. Reid

Abstract

The cell intrinsic mechanisms directing peripheral nerve regeneration have remained largely understudied, thus limiting our understanding of these processes and constraining the advancement of novel clinical therapeutics. The use of primary adult rat dorsal root ganglion (DRG) neurons cultured in vitro is well established. Despite this, these cells can be challenging to culture and have so far not been amenable to robust transfection or live-cell imaging. The ability to transfect these cells with fluorescent plasmid constructs to label subcellular structures, combined with high resolution time-lapse imaging has the potential to provide invaluable insight into how peripheral neurons coordinate their regenerative response, and which specific cellular structures are involved in this process. Here we describe a protocol that facilitates transfection and subsequent live-imaging of adult rat DRG neurons.

Key words Peripheral nerve regeneration, Transfection, Live imaging, Fluorescence microscopy, Nucleofection, Dorsal root ganglia

1 Introduction

Research into peripheral nerve regeneration has focused largely on factors external to the neuron while the intrinsic factors that control regeneration remain largely unknown [1]. Peripheral neurons possess the innate capacity to initiate regeneration in response to injury, unlike neurons of the central nervous system, yet the subcellular control of this phenomenon is poorly understood. Moreover, despite their ability to initiate a regenerative response, peripheral neurons are unable to sustain this response over prolonged periods, thereby limiting meaningful clinical recovery [1]. At present, treatment for peripheral nerve injuries relies on the surgical repair of damaged nerve ends in an effort to direct regeneration of the proximal nerve stump toward the target organ [2]. However, this treatment does not take into account the cell intrinsic mechanisms that are driving the regenerative process, and management has remained largely unchanged for over five decades.

Kazuhito Toyooka (ed.), *Neuronal Morphogenesis: Methods and Protocols*, Methods in Molecular Biology, vol. 2831,
https://doi.org/10.1007/978-1-0716-3969-6_21,

To obtain clinically relevant information on these processes, it is important to recapitulate the human response as accurately as possible. Therefore the use of an adult, mammalian system is of key importance. This is particularly relevant when considering that adult neurons display striking differences to their developmental counterparts, including having fewer than half of the same genes activated, and in deploying distinct mechanisms of cytoskeletal remodeling [3, 4]. The use of dissociated rodent dorsal root ganglion (DRG) neurons is an established system through which peripheral nerve regeneration can be studied [5]. However, these cells are challenging to culture and have thus far not been amenable to robust, long-term live-cell imaging, particularly when combined with the transfection of fluorescent plasmid constructs that facilitate visualization of specific subcellular structures. The ability to obtain precise spatial and temporal information on how neurons coordinate regeneration is likely to be transformative both for overall scientific understanding of peripheral nerve regeneration, and also for the potential of high throughput testing and validation of novel therapeutics prior to in vivo translation. Furthermore, a sound understanding of the cellular mechanisms that drive peripheral neurons to regenerate could shed light on why central neurons have lower propensity for regeneration.

Here we provide a protocol for the transfection and subsequent live imaging of adult primary DRG neurons in culture to study peripheral nerve regeneration. This protocol facilitates transfection of multiple expression constructs to study interactions between various subcellular structures to be studied in real time over long periods.

2 Materials

- Class II biological safety cabinet.
- 70% industrial methylated spirit (IMS) in spray bottle.
- Virkon disinfectant.
- P2, 200, and 1000 micropipettes.
- Sterile 10 μL, 200 μL, and 1000 μL pipette tips.
- Sterile 1.5 mL Eppendorf tubes.
- 35 mm glass bottom FluoroDishes, pre-coated with poly-D-lysine [World Precision Instruments (*see* **Note 1**)].
- Laminin from mouse Engelbreth-Holm-Swarm (EHS) sarcoma (Sigma-Aldrich).
- Dissociated adult rat primary DRG neurons (we use rats between 6 and 12 weeks of age). Dissociation and culture of DRG is covered elsewhere in this text but we use a protocol which can be found in Faroni et al. [5].

- Ham's F12 medium: a 500 mL bottle of F12 medium is supplemented with 1% penicillin streptomycin solution and 0.5% L-glutamine. Medium is stored at 4 °C and can be used for multiple experiments (*see* **Note 2**).
- N2 supplement (Gibco): N2 supplement should be aliquoted into 200 μL and stored at −20 °C.
- Lonza rat neuron nucleofector kit (Lonza): Nucleofector solution is supplemented with supplement one prior to use and stored at 4 °C.
- Lonza 2b nucleofector device [Lonza (*see* **Note 3**)].
- Water bath set to 37 °C.
- Incubator set to 37 °C and 5% CO_2.
- Brightfield phase contrast microscope.
- Widefield microscope system with environmentally controlled chamber with both temperature and CO_2 control (*see* **Note 4**).
- Lens cleaning tissues.
- Plasmids of choice. Supplied Lonza pmaxGFP® vector included with nucleofector kit may be used as a positive control.

3 Methods

3.1 Preparation of Cell Culture Dishes

For optimal imaging we use WorldPrecision instruments 35 mm glass bottom FluoroDishes, coated with Poly-D-lysine (*see* **Note 1**). Preparation of culture dishes should be carried out under sterile conditions in a class II biological safety cabinet. To prepare dishes for cell plating:

1. Place 35 mm FluoroDishes into 70 mm plastic Petri dishes with lens tissue covering the bottom of each Petri dish. This is to avoid damage to the glass bottom.
2. Dilute laminin solution 1:100 in F12 and apply 1 mL to each dish.
3. Incubate dishes at 37 °C for 2–24 h, after which the laminin solution is removed and replaced with 1 mL F12 as a wash (*see* **Note 5**).
4. Remove F12 medium. Dishes are now ready for cell plating.

3.2 Nucleofection of DRG Neurons

All cell culture steps should be carried out under sterile conditions in a class II biological safety cabinet. Care should be taken to spray all items entering the cabinet with 70% IMS to avoid contamination. Lids of cell culture dishes should be not removed when outside the cabinet.

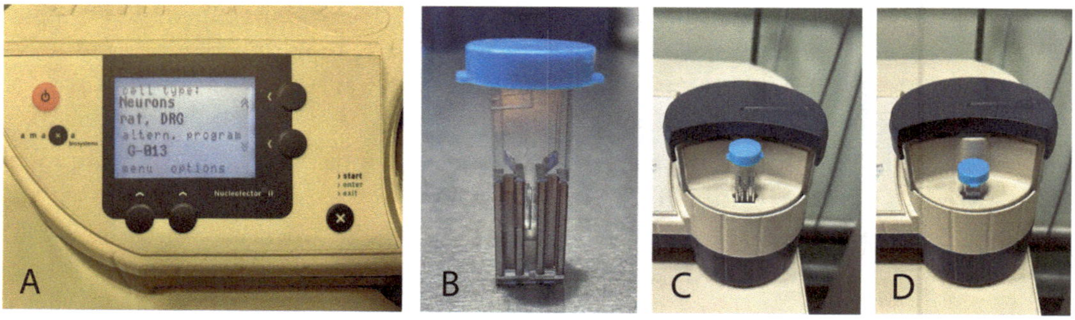

Fig. 1 Setup of Lonza 2b nucleofector device (**a**) correct program to select on the device (**b**) correct placement of cell suspension within the cuvette, (**c**) *incorrect* placement of cuvette in nucleofector, (**d**) *correct* placement of cuvette in nucleofector

1. After dissociation, cells should be counted. One million cells are required for each transfection. In our experience this equates to the harvest of all DRG from one adult rat (*see* **Note 6**).
2. After counting, centrifuge cells for 5 min at 91 × *g* then resuspend in 100 μL rat neuron nucleofector solution (*see* **Note 7**).
3. During centrifugation, prepare N2 supplemented medium. N2 should be added to pre-warmed F12 at a concentration of 1: 100 (*see* **Note 8**).
4. To this, add plasmids of interest to a total concentration of 1–2 μg/100 μL. Pipette up and down gently to mix. When using two separate constructs, we use 1 μg of each.
5. Turn on the Lonza 2b nucleofector device and select program "rat DRG" alternative program G-013 (Fig. 1a).
6. Transfer the cell/plasmid suspension to a Lonza nucleofector 2b cuvette, ensuring that the solution reaches the bottom of the cuvette (Fig. 1b). Pipette carefully to avoid introducing bubbles. Insert the cuvette into the sample carrier (*see* **Note 9**).
7. Press "X" to perform nucleofection. The sample carrier will be spun into, and then back out of, the device and a message reading "OK" should appear on the screen.
8. Remove the cuvette from the sample carrier. An indication that successful nucleofection has taken place will be the emergence of small bubbles at the top of the solution.
9. Carefully add 500 μL warmed F12 medium to the cuvette.
10. Using the supplied Lonza plastic Pasteur pipette, transfer the suspension to a 1.5 mL Eppendorf. Perform this step slowly and carefully, avoid re-pipetting the cells or shaking either the cuvette or Eppendorf.

11. Place cell suspension in a 37 °C 5% CO_2 incubator for 10 min to enhance cell recovery from the transfection.
12. In the meantime, warmed supplemented medium should be added to the pre-coated dishes ready for cell plating. For optimal cell adherence a total volume of 500 μL should be added to each dish in the first instance. Therefore, calculate the volume of cell suspension that will be required to add to each dish. For example, if using six FluoroDishes: 100 μL of cell suspension (given cells are suspended in total 600 μL of nucleofector solution plus F12) should be added to 400 μL of medium supplemented with N2 in each dish. In this example, 400 μL of F12 medium supplemented with N2 would be added to each dish. Dishes are incubated at 37 °C during cell recovery (*see* **Note 10**).
13. Following cell recovery, plate cells onto pre-prepared dishes. During this step very gentle mixing with a pipette is required to redistribute the cells in solution. After this, the above calculated volume should be added to each dish with minimal re-pipetting (*see* **Note 11**).
14. Incubate cells at 37 °C 5% CO_2 for 2–4 h to allow cells to adhere. Cell adherence can be checked using a standard bright-field microscope (*see* **Note 12**).
15. Medium should then be changed very carefully. This removes the nucleofector solution, which is toxic to cells over prolonged periods. Carefully pipette out medium and replace with 2 mL warmed N2 supplemented F12 medium (*see* **Note 8**). Incubate cells overnight at 37 °C 5% CO_2 (*see* **Note 13**).
16. The following day, and every 2 days thereafter, half of the N2 supplemented F12 medium should be removed and replaced with fresh pre-warmed medium (*see* **Note 14**).

3.3 Live-Imaging of DRG Neurons

Depending on plasmid size, transfection is evident between 24 and 36 h following nucleofection (*see* **Note 15**).

1. Perform a medium change and incubate at 37 °C 5% CO_2 in an incubator within close proximity to your microscope prior to carrying out imaging. This is to avoid unnecessary disturbance of the cells immediately prior to imaging.
2. In the meantime, ensure that the microscope incubation is turned on and set to 37 °C. Ensure the incubation chamber is fully closed to avoid heat escape. Pre-warming of the microscope incubation chamber should be carried out at least 30 min prior to imaging and should be confirmed to have reached 37 °C prior to imaging. The carrier stage should additionally be placed inside the incubation chamber to pre-warm.

3. When ready to image, move dishes into the incubation chamber to avoid cell cooling.
4. Apply a generous amount of immersion medium (approximately three drops) to the objective and mount the FluoroDish onto the stage (*see* **Notes 4** and **16**).
5. Gradually move the objective toward the dish and observe the immersion medium meet the glass interface (*see* **Note 17**). Cells can then be located in brightfield using the microscope eyepiece. Ensure the CO_2 chamber cover is mounted onto the stage.
6. Once cells are in focus, specific cells can be located for imaging using the microscope camera and imaging software (we use Zen Blue, Zeiss).
7. To avoid photobleaching it is best to locate cells using brightfield, taking snapshots only using fluorescence to identify positively transfected cells (*see* **Note 18**).
8. Record positions and the Z-stack heights for each identified cell (*see* **Note 19**).
9. Time series setup is then carried out, selecting the interval between image acquisitions and the overall length of the time-lapse depending on your experimental setup (*see* **Note 20**).
10. LEDs should be set to "gated" to minimize exposure and improve cell viability. Whole Z stacks of each channel should be acquired for each image prior to imaging the next channel to improve acquisition speed.
11. The experiment can then be started (*see* **Note 21**).
12. Upon completion of the experiment, the objective should be lowered and the stage removed from the microscope. Immersion medium should be very carefully removed from the objective using clean lens tissue.
13. Imaged dishes may then be re-incubated at 37 °C 5% CO_2 if required for further experiments or safely disposed of by adding Virkon disinfectant to the dish to kill the cells. After 24 h dishes can then be safely disposed of according to local guidelines.

3.4 Post-image Acquisition Processing

1. Post-image acquisition processing should be carried out using a deconvolution strategy to improve optical resolution of your time-lapses (*see* **Note 22**).
2. Orthogonal projections can then be created from deconvolved time-lapses by selecting in-focus Z planes of interest (Fig. 2).

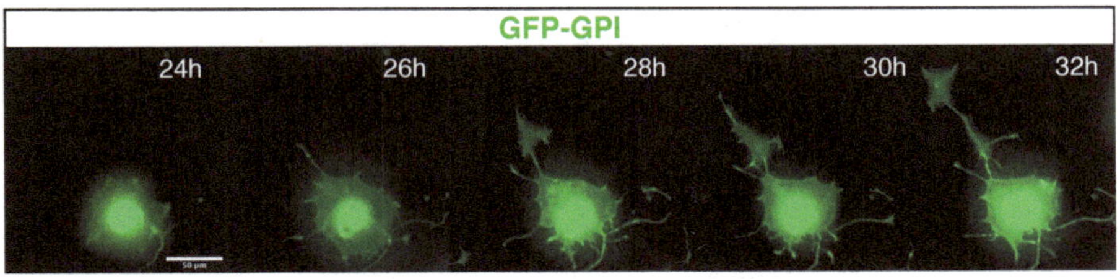

Fig. 2 Example time-series of a neuron transfected with GFP-GPI to label the cell membrane (green). This cell was imaged every 20 min for 16 h at 24 h post-transfection. Initiation of axon extension can be visualized during this time frame

4 Notes

1. Depending on your specific microscope system and experimental setup, alternative dishes may be used. However, all dishes should be glass bottom and optimized for cell imaging. If your dishes are not pre-coated with poly-D-lysine, this will require an additional step prior to laminin application. Apply 0.1 mg/mL poly-D-lysine to dishes and incubate at room temperature for 20 min. Remove Poly-D-lysine and replace with F12 as a wash. Remove F12 and proceed with subsequent laminin coating (Subheading 3.1).
2. L-glutamine is added as standard from certain suppliers of Ham's F12. If your medium already contains L-glutamine, there is no need to add any further.
3. We use the Nucleofector 2b device, however there are updated versions of nucleofector technology. Using a different device should not affect this protocol, but we recommend contacting the Lonza support team for any troubleshooting with a different device.
4. We use an Axio Observer Z.1 microscope (Zeiss) equipped with a Colibri 7 LED light source (Zeiss), Flash4 v2 sCMOS camera (Hamamatsu), and an environmentally controlled chamber. We use a 40× 1.2 NA silicone immersion objective (Zeiss) or 63× 1.4 NA oil immersion objective (Zeiss) for optimal resolution.
5. We have found that 24 h of laminin coating allows for improved cell adherence. Optimally, this should therefore be carried out a day prior to cell harvest and transfection.
6. We use an Invitrogen Countess 3 automated cell counter according to manufacturer's instructions. However, alternative methods of cell counting may be used. With optimal DRG harvest and dissociation, we have found it is no longer necessary to count cells on every occasion, since the DRG from one whole rat equates to in the region of one million cells. A further centrifugation step is also necessary after cell counting, which may affect cell viability.

7. Avoid leaving cells in nucleofector for prolonged periods as this will be toxic to cells. Nucleofector solution should also be removed from the fridge approximately 30 min before use to reach room temperature.
8. F12 supplemented with N2 should be prepared fresh for each use. Calculate the volume required for each use. After initial cell plating, we use 2 mL of N2 supplemented F12 per FluoroDish.
9. Ensure the cuvette is inserted into the device such that it does not stand too proud of the carrier (the cuvette will physically fit in two different orientations but will stand too proud of the carrier in one orientation and nucleofection will not be successful. Rotate the cuvette 90° to identify the most secure fit) (Fig. 1c, d).
10. We have found that the cells from one nucleofection will have optimal cell density for live imaging when plated onto six to eight FluoroDishes. Optimal cell density and volumes would need to be adjusted if using alternatively sized dishes.
11. Be very careful when pipetting cells at this stage. Pipette carefully up and down to redistribute cells in solution then avoid re-pipetting cells too frequently when plating. Be very careful not to shake the tubes or dishes containing cells, particularly immediately post-transfection as this is likely to cause cell death.
12. A standard benchtop microscope can be used to monitor cells prior to live imaging. Cells should be checked using brightfield phase contrast imaging after cell plating at 2–4 h to ensure an adequate number of cells have adhered to the dish. It is normal to have some non-viable cells and cell aggregates, these will be removed during the medium change. Avoid leaving cells to adhere for greater than 4 h, since the remaining nucleofector solution may induce toxicity.
13. It can be helpful to recheck cells using brightfield microscopy following the initial medium change, to ensure that cells remain adhered. However, be very careful not to shake/unnecessarily disturb culture dishes while doing so.
14. Only half of the medium is removed and replaced so as to not disturb cells in culture, which could cause cell detachment and invalidate downstream experiments.
15. Larger plasmids will take longer for transfection to become evident. This should be factored into specific experimental design. However, if no fluorescence is visualized within 36 h, it is likely that the transfection has been unsuccessful.

16. The dish should fit securely onto the stage to ensure cells remain in focus throughout image acquisition. It is sometimes necessary to use Blu Tack around the edges of the dish to mount it securely onto the stage. When mounting and removing the dish, make sure to avoid Blu Tack coming into contact with the objective.
17. When moving the objective in Z, do so carefully so as not to damage the dish glass or the objective. With brightfield switched on, you should witness a sudden "spreading" of the immersion medium onto the glass. Cells should then be visible with minimal further adjustment.
18. Snapshots should be used to set LED intensity and exposure time. To limit photobleaching and cell death these should be set as low as possible while still maintaining adequate optical resolution. We typically use a 20 ms exposure time and 30–50% LED intensity.
19. Avoid imaging too many cells in any one dish as this can lead to photobleaching, cell toxicity, and death. We usually image a maximum of 12 cells per dish. Z-stack height and interval will depend upon various factors including the desired speed of image acquisition, duration of acquisition, and the specific cellular structures being examined. For time-lapses of overall cell dynamics lasting hours to days, we typically use 30 Z stacks at 0.5 μm intervals. For shorter, faster image acquisition lasting seconds to minutes we typically focus on the subcellular locus of interest using fewer Z planes in order to improve the speed of acquisition. Once X/Y and Z positions have been saved for each individual cell, it is beneficial to return to these positions prior to starting the experiment to ensure cells have remained in focus.
20. When imaging cells for over 6 h we typically perform image acquisition every 20 min to improve cell recovery after each image is taken. For some experimental applications it is necessary to perform quicker imaging, but in these cases the overall time-lapse duration should be reduced to balance speed of acquisition with photobleaching and cell death.
21. We recommend monitoring image acquisition for the first two to three frames in the time series to ensure cells have remained in focus and the set-up parameters are correct.
22. We use ZenBlue (Zeiss) processing software and have found that a fast iterative algorithm (Richardson Lucy) set to 40 iterations and 0.1% quality threshold is optimal in our experiments, but we suggest testing various parameters and optimizing for your specific experimental setup.

Acknowledgments

This work was supported by an MRC transition support fellowship ((MR/V036386/1) to RMD and an MRC Clinical Research Training Fellowship (MR/T028785/1) to AEM.

References

1. Faroni A, Mobasseri SA, Kingham PJ, Reid AJ (2015) Peripheral nerve regeneration: experimental strategies and future perspectives. Adv Drug Deliv Rev 82:160–167
2. Midha R, Grochmal J (2019) Surgery for nerve injury: current and future perspectives. J Neurosurg 130:675–685
3. Bosse F, Hasenpusch-Theil K, Küry P, Müller HW (2006) Gene expression profiling reveals that peripheral nerve regeneration is a consequence of both novel injury-dependent and reactivated developmental processes. J Neurochem 96:1441–1457
4. Stiess M et al (2010) Axon extension occurs independently of centrosomal microtubule nucleation. Science (80-) 327:704–707
5. De Luca AC, Faroni A, Reid AJ (2015) Dorsal root ganglia neurons and differentiated adipose-derived stem cells: an in vitro co-culture model to study peripheral nerve regeneration. J Vis Exp 96:52543

Chapter 22

Nerve Regeneration Using Compartmentalized System in CIPN Model

Prashant Adhikari, Khayzaran Qubbaj, and In Hong Yang

Abstract

The analysis of nerve regeneration in the chemotherapy-induced peripheral neuropathy (CIPN) model can be achieved using the compartmentalized culture system. This system enables us to isolate the cell body from the axon physically and fluidically, therefore allowing for the independent manipulation of the cell body and axons. Compartmentalized chambers mimic the human body conditions, and can be used to study axonal degeneration, disease modeling, and drug screening. This culture system is applied to the CIPN model to study and analyze axonal behavior in response to paclitaxel (PTX) with and without fluocinolone acetonide (FA) and to better understand the site-specific target of PTX. Therefore, this compartmentalized system allows for the independent treatment of chemotherapy drugs to the cell body or axonal side which enables monitoring their reaction as a result of the treatment.

Key words Photolithography, Microfluidics, Compartmentalized chambers

1 Introduction

Peripheral nerves, unlike the central nervous system, have the ability to regenerate the lost or degenerated axons from injury or chemotherapy. Various methods such as electric stimulation, stem-cell-based therapies, and optical stimulation are advancing for the treatment of nerve regeneration. The use of compartmentalized culture systems can be integrated with such research for axonal guidance as well as separation of the cell body from the axon via the chambers that are on either side of the microchannel [2]. This methodology can be a useful tool for an in vitro model of axonal regeneration and drug testing.

The compartmentalized system uses polydimethyl siloxane (PDMS) for cell culture with an easy fabrication process. The microchannels are designed and optimized based on specific experimental outcomes using photolithography [1]. The mold of PDMS is implanted on a silicon wafer containing microchannels, and the

Kazuhito Toyooka (ed.), *Neuronal Morphogenesis: Methods and Protocols*, Methods in Molecular Biology, vol. 2831, https://doi.org/10.1007/978-1-0716-3969-6_22,

holes for 2D cell culture are created using a biopsy punch. The cells are extracted from the dorsal root ganglion (DRG), and cultured using different neurotrophic factors, and extracellular matrix proteins. They are seeded on the soma side thus allowing axons to pass through the microchannels.

The compartmentalized chamber system is used in this case for the analysis of axonal regeneration in the CIPN model using paclitaxel (PTX) and fluocinolone acetonide (FA) [4]. This culture system enables us to isolate the cell body and axon and is important for the analysis of the CIPN model because different chemotherapy drugs will target one of the following such as the microtubules, mitotic spindle, and DNA. Therefore, PTX with and without FA was treated on either side of the microchannels, targeting either the cell body or axon or both to understand their site-specific target as well as the neuroprotective function of FA [3]. Imaging is done using a fluorescent microscope and the length of axons is measured using Image J software. The dosage response curve is then generated for the chemotherapeutic drugs used to understand the degradation percentage as well as the regeneration percentage as a result of FA.

2 Materials

2.1 Microfabricated Chamber Preparation

1. Master Mold: Cover silicon wafer with SU-82002 resist (Micro Chem Product Item #: Y131240), spin, and soft bake according to manufacturer's specifications. Microchannels are defined through UV light exposure using a high-resolution DPI transparency. Bake the exposed substrate and develop it according to the resist technical sheet. Repeat the process with SU-83050 (Micro Chem Product Item #: Y311075) resist and treat the master mold with trichlorosilane for 30 min.
   (i) Compartmentalized Chamber.
      - Polydimethylsiloxane (PDMS) and SYLGARD 184 Silicone elastomer (Dow Corning, Michigan) curing agent in a 10:1 ratio.
      - Spatula for mixing.
      - Desiccator.
      - Silicon mold placed in a hot oven at 85 °C for 3 h.
      - Dermal biopsy punch of 6 mm length and 4 mm width with shape modification.
2. *Plasma Bonding and Cell Preparation*
   (i) Plasma Bonding:
      - Glass slides of 0.15 mm thickness.
      - Ethanol and distilled water for cleaning.

- For plasma bonding: Pressure of 20 kPa, oxygen flow of 50 standard cubic centimeters per minute (sccm), and 70 W of oxygen-based plasma treatment for 250 sec.
- Autoclave bag (19 × 33 cm).

3. *Coating Plate*
   (i) Poly-D-Lysine solution (100 μg/mL in distilled water).
   (ii) Laminin (5 μg/mL in phosphate-buffered saline (PBS) solution).
4. *Complete Medium*
   - 48 mL Neurobasal medium (Gibco 21103-049, w/o L-gut).
   - 200 μL of glucose.
   - 0.5 mL of Glutomax (Gibco 35050-061).
   - 0.5 mL of heat inactivated fetal bovine serum (HI-FBS).
   - 200 μL of nerve growth factor (NGF).
   - 1 mL of B-27 Supplement (Gibco 17504-044).
   - 0.5 mL of Penicillin/Streptomycin.
   - 11 μL from stock: fluorodeoxyuridine (FdUR).
5. *DRG Isolation*
   - Collagenase (10 mg/mL) stored below −20 °C.
   - Trypsin (0.25) stored below −20 °C.
   - L-15 + Penicillin/Streptomycin (P/S) media stored at 2–8 °C.
   - L-15 + P/S + FBS media stored at 2–8 °C.
   - 25 mg/mL DNase stored below −20 °C.

3 Methods

1. *Silicon Wafer with Microchannels*
   - Create a master mold using a two-step photolithographic process.
   - Coat silicon wafer with SU-82002 resist, and spin, soft-bake using manufacturer's specifications to yield a thickness of 2.5 μm.
   - Expose the wafer with UV light through high-resolution dots per inch (DPI) transparency to define arrays of microchannels.
   - Bake it again to enhance polymer cross-linking.

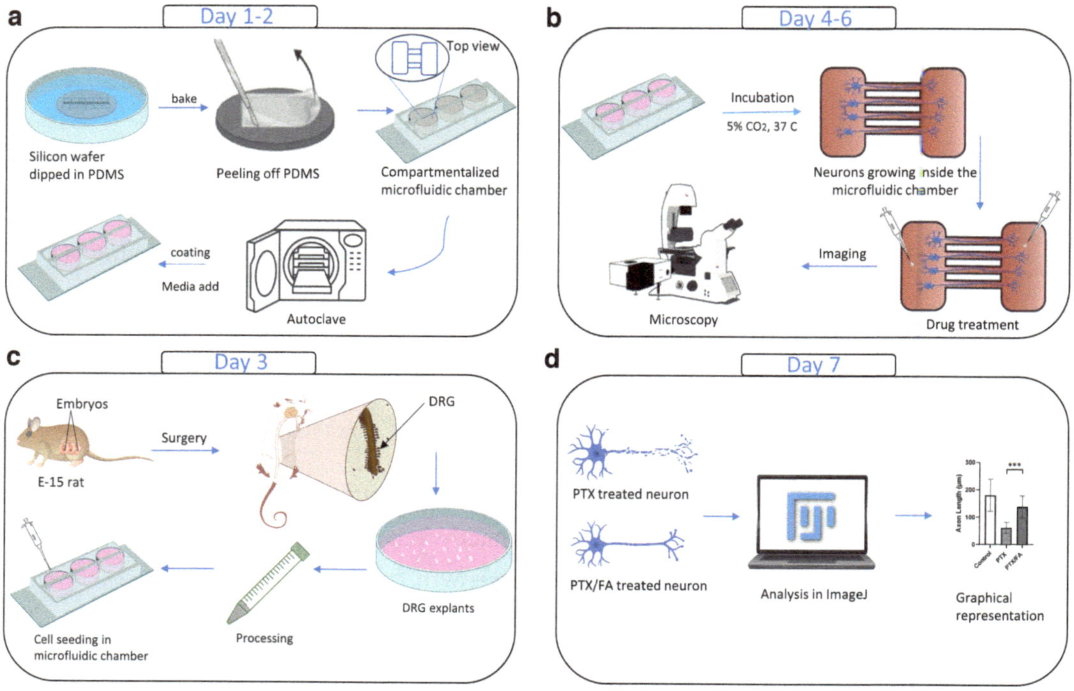

Fig. 1 Schematic of CIPN model showing the isolation, culture, and analysis of neurons. Preparation of compartmentalized microfluidic device for dorsal root ganglion (DRG) culture (**a**). E-15 Sprague Dawley rat embryonic DRG and seeding in prepared microfluidic chamber system (**b**). Cells are left to grow from day 4–6 until the axons pass through microchannels, where chemotherapy drugs are administered (**c**). Imaging of axons in the microchannels and analysis using software Image J (**d**)

- Repeat the same process with SU-83050 resist and treat the master mold with trichlorosilane for 30 min.

2. *Compartmentalized Cell Culture System*
   - Mix PDMS and curing agent thoroughly in a 10:1 ratio.
   - Remove the air bubbles using a desiccator (*see* **Note 1**).
   - Pour the mixture carefully into the silicon mold (Fig. 1a) and place it in a hot oven at 85 °C for 3 h to cure.
   - Remove all the extra mold with a blade and use a biopsy puncher to create wells of 0.6 cm × 0.6 cm on either side of the microchannels.

3. *Plasma Bonding and Cell Preparation*
   - Plasma-bond the microchannels to a glass coverslip (0.15 mm thickness) (*see* **Note 2**).
   - Autoclave the plasma-bonded slides and chambers by placing them in an autoclave bag and placing them in the autoclave machine for an hour using the "dry cycle" (Fig. 1a).

- For coating and cleaning, perform the experiment inside the biosafety hood.
- Place a drop of ethanol on one side of the chamber and then on the other side and wash the chambers three times with water and three times with PBS.
- Coat the wells with 100 μg/mL Poly-D-Lysine (PDL) and 10 μg/mL laminin and leave them in the refrigerator overnight.
- For cleaning the chambers, wash them using autoclaved distilled water and PBS three times each.
- Fill the wells with a complete media mixture and incubate them.

4. *DRG Isolation*
   1. Sterilize all dissection instruments by immersing them in 70% ethanol for 30 min or autoclaving them.
   2. Put ice-cold 10 mL L-15 with 1% P/S in two 100 mm petri plates and 3 mL in six polystyrene plates and put on ice.
   3. Euthanize 15th-day pregnant Sprague Dawley rat while in the cage via carbon dioxide (CO_2) at a flow rate of 7.5 L/min for a duration of 10 min.
   4. Use isoflurane for inhalational anesthesia.
   5. Check pain response on the tail and the change of bright red coloration in the corneal and pedal region to a dull pink to determine signs of life.
   6. Sterilize the site of the incision with 70% ethanol and make a single incision on the dorsal region of the abdomen through the skin and muscle layers to expose the abdominal cavity.
   7. Remove and detach the uterus from the cervix and ovaries, and then place the uterus in the L-15 media with 1% P/S at 4 °C.
   8. Transfer one embryo to a 3 cm dish with fresh L-15.
   9. Remove the embryo from each respective amniotic sac subsequently, expose the spinal column via a single incision along the temporal section of the cervical region of the embryo, and then remove the spinal cord from the developing spinal column.
   10. Remove the dorsal root ganglion (DRG) explants via individual plucking with a tweezer and place the DRGs into another dish of fresh 1 mL L-15 media with 1% Penicillin/Streptomycin (P/S) at 4 °C with an equal volume of 0.25% trypsin by volume for 5 min at 37 °C (Fig. 1b) (*see* **Note 4**).

5. *Cell Culture*
   1. Suspend the cells in 50,000 cells/5 mL density and seed the cells using a protein loading tip placing the petri dish at an inclined position.
   2. Seed the cells near the microchannels and at one side of the compartmentalized chamber system, which is the designated soma chamber, with allowance of the axons to extend through the 10 μm wide microchannels to the chamber on the other side (*see* **Note 3**).
   3. Leave the culture cells in the incubator at 5% CO_2 at 37 °C for the first 3–5 days.
   4. Replace 50% of the media with complete media every 3 days until day 7, after which, only the same concentration of B-27 Supplement is added to the Neurobasal media (Fig. 1b) (*see* **Note 5**).
6. *Adding of Drugs Paclitaxel (PTX) on the Axonal Side*
   1. Let the cells grow for 24 h, and check whether the axons are passed through the channels using a Leica DMi-8 Fluorescent microscope at a magnification of 20× or 40 × .
   2. Identify the location where axons are fully passed through, which can be further processed with a drug.
   3. Take the image of the axons on the microscope, which is the zero-hour image (*see* **Note 7**).
   4. Take the drug out of the freezer and put it in a hot bath at a temperature of 37 °C for a few minutes.
   5. Next, treat the well with paclitaxel (PTX) of concentration 5 μM on the axonal side. Take 70 μL of complete media out from the well and add PTX of the same volume (*see* **Note 6**).
   6. Incubate the chamber at 37 °C for 24 h. Take the image of the axon at the exact location.
   7. Repeat **steps 3–6** on different wells using PTX combined with FA (5 μM/10 nM) and take the image at 0 h and 24 h (Fig. 1c).
7. *Image Processing*
   1. Use Image J software to measure the axonal length.
   2. Compare the axon length of 0 h and 24 h (Fig. 1c).
   3. Notice the change in length using PTX alone and PTX combined with FA.
   4. Using FA and PTX both suggest the neuroprotective mechanism of FA.

4 Notes

1. Removing the air bubbles using desiccators before placing the PDMS in an oven at 80 °C, will help achieve a clear and clean set of microchannels. Ensuring the removal of air bubbles can be done when mixing the elastomer base and curing agent in the weighing boats and mixing it by hand for about 15 min and then placing the mixture in the weighing boats in the desiccator until the air bubbles are removed, and then slowly pouring it over the wafer and pouring it closer to the wafer to avoid the occurrence of air bubbles. Then place it in the desiccator to get rid of any air bubbles before placing it in the oven.
2. Before plasma bonding, make sure the microchannels are facing up, so that the upper surface is plasma bonded. Use the oxygen level around 15–20. After plasma bonding is done, flip the microchannels and attach them to the glass slide within 1 min. Make sure that microchannels are properly attached to the glass slide by slowly pushing it and heating it in the oven at 90 °C for about 10 min.
3. While seeding the cells, place the petri dish in an inclined position to increase the possibility of axons passing through the channels. Use the protein loading tip which is typically long and helps in seeding the cells near the chamber. Keep the petri dish in an inclined position for about half an hour inside the incubator.
4. An important point to note while isolating DRGs from the embryo is to not pick out the glial cells or any attached muscles. It can help to isolate the intervertebral foramen, which is adjacent to the spine and slowly pick out the DRGs by using one hand to stabilize it and the other to pick the DRGs off it.
5. When adding/replacing the media, it is important to not let the pipette tip touch the bottom of the well. It is helpful to use the pipette tip on the side of the wall and allow the media to drop down, that way the tip is not touching the bottom, as well as not agitating the cells.
6. When using the complete media with live cells, it is crucial to warm the media to room temperature before adding it to the cells as that will help avoid stressing the cells. When it comes to replacing the media for the cells, 60–70 μL of complete media is added and the chambers are immediately incubated. Therefore, it is best to avoid directly imaging the cells after adding the media.
7. Live imaging might take time in order to get good results, therefore it is necessary to provide the necessary incubation

environment during the process. That can be done by providing oxygen flow and heating during live cell imaging. If the imaging takes more than 10 min, place the chambers in an incubator for about 30 min and then image. It's really crucial to maintain an optimum environment for the cells to grow and proliferate.

References

1. Chen X, Wu H, Mao C, Whitesides GM (2002) A prototype two-dimensional capillary electrophoresis system fabricated in poly (dimethylsiloxane). Anal Chem 74(8):1772–1778
2. Park JW, Vahidi B, Taylor AM, Rhee SW, Jeon NL (2006) Microfluidic culture platform for neuroscience research. Nat Protoc 1(4): 2128–2136. https://doi.org/10.1038/nprot.2006.316
3. Tiwari AP, Tristan LJC, Albin B, Yang IH (2023) Fluocinolone Acetonide enhances anterograde mitochondria trafficking and promotes neuroprotection against paclitaxel-induced peripheral neuropathy. ACS Chem Neurosci. https://doi.org/10.1021/acschemneuro.3c00218
4. Yang IH, Siddique R, Hosmane S, Thakor N, Höke A (2009) Compartmentalized microfluidic culture platform to study the mechanism of paclitaxel-induced axonal degeneration. Exp Neurol 218(1):124–128

Chapter 23

Live Imaging Analysis of Axonal Regeneration in Human iPSC-Derived Motor Neurons Using a Microfluidic System

Katherine L. Marshall and Mohamed H. Farah

Abstract

Axonal damage is a common feature of traumatic injury and neurodegenerative disease. The capacity for axons to regenerate and to recover functionality after injury is a phenomenon that is seen readily in the peripheral nervous system, especially in rodent models, but human axonal regeneration is limited and does not lead to full functional recovery. Here we describe a system where dynamics of human axonal outgrowth and regeneration can be evaluated via live imaging of human-induced pluripotent stem cell (hiPSC)-derived neurons cultured in microfluidic systems, in which cell bodies are isolated from their axons. This system could aid in studying axonal outgrowth dynamics and could be useful for testing potential drugs that encourage regeneration and repair of the nervous system.

Key words Axonal regeneration, Motor neurons, Human iPSCs, Microfluidic devices, Live imaging, Axotomy, Outgrowth, Neuronal cytoskeletal markers

1 Introduction

Human-induced pluripotent stem-cell-derived neurons (hiPSC-neurons) cluster into circular groups and extend axons to other clumps of neurons as they grow and differentiate, which makes measurement of individual neurite outgrowth difficult. For many decades, and especially as microfabrication techniques have improved, microfluidic devices or similar compartmentalized co-culture systems have been employed to grow axons in isolation [1–7]. A key crucial feature of microfluidic devices are microchannels or microgrooves. The channels are small enough to exclude neuronal cell bodies and only allow axons to travel into connected compartments. Separation of neuronal cell bodies from their distal axons allows for efficient axotomy and discernment of individual

Supplementary Information The online version contains supplementary material available at https://doi.org/10.1007/978-1-0716-3969-6_23.

Kazuhito Toyooka (ed.), *Neuronal Morphogenesis: Methods and Protocols*, Methods in Molecular Biology, vol. 2831, https://doi.org/10.1007/978-1-0716-3969-6_23,

axons to assess regeneration [3, 6, 8–11]. Compartmentalized neuronal culture in microfluidic devices is comparable to the spatial segregation observed in vivo, where neuronal cell bodies are present in distinct nuclei, regions, or ganglia that are spatially separated from their targets, such as lower motor neuron cell bodies that reside in the anterior horn region of the spinal cord and synapse onto muscle that can be a meter away.

Separation of cell bodies from axons also is advantageous when investigating core pathological features of many neurological diseases: synaptic disconnection of axons from their targets [6, 12, 13]. In many diseases, axonal degeneration begins at the most distal part of the axon and continues toward the cell body, in a phenomenon referred to as dying-back axonopathy, resulting in loss of the functional synapse preceding the death of the neuron itself [13–21]. This makes distal axons attractive sites for early therapeutic intervention, either by maintaining synaptic connectivity and activity, which can activate pro-survival programs and inhibit apoptosis [22], or promoting axons to reconnect with their appropriate targets. Microfluidic devices that allow for specific investigation of axons are effective tools for studying axonal regeneration in addition to distal axon pathogenesis.

Axonal regeneration is dynamic, and evaluating axonal regeneration solely at fixed time points may conceal potential phenotypes related to axonal regeneration. Combining microfluidic platforms and live imaging provides the opportunity to examine dynamics of axonal regeneration of hiPSC-derived neurons. Here we describe a system for live imaging of axonal outgrowth and regeneration that can be applied to compare human neurons with disease linked-mutations and isogenic controls [7], and to test potential drugs that impact axonal regeneration.

2 Materials

All cell culture and axotomy experiments are performed in sterile biological safety cabinet, and live imaging is performed in an environmental chamber with maintenance of 37 °C temperature and 5% CO_2.

2.1 hiPSC Culture and Differentiation

1. Human iPSCs.
2. Essential 8 Medium sold together with Essential 8 Supplement.
3. Matrigel.
4. Serological pipettes.
5. 10 cm dishes.

6. Neuralization/Caudalization medium 1 (NCM1): Dulbecco's Modified Eagle Medium (DMEM)/F12 (1×), Glutamax (1×), minimal essential medium non-essential amino acids (MEM NEAA, 1×), Knock-out serum replacement (KSR 20%), β-mercapto ethanol (110 μM), SB-431542 (10 μM), LDN-193189 (0.2 μM).
7. Neuralization/Caudalization Medium 2 (NCM2): DMEM/F12 (1×), Glutamax (1×), minimal essential medium non-essential amino acids (MEM NEAA, 1×), Knock-out (KSR 20%), β-mercapto ethanol (110 μM), SB-431542 (10 μM), LDN-193189 (0.5 μM), retinoic acid (1 μM).
8. Transitional medium 1 (TM1): DMEM/F12 (1×), Glutamax (1×), minimal essential medium non-essential amino acids (MEM NEAA), Knock-out serum replacement (KSR 10%), β-mercapto ethanol (55 μM), N2 supplement (0.5 ×), heparin (1 μg/mL), penicillin-streptomycin (50 U/mL), SB-431542 (10 μM), LDN-193189 (0.5 μM), retinoic acid (1 μM).
9. Transitional medium 2 (TM2): DMEM/F12 (1×), Glutamax (1×), minimal essential medium non-essential amino acids (MEM NEAA), Knock-out serum replacement (KSR 10%), β-mercapto ethanol (55 μM), N2 supplement (0.5×), heparin (1 μg/mL), penicillin-streptomycin (50 U/mL), retinoic acid (1 μM), purmorphamine (1 μM), brain-derived neurotrophic factor (BDNF, 10 ng/mL), Ascorbic acid (0.4 μg/mL).
10. Neural Induction medium (NIM): DMEM/F12 (1×), Glutamax (1×), minimal essential medium non-essential amino acids (MEM NEAA, 1×), N2 supplement (1×), heparin (2 μg/mL), penicillin-streptomycin (100 U/mL), retinoic acid (1 μM), purmorphamine (1 μM), brain-derived neurotrophic factor (BDNF, 10 ng/mL), Ascorbic acid (0.4 μg/mL).
11. Transitional medium 3 (TM3): DMEM/F12 (0.5×), Neurobasal medium (0.5×), Glutamax (1×), minimal essential medium non-essential amino acids (MEM NEAA, 1×), N2 supplement (1×), B27 supplement (1×), N2 supplement (1×), penicillin-streptomycin (100 U/mL), heparin (1 μg/mL), brain-derived neurotrophic factor (BDNF, 10 ng/mL), glial cell-line-derived neurotrophic factor (GDNF,10 ng/mL), ciliary neurotrophic factor (CNTF, 10 ng/mL), Insulin-like growth factor (IGF, 10 ng/mL), purmorphamine (1 μM), Ascorbic acid (0.4 μg/mL), and retinoic acid (1 μM).
12. MNDM: Neurobasal medium (1×), Glutamax (1×), minimal essential medium non-essential amino acids (MEM NEAA, 1×), B27 supplement (1×), N2 supplement (1×), penicillin-streptomycin (100 U/mL), brain-derived neurotrophic factor (BDNF, 10 ng/mL), glial cell-line-derived neurotrophic factor (GDNF,10 ng/mL), ciliary neurotrophic factor (CNTF,

10 ng/mL), Insulin-like growth factor (IGF, 10 ng/mL), purmorphamine (1 μM), Ascorbic acid (0.4 μg/mL), and retinoic acid (1 μM), Compound E (125 nM),

13. Poly-L-ornithine (PLO).
14. Laminin.
15. 0.05% trypsin
16. Trypsin inhibitor.
17. Rock inhibitor Y-27632.
18. Cytosine arabinoside (Ara-C).
19. PBS: phosphate buffered saline.
20. Hemocytometer.

2.2 Microfluidic Culture and Axotomy

1. 40 mm glass bottom dishes.
2. Sylgard 184 Silicone Elastomer Kit.
3. Silane-coated, pre-patterned stainless-steel wafer molds.
4. Polydimethylsiloxane (PDMS) microfluidic devices.
5. Plasma-bonder.
6. ddH_2O.
7. Pasteur pipette.

2.3 Live Imaging

1. Inverted microscope with DIC and time-lapse imaging capabilities.
2. Environmental chamber with temperature and CO_2 control.
3. Image stitching software.
4. Tracking software.

2.4 Immunostaining in Microfluidic Devices

1. PFA: paraformaldehyde.
2. PBS: Phosphate buffered saline.
3. PBST: 0.03% Triton X-100 in PBS.
4. Goat serum.
5. Primary and secondary antibodies.
6. Hoechst 33342.

3 Methods

3.1 Differentiation of Human iPSCs into Motor Neurons

1. Generate human iPSCs from fibroblasts using a previously described four vector method [23]. Maintain and expand iPSCs on Matrigel-coated (Corning) 10 cm petri dishes as described in Taga et al. [24].

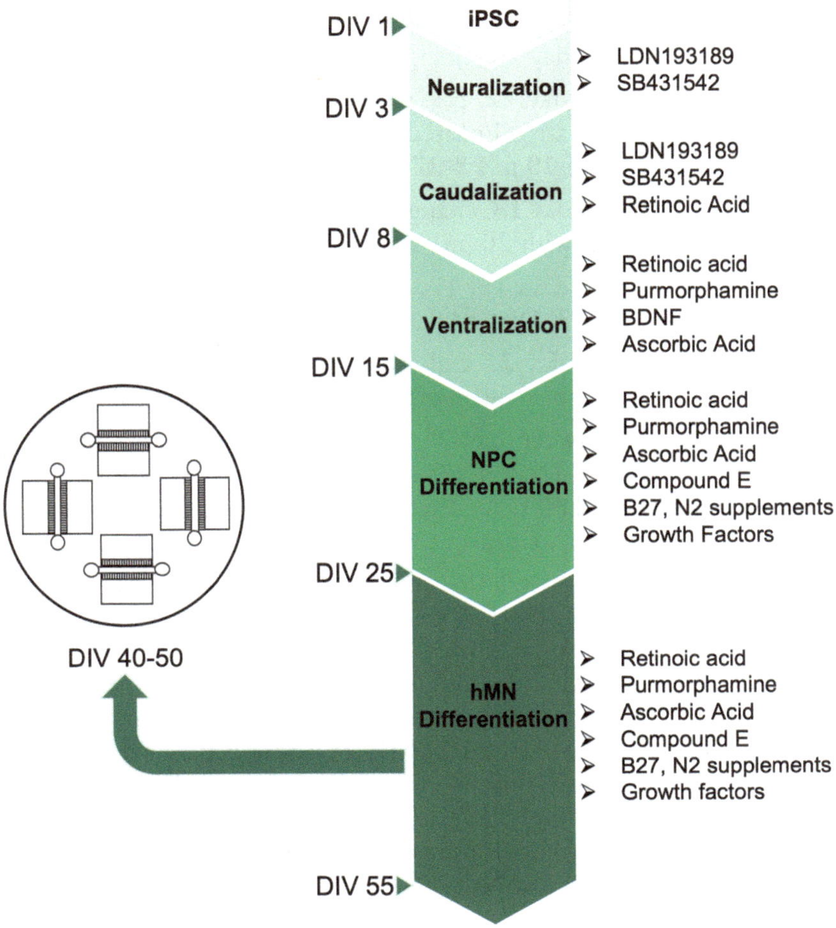

Fig. 1 Spinal motor neuron differentiation protocol. Timeline representation of differentiation protocol for human iPSC-derived spinal motor neurons adapted from references [24, 26, 27]

2. Use previously described protocols to differentiate iPSCs into spinal motor neurons [24–26], as illustrated in (Fig. 1). First, dissociate undifferentiated iPSC colonies and seed them onto a Matrigel-coated six-well plate, and culture in E8 medium supplemented with 20 μM Y-27632. When confluent, no more than 3 days following initial seeding, change medium to NCM1. Addition of the small molecules SB-431542 and LDN-193189, inhibitors of TGF-β and BMP, respectively, are used to achieve dual- SMAD inhibition, as shown in the differentiation timeline in Fig. 1. The day that cells are switched NCM1 will be considered day 1 of differentiation.
3. On day 3, change medium to NCM2. Throughout the rest of the protocol, the cells can be gently washed in their final medium to remove excess debris.

4. On day 5 and day 7, change medium with TM1.
5. On day 8, change medium to TM2.
6. On day 10, change medium to NIM1.
7. On day 12, passage cells from one well of a six-well plate to 1 Matrigel-coated T-25. Final medium is NIM1 supplemented with 20 μM Y-27632.
8. On day 13, change medium with NIM1 or NIM1 supplemented with 20 μM Y-27632 if poor survival is observed.
9. On days 15, 17, and 19, change medium and feed cells with TM3. Discard flask if cells are not confluent on day 15.
10. On day 21 and day 23, change medium and feed cells with motor neuron differentiation medium (MNDM).
11. Between day 25 and day 30, cryopreserve neural progenitor cells (NPCs) in liquid nitrogen between 25 days in vitro (DIV) and 30 DIV. Alternatively, continue with differentiation protocol by culturing NPCs in MNDM.
12. Culture NPCs in MNDM until initiation of experiments (55–60 DIV) as illustrated in Fig. 1.

3.2 Selection of Microfluidic Platform for Assessing Axon Regeneration After Injury and Microfluidic Device Assembly

1. Select microfluidic devices with large, open axonal compartments and a central cell body compartment (*see* **Notes 1** and **2**) (Figs. 2b and 3) [27, 28]. Microchannels should be no wider than 10 μm to prevent passage of cell bodies into axon compartment. Microchannels should be greater than 450 μm in length to prevent the passage of dendrites into axonal compartment (*see* **Note 3**).
2. Mix parts A and B of Sylgard 184 Silicone Elastomer Kit to generate PDMS microfluidic devices by casting the mixture onto silane-coated stainless-steel wafer molds and harden in an 80 °C oven. Alternatively, purchase PDMS microfluidic devices and proceed to **step 5**.
3. Remove hardened PDMS from wafer mold and punch holes for media entry points and axonal compartments into the molded PDMS devices.
4. Sonicate and autoclave PDMS microfluidic devices. Remove any potential remaining debris immediately prior to plasma bonding step.
5. Sonicate 40 mm glass bottom dishes, cleanse with ethanol, and visually inspect for debris.
6. Plasma treat-clean PDMS devices and round glass bottomed dishes, and bond them together following plasma-treatments (*see* **Note 4**). Plasma treatment parameters (such as normalization pressure and length of plasma treatment) must be optimized on a laboratory-by-laboratory basis.

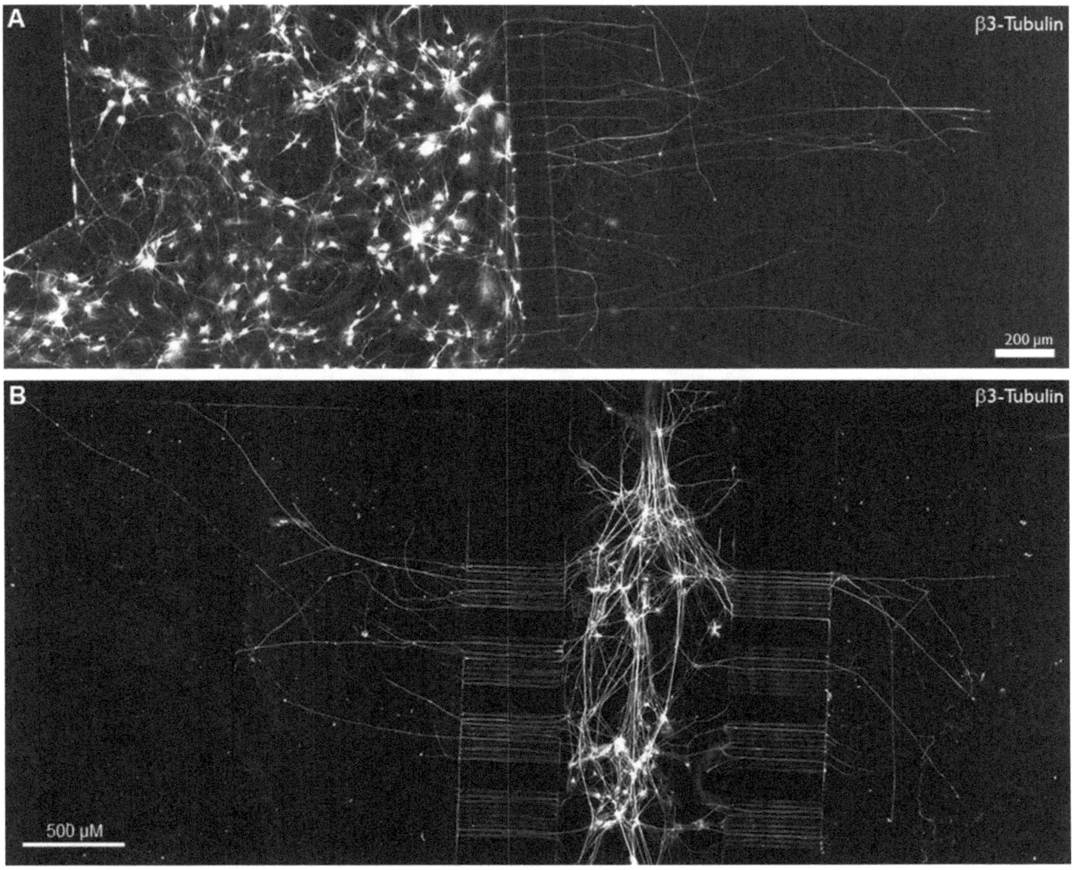

Fig. 2 hiPSC-MNs in microfluidic devices. (**a**) hiPSC-MN cultured in Xona microfluidic device immunostained for a neuronal marker, β3-tubulin. Scale bar = 200 μm. (**b**) hiPSC-MN cultured in custom microfluidic device immunostained for β3-tubulin with central cell body compartment and two axonal compartments. Scale bar = 500 μm

7. Add ddH_2O to bonded devices upon completion of plasma treatment, and store until it is time to prepare them for plating (*see* **Note 5**).

3.3 Culturing hiPSC-MNs and Plating into Microfluidic Devices

1. Thaw DIV 25–30 NPCs (neural progenitor cells) from liquid nitrogen into a 10 cm petri dish coated with poly-L-ornithine (100 μg/mL) and laminin (10 μg/mL). Thaw differentiating NPCs into MNDM media supplemented with Y-27632 (20 μM).
2. Between 34 and 37 DIV, if non-neuronal cells (e.g., glial progenitors), appearing as flat cells underneath neuronal cluster as they form, are proliferating, treat differentiating hiPSC-MN with mitotic inhibitor, cytosine arabinoside (Ara-C, 0.02 μM).

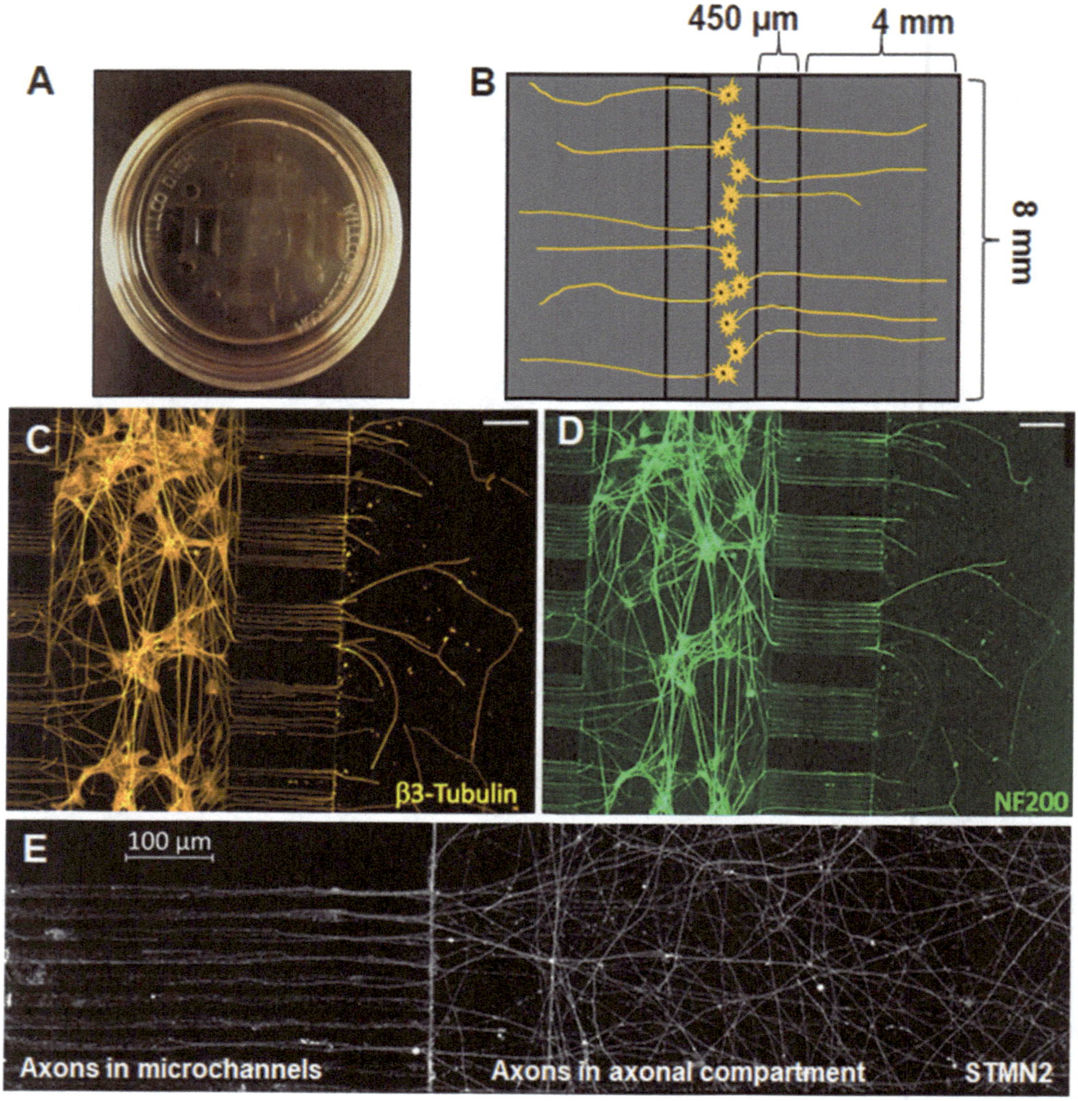

Fig. 3 hiPSC-MN axons in microfluidic device immunostained for neuronal markers. (**a**) Custom microfluidic device with 4 sectors and open axonal compartments. (**b**) Schematic diagram of one sector of the microfluidic device, depicting how neurons in a central compartment extend axons into 2 axonal compartments. (**c, d**) hiPSC-MNs at 50 DIV in custom microfluidic devices stained for β3-tubulin (**c**) and neurofilament 200 (NF200) (**d**). (**e**) hiPSC-MNs stained for axonal marker, stathmin-2 (STMN2) showing axons in microchannels and axonal compartment. Scale bars = 200 μm (**c**, **d**) and 100 μm (**e**)

3. Maintain differentiating hiPSC-MN in those dishes until 40–50 DIV, taking diligent care to not disrupt neuronal network.
4. When hiPSC-MNs reach 40–50 DIV, coat bonded microfluidic devices. Remove water and coat microfluidic device with poly-L-ornithine (100 μg/mL) and laminin (10 μg/mL) using typical incubation times (*see* **Note 6**).

5. Following laminin coat, wash device 1× with PBS. Remove PBS and refill device with MNDM.
6. Reduce media level in the axonal compartments such that media is just covering the bottom of the axonal compartment. Remove media from central compartment media entry ports. Media remains in the central compartment during this time. When prepared to plate, the central compartment should contain media (remove any bubbles), and the media entry ports should be as empty as possible without removing media from central compartment plating area.
7. Gently wash hiPSC-MNs with PBS and lift from 10 cm dish using 0.05% trypsin for 5–10 min. Gently triturate hiPSC-MNs, and halt trypsin dissociation with trypsin inhibitor (1: 10 to trypsin). Spin at 300 × *g* with neurobasal medium in 15 mL conical tube for 5 minutes.
8. Resuspend in 1 mL of MNDM+ 20 μM Y-27632, and count cells using hemocytometer while spinning cells again at 300 × *g* for 5 minutes.
9. Gently resuspend cell pellet such that there are 50,000 cells/mL in MNDM+ 20 μM Y-27632.
10. For each sector, use a P20 to pipette 3 μL of concentrated cells into the media entry port on one side of the central cell-body compartment. With the same tip, remove 3 μL from the opposite side of the cell body compartment to draw the neuronal cell bodies into the central compartment plating area (*see* **Note 7**).
11. When all four sectors are plated, place devices in 37 °C incubator for 1–2 h to allow hiPSC-MNs to adhere.
12. Add MNDM+ 20 μM Y-27632 medium to the central cell body compartments until media entry points are full (about 50 μL).
13. Add MNDM to axonal compartments. Axonal compartments should be filled such that the medium level is above the cell body compartment.
14. Return devices to incubator, with care to ensure adequate humidity levels to prevent media evaporation (*see* **Note 8**).
15. The day after plating, medium in the cell body compartment should be changed to MNDM without Y-27632.
16. Monitor axon outgrowth using an inverted light microscope (*see* **Note 9**). Usually, axons emerge from microchannels between 2 and 4 days after plating into a microfluidic device with 450 μm long microchannels.
17. hiPSC-MNs are maintained in MNDM throughout experiments. Perform media changes on cell body compartments three times per week. Medium changes on the central cell

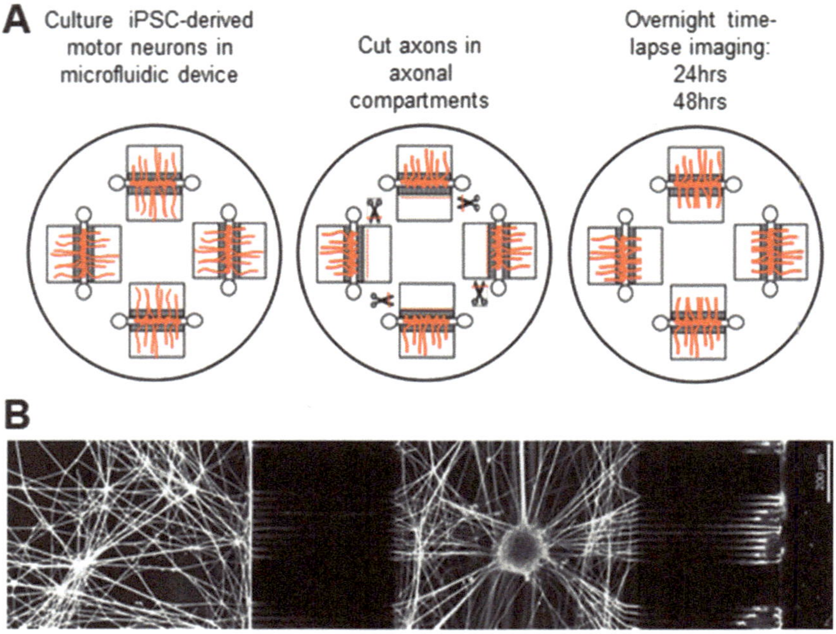

Fig. 4 Schematic diagram illustrating axonal regeneration experiments. (**a**) hiPSC-MNs were cultured in microfluidic devices and allowed to extend axons into the axonal compartments for at least 7 days following plating. Using time-lapse imaging, regenerating axons were monitored for 48 h post-axotomy. (**b**) hiPSC-MNs in microfluidic device immediately post axotomy in the right-side axonal compartment

body compartment are performed by removing medium little by little (5–10 μL each time) from both entry ports in an alternating fashion. This prevents quick flow from port-to-port through the central cell body compartment, which may cause mechanical stress to neuronal cell bodies.

18. Perform half-volume media changes on all axonal compartments when changing medium. Delicate axons in axonal compartments can be easily broken or damaged by fully removing medium from the axonal compartment.

3.4 Axotomy in Microfluidic Devices

1. Axotomy experiments in hiPSC-MNs are initiated between DIV 55 and 60, when neurons have been plated in devices for at least 7 days (Fig. 4).
2. Axotomies are achieved through vigorous vacuum aspiration of the axonal compartment with a glass Pasteur pipette. This severs axons at the junction of microchannel exit and the axonal compartment (*see* **Note 10**).
3. Remaining debris is cleared from microchannel exit using a bent 10 μL pipette tip, taking care not to scratch bottom matrix (*see* **Note 11**).
4. Replace media in empty axon compartment and repeat vacuum aspiration three times to maximize debris clearance (*see* **Note**

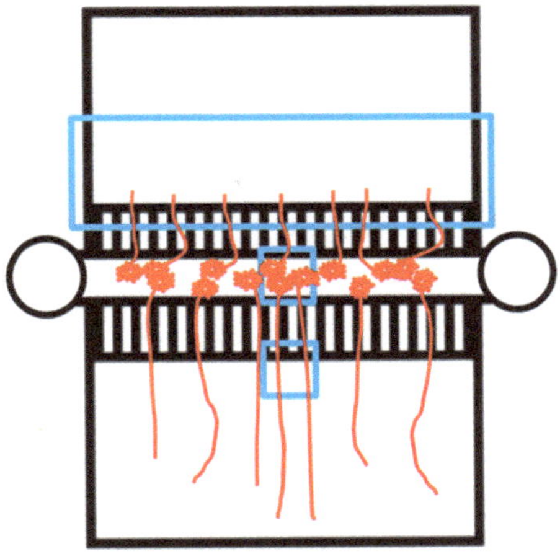

Fig. 5 Diagram of imaging area for time-lapse imaging of regenerating hiPSC-MN axons. Blue boxes indicate the areas of the microfluidic device that were captured via time-lapse imaging. All four sectors on the device were imaged overnight

12). During and following axotomy, inspect devices visually under a light microscope to verify microchannel integrity and complete severing (not dragging) of axons and removal from axonal compartment. This results in complete axotomy, which can be demonstrated by neurofilament staining immediately post-axotomy. In the axonal compartment where axons were cut, neurofilament staining is not present in the axonal compartment (Fig. 4b, right side).

3.5 Time-Lapse Imaging and Analysis of Regenerating Axons

1. Time-lapse images can be acquired on an inverted microscope with DIC capabilities, such as the Zeiss AxioObserver microscope.
2. The areas imaged are demonstrated in Fig. 5. To capture regenerating axons, axons within microchannels can be used to focus the microscope on the latter portion of the microchannels, exit into the axonal compartment, and initial portion of the axonal compartment. Time-lapse images of cell bodies and uncut axons are simultaneously acquired to monitor their status following axotomy and verify viability of the proximal axons.
3. Acquire images of regenerating axons and control areas at regular time intervals (*see* **Note 13**). These images will be stitched together to generate time-lapse videos for analysis (Videos 1 and 2).
4. Pause imaging to refocus the microscope between 24-h and 48-h time points (*see* **Note 14**). Single time point images from time-lapse videos are in Fig. 6.

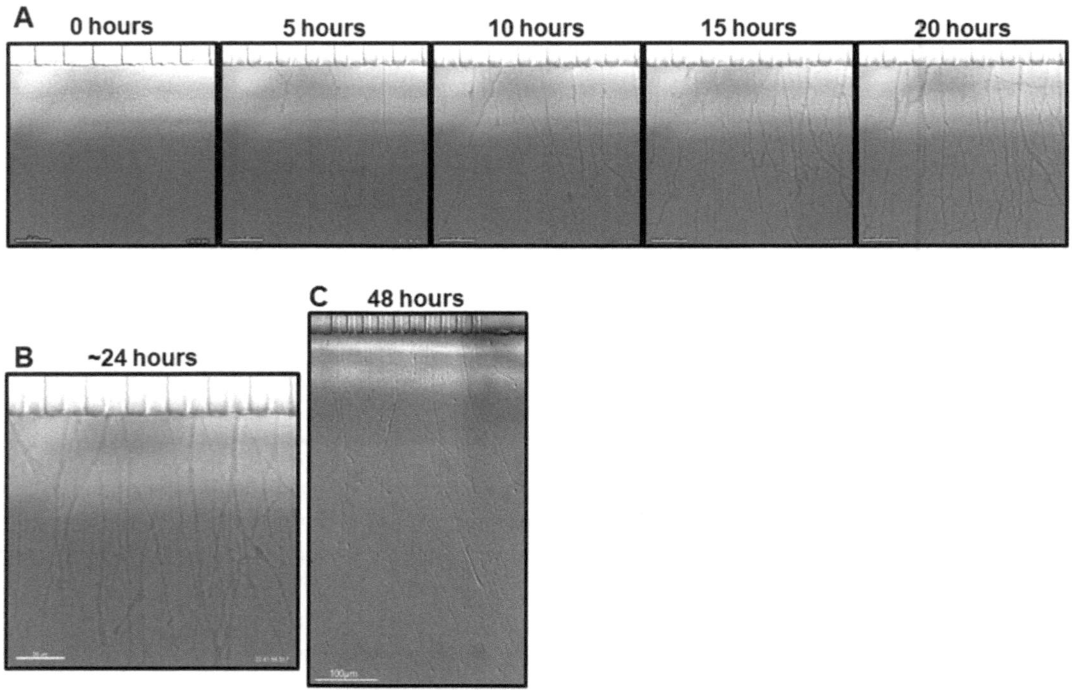

Fig. 6. Still-frames from time-lapse videos taken post axotomy. Images of the cell body compartment (not shown) and axonal chambers were acquired between 9.5 and 10 min for 48 h post-axotomy. (**a**) Individual frames from one set of microchannels take from 0, 5, 10, 15, and 20 h post axotomy. (**b**) Individual frame from microchannels about 24 h post axotomy. (**c**) Individual frame of another set of microchannels about 48 h post axotomy

5. Following time-lapse imaging, stitch large, tiled areas using Zen Blue software.
6. Upload and convert stitched files to Imaris-compatible files in Imaris file converter software.
7. Determine regeneration rate using the "spots" feature on Imaris, which has particle tracking capabilities. Inspect time-lapse videos frame-by-frame and specifying the location of regenerating growth cones throughout video, designating each regenerating axon as an individual track (*see* **Note 15**). Axon lengths at individual time points can also be determined by tracing full axons using "filament tracer" feature on Imaris.

3.6 Immunofluorescence Staining of Axons in Microfluidic Devices

1. To fix hiPSC-MNs in the microfluidic devices, first dilute cell-culture media MNDM in axon compartments using half-volume media changes until clear using PBS. This should be performed gently, and fixation should begin within 15 min of beginning this process. Wash central cell body compartment in PBS.
2. Fix hiPSC-MNs in 2% PFA in microfluidic devices at desired time points post-axotomy.

3. During staining, preserve axons in axonal compartments by performing all liquid changes at half-volume. Achieve appropriate concentrations of antibodies etc. by adding solutions at 2× concentration to remaining liquid.
4. Permeabilize and block cells with 0.3% Triton X-100 in 1× PBS (PBS-T) and 5% goat serum for 1 h at room temperature.
5. Incubate devices with primary antibodies, such as neuronal markers smi-31, β3-tubulin or stathmin-2, at 4 °C overnight.
6. Perform six half-volume washes with PBS-T.
7. Add appropriate fluorophore-conjugated secondary antibody, in addition to Hoechst 33342 for nuclear staining, for 30 min at room temperature.
8. Wash devices six times with PBS.
9. Perform imaging following immunostaining for neuronal markers. Immunostaining for β3-tubulin at various time points post-axotomy is shown in Fig. 7.

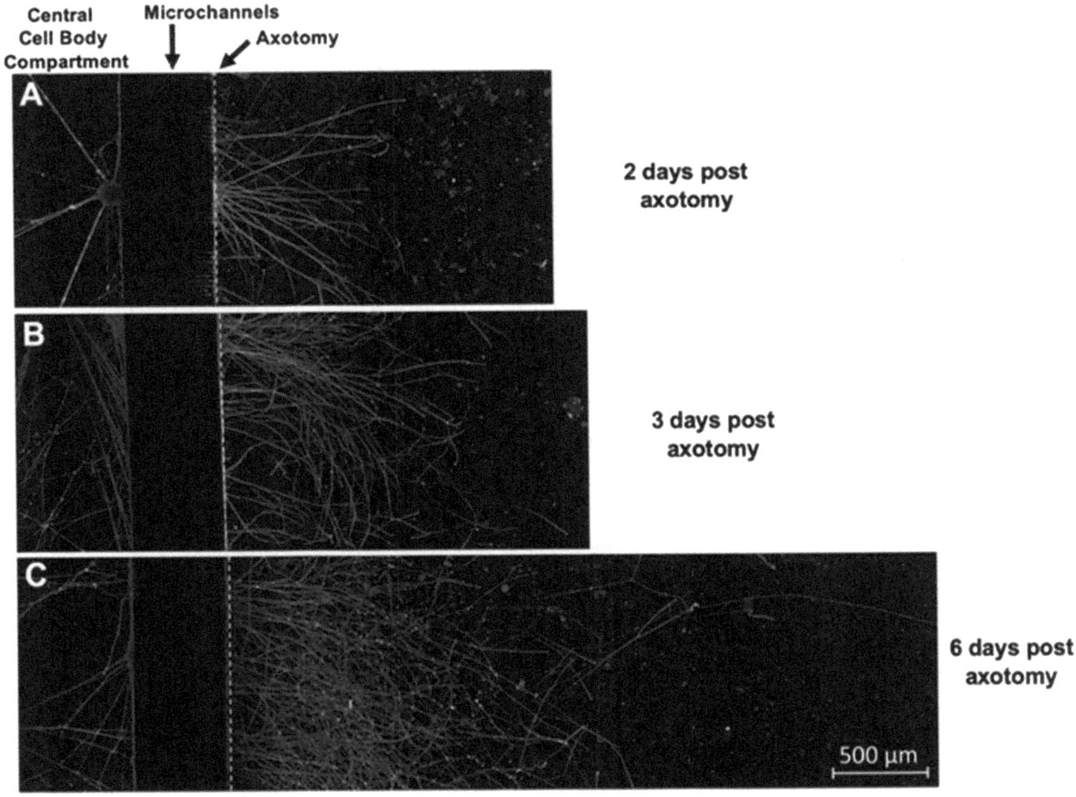

Fig. 7 Robust axonal regeneration of control hiPSC-MNs. hiPSC-MNs were cultured in the central compartment of microfluidic devices, and axons in one compartment were severed at the interface between the microchannel exit and axonal compartment using vigorous vacuum aspiration. Axons were allowed to re-extend under normal culture conditions, fixed, and stained for β3-tubulin at (**a**) 2 days post-axotomy, (**b**) 3 days post-axotomy, and (**c**) 6 days post-axotomy

3.7 Analysis

1. Consider each sector, one of the four on each microfluidic device, one biological replicate (*see* **Note 16**).
2. Frequency distribution plots, where axon lengths are binned, are useful for comparing regeneration of axon populations.
3. To ensure consistent sampling, consider using groups of microchannels as a reference, consistently tracing the same number of axons per group of microchannels.

4 Notes

1. The microfluidic platform used in our experiments has four separate microfluidic sectors on one dish. Each has a central cell body compartment and two axonal compartments. The cell body compartment is separated from the axonal compartment by 450 μm-long and 10 μm-wide microchannels. Both sets of microchannels that separate the central compartment from the axon compartments have 20 groups of 10, 10 μm-wide microchannels that are 450 μm in length. The large (8 mm wide × 4 mm long), uncovered axonal compartments allow access to axons for mechanical manipulation and can potentially facilitate addition of other cell-types, such as Schwann cells or myofibers. Because there each sector has two axonal compartments, it is possible to use one axonal compartment as a control condition with uncut axons.
2. When selecting a microfluidic platform for regeneration experiments, we first tested our hiPSC-MNs in commercial microfluidic devices from Xona Microfluidics (Fig. 2). Xona devices, with one cell body compartment and one axonal compartment separated by microchannels, were plasma-bonded to glass coverslips. hiPSC-MN culture in Xona devices resulted in separated neuronal cell bodies from their axons (Fig. 2a). Growing axons rapidly reached the limits of the axonal compartment, and we reasoned that devices with a large axonal compartment was more suitable for assessing regenerating axons for prolonged time periods post-axotomy. In addition, adding other cell-types (such as myofibers) to the axonal compartment may be more technically challenging in these Xona devices because the axonal compartment is covered with PDMS, making this compartment accessible only by media entry ports on either side. The ability to incorporate and manipulate different cell-types within compartmentalized culture systems is relevant for assessing hiPSC-MN regeneration, sprouting, and neuromuscular junction (NMJ) reinnervation. Based on these observations, we sought a microfluidic platform with ample room in the axonal compartment, such as the device described in **Note 1**.

3. It is worth considering that microchannels shorter than 450 μm may allow passage of both dendrites and axons. MAP 2 immunostaining can be used to identify dendrites in microfluidic devices [29].
4. In our attempts to culture hiPSC-MNs in microfluidic devices, we found plasma-bonding to be critical. In devices that are not plasma-bonded to the coverslips, while plating neurons is feasible, long-term culture is difficult to achieve because the seal between PDMS and glass is easily disrupted.
5. Bonded devices that are filled with water can be stored for months at a time if they are not allowed to dry. We wrap the microfluidic device dishes in parafilm to prevent water evaporation. If water has completely evaporated from device, and air is introduced to microchannels, it cannot be used.
6. Following plasma bonding step, take care not to void central cell body compartment. This can introduce bubbles to the compartment which interfere with flow of media and plating of hiPSC-MNs. If a bubble is introduced during cell plating, we find the best course of action is to forcefully introduce media into one entry port of the central compartment to push the air from the compartment. Once the bubble has been removed, plating can be resumed after removing media from entry ports. We found that bubbles are often introduced by using vacuum aspiration to remove media from the central compartment, so we use a P200 pipette to remove media during cell plating.
7. Immediately after plating, cell bodies should appear densely and regularly packed into the cell body compartment. If hiPSC-MNs appear too sparse or are accumulating in media entry ports for the central compartment, it is likely that there is excess media in the central compartment. This can happen when not enough media is removed from media entry ports. When preparing to plate, the central compartment should contain media (no bubbles), and the media entry ports should be as empty as possible. Alternatively, there may be a slow leak, from incomplete bonding or cracks introduced while punching out device features, in the sector that is introducing media from axonal compartment.
8. We found that maintaining humidity is critical for maintenance of microfluidic devices. Low media volumes in microfluidic devices make cultures particularly susceptible to evaporation and concentration of media. We kept 3–4 microfluidic devices in one 15 cm dish, that also contained an extra 35 mm water dish.
9. In our hands, hiPSC-MN axons typically grow through microchannels and into the axonal compartment within 2–4 days following plating with no supplementation of neurotrophic

factors into the axonal compartment to encourage axonal outgrowth through the microchannels. Other researchers create a concentration gradient of neurotrophic factors like BDNF and GDNF to encourage axon outgrowth through microchannels and into the axon compartment [10, 11, 30].

10. Care must be taken while performing axotomy to prevent damage to microchannels. During visual inspection of microfluidic devices to check for removal of axonal debris, extra attention should be paid to area of microchannel exit. If axons looked like they have been dragged underneath area in between microchannels or there is visible damage to microchannels, the device seal has been disrupted. Bond disruption can block the path of regenerating axons.
11. Care must be taken to avoid scratching the basement matrix while clearing remaining debris with a bent pipette tip. Damage to matrix coating can block or redirect the path of regenerating axon, making tracing of individual axons difficult and potentially confounding results.
12. We and others find vacuum axotomy to be reliable, but other methods of axotomy are possible within microfluidic systems including enzymatic digestion of axons (trypsin or accutase) or combined mechanical and chemical axotomy [10, 11].
13. Keep in mind that the length of time and interval of imaging will impact the captured axon dynamics. We captured images on a Zeiss AxioOberserver microscope every 9.5–10 min for 48 h post-axotomy. The time interval of image acquisition can depend on the intended axon dynamics to capture, number of areas to image, and speed of the microscope. If time-lapse imaging is to be based on a fluorescent marker, phototoxicity and bleaching should be considered when determining time interval.
14. Imaging drift is an inevitable aspect of live-cell imaging over longer periods and when imaging large areas. We recommend taking Z-stacks at each position to account for this. During manual tracking, this allows us to change the displayed Z-slice at different time points to improve growth cone visibility. We captured three Z-stack images 5 μm apart at each position using DIC on the 10× objective for each image. When arranging the large area to be captured in the axonal compartment, we found it useful to distribute focus points regularly throughout.
15. We used Imaris for manual tracking of regenerating axons. Imaris software is more optimized for fluorescent images, and it is likely that axon tracing can be automated to some degree using Imaris wizard. Automated tracking of axons may be further enabled as artificial intelligence-powered tools are increasingly developed and incorporated into analysis software.

16. Given the variation that is inherent to the use of human cells and axonal regeneration in general, we find it useful to isogenic cell lines to conduct experiments.

Acknowledgments

The authors thank Aundree Hermawan, Arens Taga, Maria Majid, and Erin Beebe for their technical assistance. We also thank Michele Pucak and the Neuroscience Multiphoton Imaging Core for their imaging assistance.

Grants Maryland Stem Cell Research Fund: MSCRFD-5093 (M.H.F.), 2022-MSCRFF-5833 (K.L.M.); NIH: R21NS130900 (M.H.F.)

References

1. Campenot RB (1977) Local control of neurite development by nerve growth factor. Proc Natl Acad Sci USA. https://doi.org/10.1073/pnas.74.10.4516
2. Kimpinski K, Campenot RB, Mearow K (1997) Effects of the neurotrophins nerve growth factor, neurotrophin-3, and brain-derived neurotrophic factor (BDNF) on neurite growth from adult sensory neurons in compartmented cultures. J Neurobiol 33(4):395–410. https://doi.org/10.1002/(sici)1097-4695(199710)33:4<395::aid-neu5>3.0.co;2-5
3. Taylor AM, Blurton-Jones M, Rhee SW, Cribbs DH, Cotman CW, Jeon NL (2005) A microfluidic culture platform for CNS axonal injury, regeneration and transport. Nat Methods. https://doi.org/10.1038/nmeth777
4. Millet LJ, Gillette MU (2012) Over a century of neuron culture: from the hanging drop to microfluidic devices. Yale J Biol Med 85(4):501–521
5. Neto E, Leitão L, Sousa DM, Alves CJ, Alencastre IS, Aguiar P et al (2016) Compartmentalized microfluidic platforms: the unrivaled breakthrough of in vitro tools for neurobiological research. J Neurosci. https://doi.org/10.1523/JNEUROSCI.1748-16.2016
6. Marshall KL, Farah MH (2021) Axonal regeneration and sprouting as a potential therapeutic target for nervous system disorders. Neural Regen Res 16(10):1901–1910. https://doi.org/10.4103/1673-5374.308077
7. Marshall KL, Rajbhandari L, Venkatesan A, Maragakis NJ, Farah MH (2023) Enhanced axonal regeneration of ALS patient iPSC-derived motor neurons harboring SOD1 (A4V) mutation. Sci Rep 13(1):5597. https://doi.org/10.1038/s41598-023-31720-7
8. Klim JR, Williams LA, Limone F, Guerra San Juan I, Davis-Dusenbery BN, Mordes DA et al (2019) ALS-implicated protein TDP-43 sustains levels of STMN2, a mediator of motor neuron growth and repair. Nat Neurosci. https://doi.org/10.1038/s41593-018-0300-4
9. Melamed Z, López-Erauskin J, Baughn MW, Zhang O, Drenner K, Sun Y et al (2019) Premature polyadenylation-mediated loss of stathmin-2 is a hallmark of TDP-43-dependent neurodegeneration. Nat Neurosci. https://doi.org/10.1038/s41593-018-0293-z
10. Stoklund Dittlau K, Krasnow EN, Fumagalli L, Vandoorne T, Baatsen P, Kerstens A et al (2021) Human motor units in microfluidic devices are impaired by FUS mutations and improved by HDAC6 inhibition. Stem Cell Rep 16(9):2213–2227. https://doi.org/10.1016/j.stemcr.2021.03.029
11. Garone MG, Birsa N, Rosito M, Salaris F, Mochi M, De Turris V et al (2021) ALS-related FUS mutations alter axon growth in motoneurons and affect HuD/ELAVL4 and FMRP activity. Commun Biol 4(1). https://doi.org/10.1038/s42003-021-02538-8
12. Henstridge CM, Pickett E, Spires-Jones TL (2016) Synaptic pathology: a shared mechanism in neurological disease. Ageing Res Rev. https://doi.org/10.1016/j.arr.2016.04.005

13. Tallon C, Russell KA, Sakhalkar S, Andrapallayal N, Farah MH (2016) Length-dependent axo-terminal degeneration at the neuromuscular synapses of type II muscle in SOD1 mice. Neuroscience 312:179–189. https://doi.org/10.1016/j.neuroscience.2015.11.018
14. Tampellini D (2015) Synaptic activity and Alzheimer's disease: a critical update. Front Neurosci. https://doi.org/10.3389/fnins.2015.00423
15. Moloney EB, de Winter F, Verhaagen J (2014) ALS as a distal axonopathy: molecular mechanisms affecting neuromuscular junction stability in the presymptomatic stages of the disease. Front Neurosci. https://doi.org/10.3389/fnins.2014.00252
16. Fischer LR, Culver DG, Tennant P, Davis AA, Wang M, Castellano-Sanchez A et al (2004) Amyotrophic lateral sclerosis is a distal axonopathy: evidence in mice and man. Exp Neurol 185(2):232–240. https://doi.org/10.1016/j.expneurol.2003.10.004
17. Han I, You Y, Kordower JH, Brady ST, Morfini GA (2010) Differential vulnerability of neurons in Huntington's disease: the role of cell type-specific features. J Neurochem 113(5):1073–1091. https://doi.org/10.1111/j.1471-4159.2010.06672.x
18. Valadão PAC, Gomes MPSM, Aragão BC, Rodrigues HA, Andrade JN, Garcias R et al (2018) Neuromuscular synapse degeneration without muscle function loss in the diaphragm of a murine model for Huntington's disease. Neurochem Int. https://doi.org/10.1016/j.neuint.2018.03.007
19. Chu Y, Morfini GA, Langhamer LB, He Y, Brady ST, Kordower JH (2012) Alterations in axonal transport motor proteins in sporadic and experimental Parkinson's disease. Brain 135(7):2058–2073. https://doi.org/10.1093/brain/aws133
20. Dadon-Nachum M, Melamed E, Offen D (2011) The "dying-back" phenomenon of motor neurons in ALS. J Mol Neurosci 43(3):470–477. https://doi.org/10.1007/s12031-010-9467-1
21. Tallon C, Marshall KL, Kennedy ME, Hyde LA, Farah MH (2020) Pharmacological BACE inhibition improves axonal regeneration in nerve injury and disease models. Neurotherapeutics. https://doi.org/10.1007/s13311-020-00852-3
22. Bell KFS, Hardingham GE (2011) The influence of synaptic activity on neuronal health. Curr Opin Neurobiol. https://doi.org/10.1016/j.conb.2011.01.002
23. Takahashi K, Tanabe K, Ohnuki M, Narita M, Ichisaka T, Tomoda K et al (2007) Induction of pluripotent stem cells from adult human fibroblasts by defined factors. Cell. https://doi.org/10.1016/j.cell.2007.11.019
24. Taga A, Habela CW, Johns A, Liu S, O'Brien M, Maragakis NJ (2021) Establishment of an electrophysiological platform for modeling ALS with regionally-specific human pluripotent stem cell-derived astrocytes and neurons. J Vis Exp 174. https://doi.org/10.3791/62726
25. Boulting GL, Kiskinis E, Croft GF, Amoroso MW, Oakley DH, Wainger BJ et al (2011) A functionally characterized test set of human induced pluripotent stem cells. Nat Biotechnol. https://doi.org/10.1038/nbt.1783
26. Roybon L, Lamas NJ, Garcia-Diaz A, Yang EJ, Sattler R, Jackson-Lewis V et al (2013) Human stem cell-derived spinal cord astrocytes with defined mature or reactive phenotypes. Cell Rep. https://doi.org/10.1016/j.celrep.2013.06.021
27. Hosmane S, Yang IH, Ruffin A, Thakor N, Venkatesan A (2010) Circular compartmentalized microfluidic platform: study of axon–glia interactions. Lab Chip 10(6):741. https://doi.org/10.1039/b918640a
28. Marshall KL, Labchan R, Venkatesan A, Maragakis NJ, Farah MH (2023) Enhanced axonal regeneration of ALS patient iPSC-derived motor neurons harboring SOD1(A4V) mutation. Sci Rep. https://doi.org/10.1038/s41598-023-31720-7
29. Virlogeux A, Moutaux E, Christaller W, Genoux A, Bruyère J, Fino E et al (2018) Reconstituting corticostriatal network on-a-Chip reveals the contribution of the presynaptic compartment to Huntington's disease. Cell Rep 22(1):110–122. https://doi.org/10.1016/j.celrep.2017.12.013
30. Nijssen J, Aguila J, Hoogstraaten R, Kee N, Hedlund E (2018) Axon-Seq decodes the motor axon transcriptome and its modulation in response to ALS. Stem Cell Rep. https://doi.org/10.1016/j.stemcr.2018.11.005

Chapter 24

In Vivo Neural Tract Tracing Procedures: Unravelling Neural Hodology Using Fluorescent and Non-fluorescent Neural Tract Tracers, Optogenetic Approach, and Diffusion Tensor Neurotractography Protocols

Norberto Cysne Coimbra, Paloma Molina Hernandes, Davi Henrique Salvador Pereira dos Santos, Maria de Fátima dos Santos Sampaio, Renato Leonardo de Freitas, Farhad Ullah, Carlos Ernesto Garrrido Salmon, and Rafael Carvalho Almada

Abstract

Fluorescent and non-fluorescent neural tract tracers enable the investigation of neural pathways in both peripheral and central nervous systems in laboratory animals demonstrating images with high resolution and great anatomic precision. Anterograde and retrograde viral tracers are important cutting-edge tools for neuroanatomical mapping. The optogenetic consists of an advanced alternative for in vivo neural tract tracing procedures, fundamentally considering the possibility to dissect and modulate pathways either exciting or inhibiting neural circuits in laboratory animals. The neurotractography by diffusion tensor imaging in vivo procedures enables the study of neural pathways in humans with reasonable accuracy. Here we describe the procedure of classical anatomic neural tract tracing and modern optogenetic technique performed *in anima vili* in addition to different diffusion tensor neurotractography performed in anima nobili.

Key words Neural tract tracers, Biotinylated dextran amine, Texas red-conjugated dextran, Optogenetics, Human neurotractography, Cortico-neostriato-nigral disinhibitory pathways, Nigro-tectal inhibitory pathways

1 Introduction

Classical neuroanatomic studies are performed for decades using fluorescent and non-fluorescent neural tract tracers, such as fast blue [1–3], true blue [4], nuclear yellow [5], Lucifer yellow [6], viral vectors [7], *Phaseolus vulgaris* Leucoagglutinin [8, 9], etc. The advantages of these neuroanatomical tools is the retrograde neural

Kazuhito Toyooka (ed.), *Neuronal Morphogenesis: Methods and Protocols*, Methods in Molecular Biology, vol. 2831, https://doi.org/10.1007/978-1-0716-3969-6_24,

tract tracing in a short survival time and the neural labeling of either neuronal perikarya (e.g., with fast blue and true blue) or neuronal nuclei (e.g., with nuclear yellow), and the possibility of double and triple neural labeling. The Lucifer yellow allows the observation of the entire cellular body in addition to dendrites and axons when microinjected inside the neuronal cytoplasm by patch clamp procedure. The disadvantage of some of these neural tract tracers is the possibility of transneuronal tracing if the peripheral/central nervous system slices are maintained at room temperature. This inconvenience can be minimized when the histological laminas are kept under low temperatures. Retrograde axonal transport is also observed by certain viruses conjugated to fluorophores spreading from one neuron to the next in a neural network (transneuronal neural tract tracing). However, biodextran amine (BDA) neural tract tracers are a good alternative for precise neural tract tracing procedures without transneuronal tracing showing precise axosomatic, axodendritic, and axoaxonic appositions suggesting synapses [10]. These neural tract tracers can be either directly microinjected into the central nervous system or through a guide cannula. The advantage of stereotaxically inserting a guide cannula is the avoidance of deposits of the neural tract tracer outside the structure of interest; the disadvantage is the mechanical lesion caused by the guide cannula. Researchers must be cautious regarding the volume of microinjections of the neural tract tracer inside the central nervous system. Considering previous evidence suggesting that the diffusion of substances microinjected in the central nervous system into the tissue surrounding the site of injections has a direct relationship to the volume given [11, 12], a volume no larger than 0.4μL (a volume of 0.5μL causes an average spread of 1.04 mm [11]) is recommended for rats. In mice, for example, volumes no larger than 0.1μL should be administered. A good alternative would be the use of iontophoretic deposits of neural tract tracers in brain tissue. Either fluorescent or non-fluorescent dextran amine deposits in the central nervous system by iontophoresis procedure is followed by precise neural tract tracing with deposits of small amount of drug inside a given neuronal nucleus or in a discrete division of a given nucleus [3, 13]. However, the neural tract tracer molecule must be polarized to allow the efficiency of that technique.

The use of viral tools in neuroscience is highly interdisciplinary and viral tracers have been important tools for neuroanatomical mapping and genetic payload delivery. Genetically modified viruses allow for cell-type-specific targeting and overcome many limitations of non-viral tracers. In this context, new viral-genetic tracing tools such as recombinant rabies viral vectors are efficient retrograde labeling and monosynaptic tracing that allows detailed synaptic circuit inputs [14]. The use of viral vectors in combination with optical stimulation methods made it possible to analyze the structure and function of the brain neural network.

The optogenetic manipulation of the neuronal activity is of key importance for mapping functional connectivity and dissecting the neural circuitry underlying behaviors once that enables to analyze the neuronal network both in vivo and in vitro with precise spatio-temporal resolution [15, 16]. This technique is based on the exogenous expression of light-sensing proteins in neurons allowing manipulation of membrane potential and thus the activation and/or inhibition of a neural network by light [15, 17]. Another advantage of optogenetics is the manipulation of defined cell groups or projections promoted by the development of viral vectors with specially engineered properties, such as axonal transduction and/or retrograde propagation, cell type-specific promoter dependence and/or recombinase dependence (e.g., Cre, Flp, and Dre recombinases, which can be targeted to specific cell types using recombinase-driver rodent lines or recombinase expressing viruses) [18].

Diffusion tensor imaging (DTI) is a more suitable procedure for in vivo neural tract tracing in human participants in comparison to classical neural tract tracing performed in laboratory animals. The disadvantages of this technique are the low resolution and the lack of precision, in addition to the fact that DTI is based on a probabilistic existence of a given tract in the human brain. Another limitation of DTI is the size of the connected structures to be investigated, although recent advances in this field are providing more powerful brain scanners and accuracy of DTI performed for example in the brainstem [19] and spinal cord [20].

The goal of this approach is to compare data obtained by the following procedures: (a) classical neuroanatomical techniques for neural tract tracing in the central nervous system of Wistar rats; (b) optogenetics procedure performed in mice; (c) human neurotractography by diffusion tensor imaging.

2 Material

2.1 Neural Tract Tracing In Anima Vili

Fluorescent Neural Tract Tracers

(a) Male Wistar rats (*Rattus norvergicus*; Rodentia; Muridae) weighing 200–250 g.

(b) Plexiglas cages with free access to food and water throughout the experiment.

(c) Animal Facility where laboratory animals are maintained on a 12 h light/12 h dark cycle (lights on at 7:00 a.m.) at 23–25 °C.

(d) Anesthetic drugs: ketamine and xylasine. Antibiotics: penicillin G-benzathine. Analgesic drug: flunixin meglumine. Neural tract tracer: AlexaFluor Texas red-conjugated dextran; 3000 MW.

(e) A non-steroidal analgesic and anti-inflammatory drug (Meglumine; 2.5 mg/kg; subcutaneous injection).

(f) Antibiotics (penicillin G-benzatine; 120,000 UI; 0.1 mL; intramuscular injection).

(g) Stainless steel guide cannulae (0.6 mm outside diameter, 0.4 mm inside diameter).

(h) Liquid and powder polymerizing acrylic resin (liquid acrylic emulsion polymer; mixture based on methyl methacrylate monomer), used to anchor the guide cannula to the skull of laboratory animal.

(i) Stainless steel wire to protect the guide cannula from obstruction (28 gauge for *Rattus norvegicus/Cavia porcellus* and 33-gauge for *Mus musculus*, cut to extend 0.1 mm blow the guide cannula).

(j) Stereotaxic frame (David Kopf Instruments, Tujunga, CA, USA).

(k) Drug infusion system: infusion pump (Stoelting, Kiel, Wisconsin, USA); polyethylene tube (PE10) attached to a dental needle.

(l) Perfusion solution: physiological saline followed by 4% paraformaldehyde (PFA, Sigma) dissolved in 0.1 M phosphate buffer pH 7.4 (9 g NaCl, 5.52 g 23 mM NaH_2PO_4, and 10.93 g 77 mM Na_2HPO_4 for 1000 mL of solution; adjustment of pH may be performed with either HCl or NaOH).

(m) Crioprotection: 10% and 20% sucrose dissolved in 0.1 M sodium phosphate buffer, pH 7.3, at 4 °C. HCl or NaOH may be used to adjust pH.

(n) Material for histology: 2-methylbutane (Sigma), dry ice, Tissue-Tek, cryostat (CM 1950, Leica, Wetzlar, Germany); Fluoromount with the blue fluorescent DNA stain 4′,6-diamidino-2-phenylindole (DAPI) (Electron Microscopy Sciences, Hatfield, Pennsylvania, USA); silanized slides.

(o) AxioImager Z1 with APOTOME II (Zeiss, Oberkochen, Germany).

Iontophoretic Deposits of Biotinylated Dextran Amines in Brain Tissue

(p) Laboratory animals, anesthetic and analgesic drugs, stereotaxic frame, perfusion pump, cryoprotective solutions and material for histology as described above; cyanoacrylate glue and dental cement powder; perfusion solution (2% paraformaldehyde and 2% glutaraldehyde in 0.1 M phosphate buffer, pH 7.4); ABC kit (Vector laboratories), 3,3′-diaminobenzidine (Sigma-Aldrich, USA); counterstaining dyes: either hematoxylin-eosin or Nissl method.

(q) Glass capillary; Narishige (Japan) micropipette puller; micropipettes with a tip diameter with either 10–15μm or 40–50μm (depending on the dimensions of the structure) back-filled with the BDA solution.

(r) Iontophoresis system: iontophoresis device; electrode tip positive, DC current pulses 5μA, 7s on, 7s off, over a period of 15 min.

Optogenetics

(s) Laboratory animals (C57BL/6 N mice).

(t) Anesthetic drug: isoflurane (Forene®, Abbott, Germany, under induction at 4%, maintained at 1.5%).

(u) Stereotaxic frame.

(v) viral construct AAV5-hSyn-Jaws-GFP-ER2 (350 nL).

(w) Hamilton syringe, polyethylene tube PE-10 and Mizzi-like needles.

(x) Optic cannulas (CFML12L10, Ø200 μm, NA 0.39, cut to 3 mm length).

(y) Optical stimulation of ChR2: pulsed (5 ms at 20 Hz) laser light (Omicron-Laserage, Rodgau-Dudenhofen, Germany, LightHUB-4) of 460 nm.

2.2 Diffusion Tensor Imaging (DTI) Neurotractography in Anima Nobili

(a) Human participants (ethical approval and informed consent must be provided in accordance with local Ethics Committee rules).

(b) Either Siemens (Erlangen, Germany) Trio 3T scanner with a 12-channel head coil or 3 Tesla Philips Achieva magnetic resonance imaging (MRI) scanner.

(c) Diffusion toolbox (FDT) from FMRIB Centre in the FSL software package (http://www.fmrib.ox.ac.uk/fsl/).

(d) Diffusion Toolkit [21] for tract reconstruction.

(e) TrackVis visualization software [21].

3 Methods

3.1 Unravelling Neural Hodology Using Fluorescent and Non-fluorescent Neural Tract Tracers

1. *Neural Tract Tracing of Cortico-neostriato-nigro-tectal Pathways with a Fluorescent Dextran*

Animals

(a) Male Wistar rats (*Rattus norvergicus*; Rodentia; Muridae) aged 9–11 weeks (weighing 200–250 g, $n = 8$ per group) at the beginning of the experiment, from the animal facility of the Ribeirão Preto School of Medicine of the University of São Paulo (FMRP-USP), were used.

(b) These animals were housed four in a group in Plexiglas cages and provided free access to food and water throughout the experiment. They were kept in the experimental room for 48 h prior to the experiment and were maintained on a 12 h light/12 h dark cycle (lights on at 7 a.m.) at 23–25 °C.

(c) The study protocols complied with the Ethical Commission in Animal Experimentation of the FMRP-USP recommendations, which fulfils the principles of ethics for animal research adopted by the National Council for Control of Animal Experimentation (CONCEA) and were approved by the Commission of Ethics in Animal Research (FMRP-USP CEUA process 0074/2015).

Stereotaxic Surgery

(a) The animals were anesthetized with 92 mg/kg ketamine and 9.2 mg/kg xylazine, and submitted to a stereotaxic surgery as previously described [22] for unilateral implantation of two stainless steel guide cannulae (0.6 mm outside diameter, 0.4 mm inside diameter) in the midbrain aimed at the substantia nigra pars reticulata (SNpr) and in the telencephalon, aiming the caudate nucleus and putamen (CPu), according to the following coordinates (with Bregma as the reference): SNpr (anterior/posterior (AP): -6.36 mm; medial/lateral (ML): ± 1.2 mm; dorsal/ventral (DV): 3.6 mm) and CPu (AP: -1.80 mm; ML: ± 3.8 mm; DV: 4.8 mm).

(b) The cannulae were fixed to the skull with acrylic resin and two stainless-steel screws. Each guide cannula was sealed with stainless steel wire to protect it from blockage.

(c) Each rat was treated with an intramuscular injection of penicillin G-benzathine (120,000 UI; 0.2 mL) followed by intramuscular injection of the analgesic and anti-inflammatory drug flunixin meglumine (2.5 mg/kg). Afterward, the rats were allowed 5 days to recover from the surgical procedure.

Neural Tract Tracing

(a) A bidirectional neural tract tracer (AlexaFluor Texas red-conjugated dextran; 3000 MW; Molecular Probes, Eugene, OR, USA) was microinjected either into the CPu or in the SNpr in independent groups of animals in a volume of 0.3μL over the course of 5 min.

(b) Infusions were delivered using an infusion pump (Stoelting, Kiel, Wisconsin, USA) through a polyethylene tube (PE10) attached to a dental needle.

(c) The dental needle was left in place for 2 min after the end of each microinjection to allow local drug diffusion. After completing the microinjection procedure, the dental needle was removed, and the skin was sutured.

(d) Then, 7 days after the neural tract tracer microinjections, mice were deeply anesthetized with intramuscular injections of 92 mg/kg ketamine (Ketamina®) and 10 mg/kg xylazine (Dopaser®) and perfused intracardially with physiological saline followed by 4% paraformaldehyde (PFA, Sigma) dissolved in 0.1 M phosphate buffer (pH 7.4).

(e) The encephalon was removed, post-fixed in PFA for 4 h, and then transferred to 10% and 20% sucrose dissolved in 0.1 M sodium phosphate buffer, pH 7.3, at 4 °C for at least 12 h in each solution.

(f) The nervous tissue was immersed in 2-methylbutane (Sigma), frozen in dry ice (30 s), embedded in Tissue-Tek, sectioned (20 mm thick) using a cryostat (CM 1950, Leica, Wetzlar, Germany) at 20 °C and mounted with Fluoromount with DAPI (Electron Microscopy Sciences, Hatfield, Pennsylvania, USA) on silanized slides.

(g) After this procedure, the forebrain slices with the bidirectional neural tract tracer were mounted on glass slides and coverslipped with mineral oil, and both the SNpr and CPu were observed under fluorescence microscopy (AxioImager Z1 with APOTOME II, Zeiss, Oberkochen, Germany).

2. *Results and Discussion: Labeling of Cortico-neostriatal, Cortico-nigral, Neostriato-nigral, and Nigro-tectal Pathways*

Microinjection of the neural tract tracer into the neostriatum showed Texas red-Dextran labeled neurons in the deep pyramidal layer of the motor cortex (Fig. 1a), axonic fibers, and terminal buttons located in the reticulate division of the substantia nigra (SNpr), as shown in Fig. 1b, on the left, and Texas red-Dextran labeled cell bodies located in the substantia nigra pars compacta (SNpc), as shown in Fig. 1b (upper right corner). Fluorescent neural tract tracer microinjections in SNpr showed the presence of Texas red-Dextran labeled cell bodies located in the neostriatum (Fig. 1c), Texas red-Dextran labeled fibers and terminal buttons located in the deep layers of the superior colliculus (Fig. 1d) and the dorsal columns of the periaqueductal gray matter (dlPAG) (data not shown), in addition to neurons (Fig. 1e, f) and fibers (Fig. 1f) located in the deep pyramidal layer of the prelimbic cortex. These data correspond to the tractography of the cortico-neostriado-nigro-tectal pathway, which recruits brain and brainstem intermediate relé, and of the direct cortico-nigral pathway.

The mesencephalic tectum is involved with the processing of aversive stimuli both in humans [23, 24] and in other animals [25] and is recruited for the elaboration of panic attack-like behavior [26, 27]. Nevertheless, there are also connections between the mesencephalic tectum and the anterior

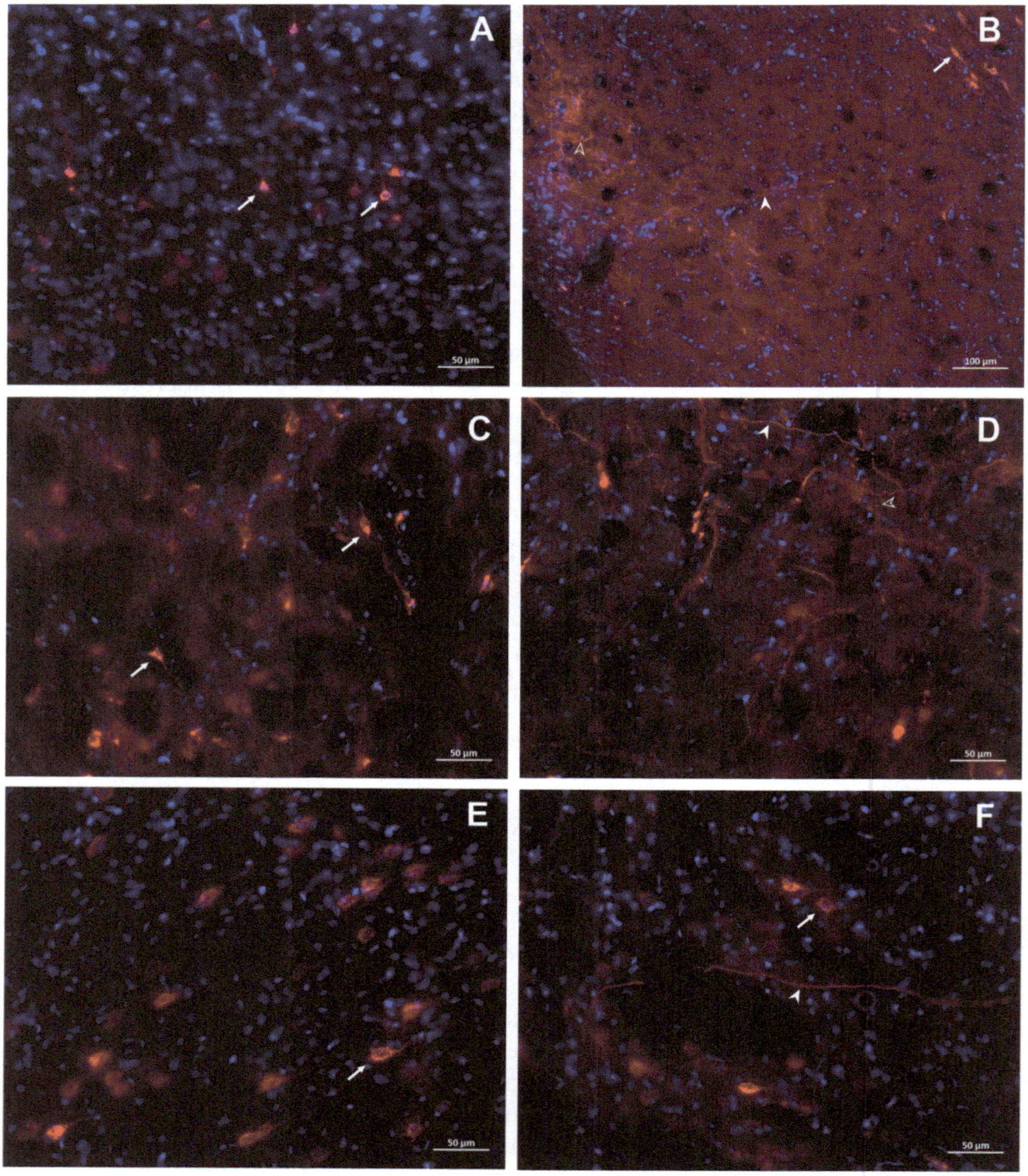

Fig. 1 Photomicrographs of coronal sections of the motor cortex (**a**), substantia nigra (**b**), neostriatum (**c**), deep layers of the superior colliculus (**d**), and prelimbic cerebral cortex (e and f), showing cell bodies (white arrows), axons (closed arrowheads), and varicosities suggestive of synapses (open arrowheads) labeled by the Texas Red-conjugated dextran bidirectional neural tract tracer, microinjected either in the neostriatum or in the reticulate division of the substantia nigra, which resulted in the labeling shown in **a** and **b**, and from **c** to **f**, respectively

[28] and medial hypothalamus [26, 29], probably recruited at different times during the elaboration of defense behavior, when there is a need to change the characteristics of non-oriented and oriented escape behavior toward safe places [30, 31]. Recruitment of endocannabinoid and endovanilloid neuromodulation in both the anterior [32] and medial hypothalamus [33] is relevant in modulating panic attack-like responses. The SNpr also consists of an important relay of the cortico-neostriado-pallido-nigro-tectal pathway [3, 22], and consists in a ventral midbrain structure connected to cpCS and SCPdl, where Gamma-aminobutyric acid (GABA)ergic afferents from the striatum-nigral projection may also be modulated by endocannabinoids [30, 31, 34]. Cortico-nigral connections, such as those demonstrated in the present work, may appear as a direct pathway between the pre-limbic and motor cortexes and the ventral midbrain, directly stimulating neuronal bodies of the nigro-tectal GABAergic pathway, which modulates the expression of innate fear-related defensive responses and panic attacks [3, 35]. The study of the connections demonstrated in the present work could open new horizons for understanding of defense behavior and treatment of the panic syndrome.

3. *Neural Tract Tracing with Biotinylated Dextran Amine Deposited by Iontophoresis in the Dorsal Midbrain*
   1. Stereotaxic Surgery
      - (a) The animals ($n = 6$) were anesthetized with ketamine at 92 mg/kg and xylazine at 9.2 mg/kg and fixed in a stereotaxic frame (David Kopf, Tujunga, California, USA).
      - (b) Biotinylated dextran amine (BDA: 3.000; Molecular Probes; USA) was deposited in the dorsolateral columns of the periaqueductal gray matter (dlPAG), by means of a micropipette implanted in the dorsal midbrain.
      - (c) The upper incisor bar was set at 3.3 mm below the interaural line so that the skull was horizontal between bregma and lambda.
      - (d) The micropipette was introduced vertically using the following coordinates, with the bregma serving as the reference for each plane: AP: −6.72 mm; ML: 0.8 mm, DV: 5.2 mm, considering the dlPAG.
      - (e) Micropipettes were made with a tip diameter of 40–50μm filled with 10% BDA dissolved in 0.01 M phosphate buffer, pH 7.4.

(f) BDA was deposited by iontophoresis: electrode tip positive, DC current pulses 5μA, 7 s on, 7 s off, over a period of 15 min.

(g) After the injection was completed, a protective layer of sterile gelatin foam was put in the bone cavity above the mesencephalon and the bone was closed with cyanoacrylate glue thickened with dental cement powder.

(h) Four days after surgery, the animal was deeply anesthetized with ketamine at 92 mg/kg and xylazine at 9.2 mg/kg and perfused transcardially at a rate of 3.5 mL/min with a 20 mL physiological saline followed by 100–200 mL of 2% paraformaldehyde and 2% glutaraldehyde in 0.1 M phosphate buffer, pH 7.4.

(i) After fixation, the midbrain was sectioned and mesencephalic tissues were rinsed in 10% and 20% sucrose dissolved in 0.1 M phosphate buffer, pH 7.4, at 4 °C for at least 12 h, in each solution.

(j) Tissue fragments were immersed in 2-methylbutane (Sigma), frozen on dry ice (30 s), embedded in Tissue Tek O.C.T. and cut with a cryostat (HM 505 E, Microm).

(k) During the processing of non-fluorescent BDA-labeled brainstem tissue, endogenous peroxidase activity was blocked by pre-incubation in 50, 70, and 50% ethanol solutions for 15, 20, and 15 min, respectively.

(l) BDA labeling was visualized using the avidin-biotin method (ABC standard Elite kit; Vector Laboratories) with nickel-enhanced DAB reactions [36].

(m) The sections were then washed thoroughly with 0.1 M phosphate buffer, pH 7.4, mounted on glass slides, counterstained with either hematoxylin-eosin or Nissl method and cover-slipped with Canada balsam and observed in the AxioObserver microscope (Zeiss, Germany).

2. Results and Discussion: Labeling of Intratectal and Nigrotectal Pathways

Iontophoretic deposits of BDA in the dlPAG (Fig. 2a) showed BDA-labeled neuronal perikarya, axonal fibers, and appositions suggesting terminal buttons in the mesencephalic reticular formation (Fig. 2b). It shows a reciprocate connections between these two structures, BDA labeled perikarya connected to dlPAG and ascending nigro-tectal connections in the deep layers of the superior colliculus (Fig. 2c). These data comprise part of the neural bases of

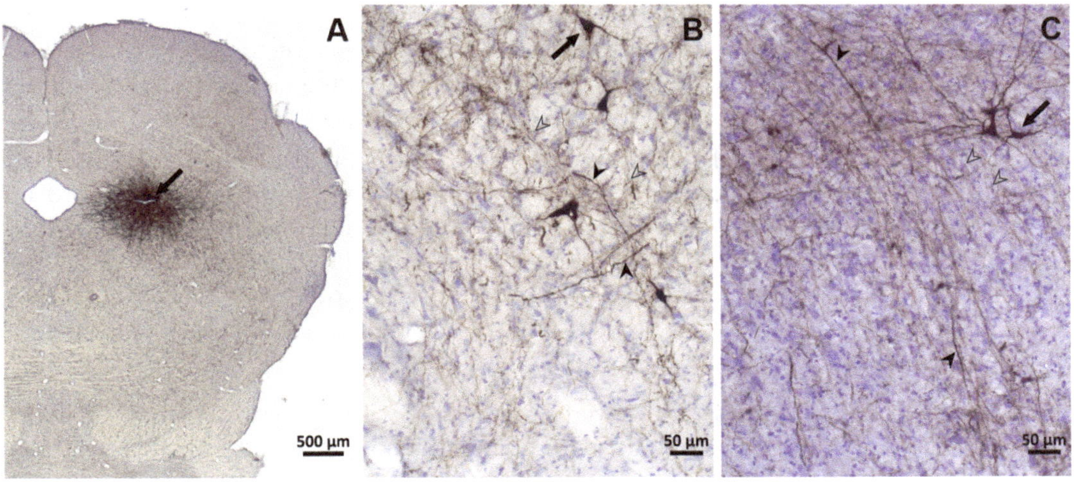

Fig. 2 Photomicrographs of transversal sections of Wistar rats mesencephalon, showing a representative site of iontophoretic deposit of biotinylated dextran amine (BDA) neural tract tracer (black arrow) in the dorsolateral column of the periaqueductal gray matter (**a**). BDA-labeled perikarya (black arrows) were found in both the mesencephalic tegmentum (**b**), at the level of the ascending activating reticular formation, and the midbrain tectum (**c**), at the level of the deep layers of the superior colliculus (dlSC). BDA-labeled axonal fibers (black arrowheads) and terminal buttons (open arrowheads) were also found in both the deep mesencephalon (**b**) and in the dlSC (**c**). Verticalized BDA-labeled axons found in the dlSC (**c**) seem to reach the dorsal midbrain from the substantia nigra pars reticulata, while horizontally oriented fibers also found in dlSC seem to reach the lateral regions of the midbrain tectum form neurons situated in the dlPAG

attentional posture displayed by the prey during threatening situations [30, 31] in which dlPAG neurons can activate the ascending reticular activation system in addition to the endogenous pain modulatory system nuclei recruited during the elaboration of unconditioned fear-induced antinociception [10]. Dorsal midbrain structures seem to be modulated by nigro-tectal ascending pathways [3].

3.2 Unraveling Functional Neuroanatomy by Optogenetic Procedure

1. *Neural Tract Tracing of Nigro-collicular Pathways by AAV5-hSyn-Jaws-GFP-ER2*
   (a) Experiments were performed with male C57BL/6 N mice purchased from Charles River (Bad Sulzfeld, Germany) aged 8 to 14 weeks.
   (b) To identify the origin of SNpr inputs to SC, mice were anesthetized with isoflurane (Forene®, Abbott, Germany, under induction at 4%, maintained at 1.5%), and head fixed in a stereotaxic frame (Leica Biosystems, Nussloch, Germany, AngleTwo). Craniotomies were made bilaterally and the stereotaxic coordinates used for SNr were AP −3.2 mm, ML ± 1.5 mm, DV −4.2 mm, from the skull surface.

(c) 350 nL of the viral construct AAV5-hSyn-Jaws-GFP-ER2 (University of North Carolina Vector Core, Chapel Hill, NC, USA) was one side injected using a Hamilton syringe (80 nL/min) into the SNpr.

(d) Four weeks after the injection, the animals were sacrificed with an overdose of isoflurane immediately transcardially perfused with ice cold PBS followed by 4% paraformaldehyde. Brains were post-fixed overnight at 4 °C in paraformaldehyde and then cryoprotected with 30% sucrose in PBS. The brains were sliced at 40 mm with a cryostat.

2. *Optogenetics Experiments*

(a) Mice were anesthetized with isoflurane (Forene®, Abbott, Germany, under induction at 4%, maintained at 1.5%), and head fixed in a stereotaxic frame (Leica Biosystems, Nussloch, Germany, AngleTwo). Craniotomies were made bilaterally and the stereotaxic coordinates used for SNpr were AP −3.2 mm, ML ± 1.5 mm, DV −4.2 mm, from the skull surface.

(b) Either 350 nL of AAV5-hSyn-ChR2(H134R)-mCherry (ChR2; n = 10) or the control vector AAV5-hSyn-mCherry (mCherry; n = 8) was bilaterally injected using a Hamilton syringe (80 nL/min) into the SNpr and mice were randomly assigned to ChR2 or mCherry groups. All viral aliquots were obtained from the University of North Carolina Vector Core (Chapel Hill, NC, USA).

(c) Six weeks later, fiber optic cannulas (Thorlabs, Dachau/Munich, Germany, CFML12L10, Ø200 μm, NA 0.39, cut to 3 mm length) were implanted targeting the dlSC (angle: 12°, AP: −3.9 mm, ML: ± 1.1 mm, DV: −2.0 mm). The mice were allowed to recover for 2 weeks after the fiber implantation.

(d) For optical stimulation of ChR2, pulsed (5 ms at 20 Hz) laser light (Omicron-Laserage, Rodgau-Dudenhofen, Germany, LightHUB-4) of 460 nm was applied. The laser output power was set to measure 7.5 mW at the single fiber tip. Bilateral optical stimulation was achieved using a fiber-optic rotary joint with two output ports (Doric Lenses Inc., Quebec, Canada, FRJ_1x2i_FC-2FC).

(e) For behavioral analyses the beetle mania task (BMT) and open field test (OFT) were used. The OFT session was divided into three 3-min epochs with alternating laser manipulation (OFF–ON–OFF; [37]). In the BMT, laser stimulation was performed only in the test phases. In Experiment 1, the laser was activated when the robo-beetle was in the same segment or the adjacent segment as the mouse. The offset of the stimulation was triggered if either the robo-beetle or the mouse left the "critical segments." The number of laser events was recorded. In Experiment 2, the laser was activated during the whole test phase. See findings elsewhere [17].

(f) The histological processing was performed as described in the experiment above. Placements of injection and implantation sites were verified at the end of the study, and only mice with adequate placement of the injection sites and optical fibers were included into the analyses.

3. *Results and Discussion: Nigro-tectal Projections and Functions*

To investigate the origin of the nigro-tectal pathway for optogenetic experiments, the AAV5-hSyn-Jaws-GFP-ER2 was microinjected in the SNpr (Fig. 3a). The transfection was robust and allowed the identification of axonal fibers from the SNpr in the dlSC (Fig. 3b). This indicates a homogeneous distribution of dlSC-projecting neurons within the SNpr. Using optogenetics in combination with the established etho-behavioral beetle mania task (data not shown), we provide first direct evidence that activation of the nigro-tectal pathway at the level of the SC reduces threat recognition and active avoidance, while increasing tolerance and approach behavior toward an erratic moving robo-beetle, without affecting exploratory behavior in an open field [17].

3.3 Human Brain Connectivity Revealed by Diffusion Tensor Neurotractography

1. *Data Acquisition (Protocol 1)*

(a) The images were acquired on a Siemens (Erlangen, Germany) Trio 3T scanner with a 12-channel head coil.

(b) Diffusion-weighted images were acquired in the axial plane using an echo planar imaging sequence (3 acquisitions of 60 directions with 5 non-diffusion-weighted images, b-value 1000 s mm −2, voxel size 1.5 × 1.5 × 1.5 mm, 100 slices).

(c) Encephalic areas incorporated the whole brain including brainstem. Cardiac gating was performed to minimize artifacts from pulsatile flow of the cerebrospinal fluid of which the brainstem is particularly sensitive.

(d) Brainstem optimization has resulted in a unique high-quality dataset that is not available on public databases. Each subject also had a T1-weighted high-resolution (1 × 1 × 1 mm voxels) structural image acquired to aid registration.

2. *Diffusion MRI Data Preprocessing.*

(a) Preprocessing was performed using diffusion toolbox (FDT) from the University of Oxford Functional

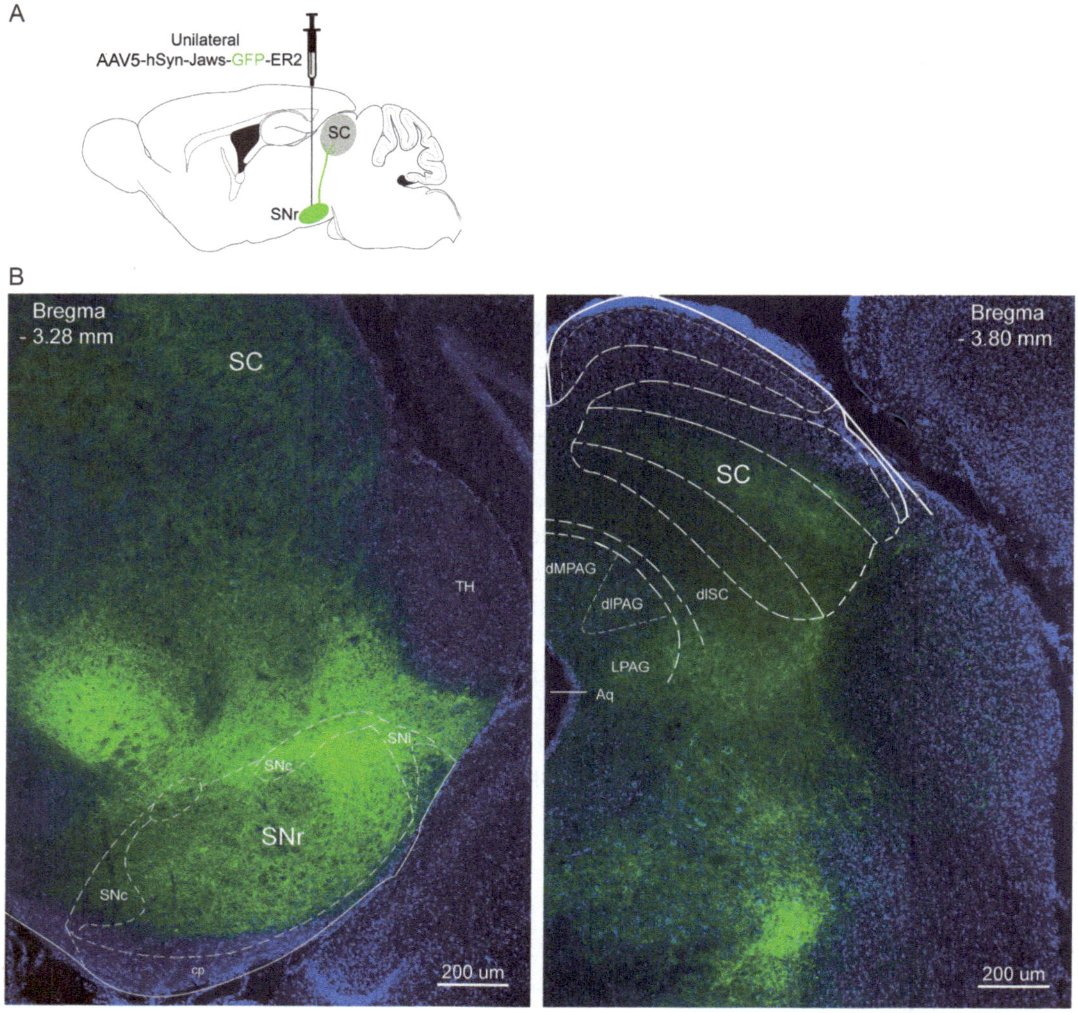

Fig. 3 Infusion of viral vectors into the substantia nigra pars reticulata (SNr). (**a**) Schematic illustration of the viral vector injection. (**b**) Representative coronal photomicrographs showing the expression of GFP in SNr somata, as well as in SNr terminals within the periaqueductal gray (PAG) matter and superior colliculus (blue: DAPI, green: GFP). SC: Superior colliculus; TH: metathalamus (medial geniculate nucleus); SNr: Substantia nigra pars reticulata; SNc: Substantia nigra pars compacta; SNl: Substantia nigra pars lateralis; dmPAG: Dorsomedial column of the periaqueductal gray matter; dlPAG: Dorsolateral column of the periaqueductal gray matter; LPAG: Lateral column of the periaqueductal gray matter; Aq: Aqueductus Sylvii

Magnetic Imaging of the Brain (FMRIB) Center FSL (FMRIB Center Software Library) software package (http://www.fmrib.ox.ac.uk/fsl/). This included extraction of non-brain tissue using brain extraction tool (BET) [38] and affine registration to a reference volume to correct for eddy currents and head motion using a new tool to correct for eddy current-induced distortions and subject movements [39].

(b) The data from the three acquisitions for each subject were averaged to improve the signal to noise ratio and voxel-wise estimates of fiber orientation and uncertainty was carried out using BEDPOSTX (the second step in the DTI processing pipeline that sets the stage for probabilistic tractography; BEDPOSTX stands for Bayesian Estimation of Diffusion Parameters Obtained using Sampling Techniques. The X stands for modelling Crossing Fibres [40, 41].

3. *Image Registration*

(a) After preprocessing, the participant's diffusion-weighted scans were registered to the MNI152 1 mm standard space (average T1 brain image constructed from 152 normal subjects at the Montreal Neurological institute, Montreal, QC, Canada).

(b) Registration was performed as a three-step procedure via the high-resolution T1-weighted structural image with linear registration using the FMRIB Center-produced Linear Image Registration Tool (FLIRT), a fully automated robust and accurate tool for linear (affine) intra- and inter-modal brain image registration and non-linear registration using FNIRT, a non-linear registration method for FSL (a comprehensive library of tools for fMRI, MRI and diffusion brain imaging-related data), which outputs a deformation or coefficient field [39, 42].

4. *Definition of Seed and Target Masks (Structures of Interest)*

(a) Seed masks for the regions of interest for tractography were drawn for one brain hemisphere for the: (i) caudate nucleus, (ii) substantia nigra. This was done with fslview in FSL (http://www.fmrib.ox.ac.uk/fsl) for only one subject.

(b) These regions were chosen based on their involvement in the control of unconditioned fear-related behavior [3, 22, 30, 31].

(c) Duvernoy's atlas of the human brain [43] was used to locate the caudate nucleus, the substantia nigra, and the periaqueductal gray matter.

(d) The volunteers were screened for magnetic resonance imaging (MRI) compatibility and scanned with ethical approval and informed consent in accordance with the Oxfordshire Clinical Research Ethics Committee. (Data were obtained by Oxford University FMRIB Centre data collection. No new brain was scanned at this time.)

5. *Probabilistic Tractography*

(a) Probabilistic tractography was carried out for the participant using previously described methods [40, 41], with

FMRIB Center Diffusion Toolbox (FDT) (http://www.fmrib.ox.ac.uk/fsl/) with 10,000 samples per voxel. FDT is part of FSL software, consisting in an easy-to-use graphical user interface (GUI) and its component programmes can also be run from the command line.

(b) Estimates of the connections between each voxel in the seed regions and every voxel of the whole brain were then calculated. This generates a connectivity profile for each seed voxel and is derived from the number of samples that arrive at each target voxel. To reduce the false-positive connections, the path distribution estimates were thresholder to a connection probability of $P < 0.0003$.

(c) A high sample number and low threshold was chosen to improve identification of small cortical tracts.

(d) A cross-correlation matrix between the connectivity profiles of all voxels in the seed mask was then calculated [44].

6. *Results and Discussion: Neurotractography of Neostriato-nigrotectal Pathways*

Our findings showed connections between the neostriatum and structures situated in at least two brainstem main divisions (mesencephalon and pons) also reaching the cerebellum, as shown in Fig. 4a. However, our data highlight connections between the neostriatum (caudate nucleus) and the substantia nigra, reaching the ventral midbrain through the ventromedial division of the internal capsule, and these connections involve the ventromedial two-thirds of the substantia nigra (Fig. 4b). In addition, high probabilistic connections were identified in the mesencephalic tegmentum projecting from the substantia nigra toward the dorsal midbrain, reaching the periaqueductal gray matter (Fig. 4c). The neostriato-nigrotectal pathways showed by DTI procedure demonstrate that similar connections between the caudate-putamen, the SNpr, and the dlSC/dlPAG continuum described in laboratory animals [3, 30] are also found in humans. These findings support the hypothesis that similar control of midbrain tectum neurons by SNpr inhibitory neurons that are under a disinhibitory projection from neostriatum and pallidum, as described elsewhere in laboratory animals [22, 30], seems to be also present in our species.

7. *Data Acquisition Protocol 2*

(a) Magnetic resonance images were acquired in a 3 T Philips Achieva MRI scanner; Diffusion-weighted imaging (DWI), a form of magnetic resonance imaging based upon measuring the random Brownian motion of water molecules within a voxel of tissue: The MRI sequence, a number of radiofrequency pulses and gradients that result

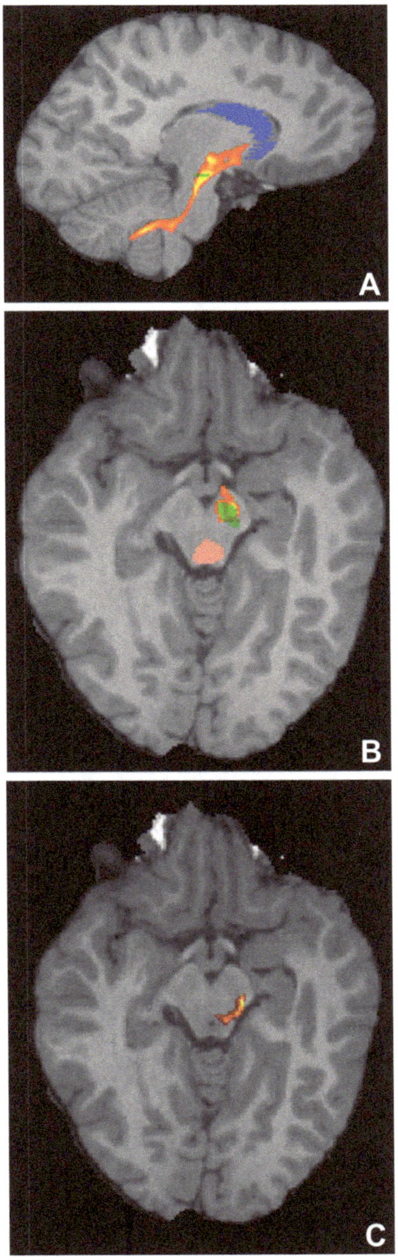

Fig. 4 Neurotractography by diffusion tensor imaging of neostriato-nigral pathways (**a**) reaching the substantia nigra in the ventral midbrain (**b**), and the nigro-tectal pathways (**c**) connecting the substantia nigra to the periaqueductal gray matter. Non-published data obtained by N.C.Coimbra, S. Jbabdi, and I. Tracey (2006). Reproduced with permission by Dr. Saad Jbabdi from Wellcome Centre for Integrative Neuroimaging, FMRIB, Wolfson Building, Nuffield Department of Clinical Neuroscience, John Radcliffe Hospital, University of Oxford, Oxford, OX3 9DU, UK

in a set of images with a particular appearance, was combined with echo planar imaging (EPI) acquisition, the fastest acquisition method in MRI (100 ms / slice), but with limited spatial resolution, was performed with the following parameters: TR/TE = 9300/54 ms, voxel size 2 mm isotropic, EPI factor = 67, FOV = 256 × 256 mm2, 60 slices, 33 diffusion gradients, being 32 with b = 1000 s/mm2 and one with b = 0, overplus = no. The PAR/REC images collected from the scanner were converted to NIfTI format using the dcm2nii extension MRIcroGL (http://www.mccauslandcenter.sc.edu/mricrogl/).

8. *DWI Processing*

   (a) Images were motion corrected by standard Philips processing in the scanner, and a preprocessing pipeline was performed on the DWI data.

   (b) The preprocessing pipeline in FSL software (Oxford University FMRIB Centre, Oxford, United Kingdom) [45] includes a correction for eddy current distortions and motion [46], brain extraction [38], bias field corrections, tensor model fitting using ordinary least squares, and FA and MD scalar map calculation using FDT toolbox [45].

   (c) Tract reconstruction was performed using the Diffusion Toolkit [21].

9. *Results and Discussion: Neurotractography of Cortico-hypothalamic, cortico-midbrain Pathways, and Neural Links Between Hypothalamic Nuclei and the Midbrain Tectum*

   Neural pathways were found connecting the rostral division of the frontal lobe cortex, such as the medial part of the superior frontal gyrus (Fig. 5a, b) and the anterior region of the cingulate gyrus (Fig. 5a, c) with the ventral midbrain (Fig. 5a and, respectively), where the substantia nigra is located, and also with the posterior hypothalamus (Fig. 5b, c, respectively), as well as connections between nuclei from medial hypothalamus (Fig. 5a) and the anterior hypothalamus (Fig. 5b, c) with the dorsal periaqueductal gray matter (Fig. 5a–c). These data suggest a connection between forebrain areas and human diencephalic and mesencephalic structures related to the elaboration of panic attack-like behavior both in humans [48] and in laboratory animals [22, 28, 30, 49].

   The rostral part of the frontal lobe, also known as the medial prefrontal cortex, has been reported to contribute to the elaboration of defensive behavior. Indeed, serotonergic and γ-aminobutyric acid (GABA)ergic neurotransmission in the medial prefrontal cortex are implicated in the modulation of conditioned fear [34].

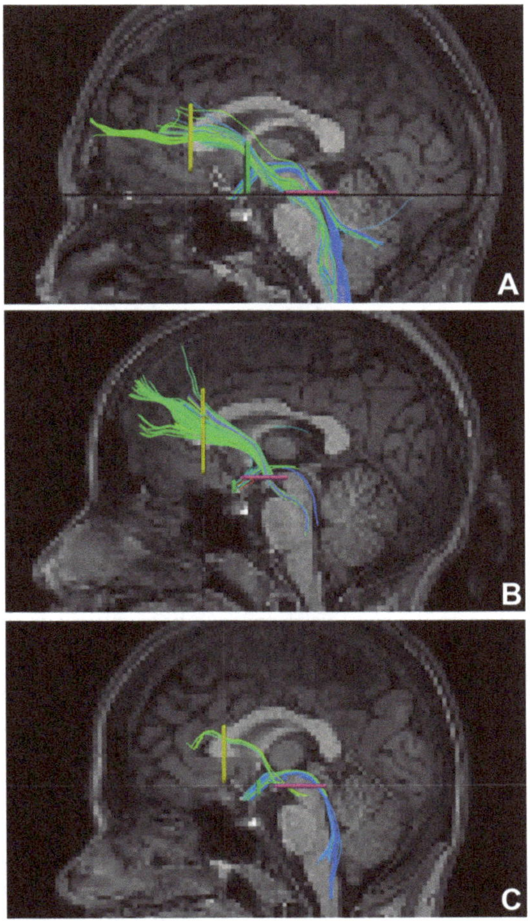

Fig. 5 Photomicrographs of parasagittal sections of the human brain, showing in artificial green and purple colors and indicated by the yellow bar, pathways that connect the rostral division of the frontal lobe cortex, such as the medial part of the superior frontal gyrus (**a** and **b**) and the anterior region of the cingulate gyrus (**a** and **c**) with the ventral midbrain (**a** and **c**, respectively), where the substantia nigra is located, and also with the posterior hypothalamus (**b** and **c**, respectively), as well as connections between the nucleus of the medial hypothalamus (**a**) and anterior hypothalamus (**b** and **c**) and the dorsal periaqueductal gray matter (**a**–**c**), in blue and green color, indicated by the green bar

Our team has demonstrated in laboratory animals that glutamatergic connections between the anterior part of the cingulate gyrus and the posterior hypothalamus modulate both the affective defense responses elicited by microinjection of N-methyl-D-aspartate (NMDA) into the posterior hypothalamus and the antinociception induced by unconditioned fear [49]. Similarly, inputs from the dorsomedial division of the ventromedial hypothalamus to the dorsal periaqueductal gray matter (dPAG) exert a role during the orchestration of interchanging between oriented escape elaborated by hypothalamic neurons to non-oriented/explosive escape behavior and freezing responses, organized by dlPAG neurons [26]. Nuclei of the medial hypothalamus are effectively involved in defense, such as panic attack-like responses displayed by laboratory animals [33] and panic symptoms reported by humans [48]. Nevertheless, the cannabinoid system of the anterior nucleus of the hypothalamus also plays an important role in modulating defensive behavior [32] in threatening situations. Interestingly, it was demonstrated by our team, also in laboratory animals, that glutamatergic projections from the 24b area of the cingulate gyrus modulate defensive antinociception mediated by nitric oxide in the anterior nucleus of the anterior hypothalamus [50]. It is also interesting to point out that responses such as panic attacks and antinociception induced by innate fear caused by the endogenous increase of nitric oxide in the anterior nucleus of the hypothalamus are modulated by glutamatergic pathways that connect the anterior hypothalamus to the dlPAG [28].

Evidence for pathways connecting the medial prefrontal cortex and the rostral part of the cingulate gyrus with hypothalamic nuclei and with the ventral midbrain and between hypothalamic nuclei and dorsal midbrain in humans corroborates our recently published data. These findings further expand current knowledge of the complex neural network that elaborates and modulates conditioned and unconditioned fear-related behavior and defensive antinociceptive processes.

4 Notes

1. The fluorescent neural tract tracer must be handled in an environment protected from light [3, 30].
2. Both dextran amine and biotinylated dextran amine neural tract tracers can be stored at −80 °C in the freezer in aliquots for further experimental use and are prepared for delivery as a 10% solution in 0.1 M sodium phosphate-buffered saline (pH 7.4) [3].
3. During the processing of non-fluorescent BDA- labeled brainstem tissue, endogenous peroxidase activity must be blocked by pre-incubation in 50, 70, and 50% ethanol solutions for 15, 20, and 15 min, respectively [3, 10, 13, 36].
4. Avidin-biotin method can be or not associated with nickel-enhanced DAB reactions. VIP DAB kits are an alternative for double-labeling procedures [3].

5. For beetle mania task an open field test was used (data not shown) [17]. However, several other psychobiological experiment could be done for different purposes [10].
6. The processing occurs in two steps: (1) reconstruction-diagonalization of diffusion data, in which the input data are the corrected 4D DW image, bvecs (3 rows of space-separated floating-point value) table, and value of bval (a single row of space-separated floating-point value); (2) tracking-reconstruction of whole brain fibers, considering angle threshold of 35° and FA threshold of 0.2, using fiber assignment by continuous tracking (FACT) propagation algorithm [47], with inversion of the X diffusion gradient direction.
7. Tract visualization and processing were performed with the TrackVis visualization program [21], in which track files generated in the prior step were uploaded, together with FA grayscale, FA color, and MD maps, for tract selection and analysis.
8. Several tracts were manually segmented using TrackVis, based on known anatomical information in the literature [43].

Acknowledgments

This work was supported by the Conselho Nacional de Pesquisa e Desenvolvimento Tecnológico (CNPq) (Research Grants 470119/2004-7 and 427397/2018-9), Fundação de Amparo à Pesquisa do Estado de São Paulo (FAPESP) (Research Grants 2020/15050-7 and 2021/14073-6), and a Pro-Rectory of the University of São Paulo (USP) Research Grant (NAP-USP-NuPNE; process IaPq2012-156-USP-12.1.25440.01.6). R.C. Almada was supported by CAPES (postdoctoral fellowship grant: PNPD20131680-33002029012P3-PNPD-USP/RP/MEDICINA NEUROLOGIA) and FAPESP (postdoctoral fellowship grant 2012/22681-7; Young Investigator Program research grant 2018/03898-1). Ullah was supported by The Academy of Sciences for the Developing World (TWAS)-CNPq (Scientiae Doctor Fellowships processes 190316/2010-1). R. L. de Freitas was supported by FAPESP (Scientific Initiation Scholarship 01/03752-6, M.Sc. fellowship 03/05256-1, post-doctoral fellowship 2009/17258-5, Young Researchers in Emergent Centres Program fellowship 2013/12916-0, and research grant 2014/11869-0) and CAPES (Sc.D. Fellowship). Each organization played no further role in the design of the study; the collection, analysis, and interpretation of data; the writing of the report; or the decision to submit the paper for publication. N.C. Coimbra is a researcher from CNPq (grants PQ_{1A}: 301905/2010-0 and 301341/2015-0, PQ_2: 302605/2021-5) and was a CNPq-Post-doctoral fellow (grant 200629/2005-0) in the Physiology, Anatomy and Genetics Department and the Clinical Neurology (FMRIB Centre) Department of the University of Oxford, England, United Kingdom. The authors are grateful to D.H. Elias-Filho for their expert technical assistance. D.H. Elias Filho was a recipient of scholarships sponsored by CNPq (grants 505461/2010-2 and 372838/2018-9) and FAEPA (grants 345/2009 and 185/2010).

Conflict of Interest The authors declare that there are no conflicts of interest with respect to the presented work.

References

1. Freitas RL, Ferreira CMR, Ribeiro SJ, Carvalho AD, Elias-Filho DH, Garcia-Cairasco N, Coimbra NC (2005) Intrinsic neural circuits between dorsal midbrain neurons that control fear-induced responses and seizure activity and nuclei of the pain inhibitory system elaborating postictal antinociceptive processes: a functional neuroanatomical and neuropharmacological study. Exp Neurol 191:225–242. https://doi.org/10.1016/j.expneurol.2004.10.009
2. Couto LB, Moroni CR, dos Ferreira CMR, Elias-Filho DH, Parada CA, Pelá IR, Coimbra NC (2006) Descriptive and functional neuroanatomy of locus coeruleus-noradrenaline-containing neurons involvement in bradykinin-induced antinociception on principal sensory trigeminal nucleus. J Chem Neuroanat 32:28–45. https://doi.org/10.1016/j.jchemneu.2006.03.003
3. Castellan-Baldan L, da Costa Kawasaki M, Ribeiro SJ, Calvo F, Corrêa VMA, Coimbra NC (2006) Topographic and functional neuroanatomical study of GABAergic disinhibitory striatum-nigral inputs and inhibitory nigrocollicular pathways: Neural hodology recruiting the substantia nigra, pars reticulata, for the modulation of the neural activity in the inferior colliculus involved with panic-like emotions. J Chem Neuroanat 32:1–27. Erratum in: J Chem Neuroanat 2007;33(1):65–66. https://doi.org/10.1016/j.jchemneu.2006.05.002
4. van der Kooy D, Koda LY, McGinty JF, Gerfen CR, Bloom FE (1984) The organization of projections from the cortes, amygdala, and hypothalamus to the nucleus of the solitary tract in rat. J Comp Neurol 224:1–24. https://doi.org/10.1002/cne.902240102
5. de Olmos J, Heimer L (1980) Double and triple labeling of neurons with fluorescent substances; The study of collateral pathways in the ascending raphe system. Neurosci Lett 19:7–12. https://doi.org/10.1016/0304-3940(80)90247-5
6. Takato M, Goldring S (1979) Intracellular marking with Lucifer Yellow CH and horseradish peroxidase of cells electrophysiologically characterized as glia in the cerebral cortex of the cat. J Comp Neurol 186:173–188. https://doi.org/10.1002/cne.901860205
7. Xu X, Holmes TC, Luo MH, Beier KT, Horwitz GD, Zhao F, Zeng W, Hui M, Semler BL, Sandri-Goldin RM (2020) Viral vectors for neural circuit mapping and recent advances in trans-synaptic anterograde tracers. Neuron 107:1029–1047. https://doi.org/10.1016/j.neuron.2020.07.010
8. Gerfen CR, Sawchenko PE (1984) An anterograde neuroanatomical tracing method that shows the detailed morphology of neurons, their axons and terminals: Immunohistochemical localization of an axonally transported plant lectin, Phaseolus vulgaris leucoagglutinin (PHA-L). Brain Res 290:219–238. https://doi.org/10.1016/0006-8993(84)90940-5
9. Thompson AM, Thompson GC (1988) Neural connections identified with PHA-L anterograde and HRP retrograde tract-tracing techniques. J Neurosci Methods 25:13–17. https://doi.org/10.1016/0165-0270(88)90115-X
10. Coimbra NC, De Oliveira R, Freitas RL, Ribeiro SJ, Borelli KG, Pacagnella RC, Moreira JE, Da Silva LA, Melo LL, Lunardi LO, Brandão ML (2006) Neuroanatomical approaches of the tectum-reticular pathways and immunohistochemical evidence for serotonin-positive perikarya on neuronal substrates of the superior colliculus and periaqueductal gray matter involved in the elaboration of the defensive behavior and fear-induced analgesia. Exp Neurol 197:93–112. https://doi.org/10.1016/j.expneurol.2005.08.022

11. Myers RD (1966) Injection of solutions into cerebral tissue: relation between volume and diffusion. Physiol Behav 1:171. https://doi.org/10.1016/0031-9384(66)90064-3
12. Routtenberg A (1972) Intracranial chemical injection and behavior: a critical review. Behav Biol 7:601–641. https://doi.org/10.1016/S0091-6773(72)80073-7
13. Ribeiro SJ, Ciscato JG Jr, De Oliveira R, De Oliveira RC, D'Ângelo-Dias R, Carvalho AD, Felippotti TT, Rebouças ECC, Castellan-Baldan L, Hoffmann A, Corrêa SAL, Moreira JE, Coimbra NC (2005) Functional and ultrastructural neuroanatomy of interactive intratectal/tectonigral mesencephalic opioid inhibitory links and nigrotectal GABAergic pathways: involvement of $GABA_A$ and μ_1-opioid receptors in the modulation of panic-like reactions elicited by electrical stimulation of the dorsal midbrain. J Chem Neuroanat 30:184–200. https://doi.org/10.1016/j.jchemneu.2005.07.004
14. Zhu X, Lin K, Liu Q, Yue X, Mi H, Huang X, He X, Wu R, Zheng D, Wei D, Jia L, Wang W, Manyande A, Wang J, Zhang Z, Xu F (2020) Rabies virus pseudotyped with CVS-N2C glycoprotein as a powerful tool for retrograde neuronal network tracing. Neurosci Bull 36:202–216. https://doi.org/10.1007/s12264-019-00423-3
15. Yawo H, Asano T, Sakai S, Ishizuka T (2013) Optogenetic manipulation of neural and non-neural functions. Develop Growth Differ 55:474–490. https://doi.org/10.1111/dgd.12053
16. Zhao S, Cunha C, Zhang F, Liu Q, Gloss B, Deisseroth K, Augustine GJ, Feng G (2008) Improved expression of halorhodopsin for light-induced silencing of neuronal activity. Brain Cell Biol 36:141–154. https://doi.org/10.1007/s11068-008-9034-7
17. Almada RC, Genewsky AJ, Heinz DE, Kaplick PM, Coimbra NC, Wotjak CT (2018) Stimulation of the nigrotectal pathway at the level of the superior colliculus reduces threat recognition and causes a shift from avoidance to approach behavior. Front Neural Circuits 12:1–9. https://doi.org/10.3389/fncir.2018.00036
18. Kim CK, Adhikari A, Deisseroth K (2017) Integration of optogenetics with complementary methodologies in systems neuroscience. Nat Rev Neurosci 18:222–235. https://doi.org/10.1038/nrn.2017.15
19. Surgent O, Riaz A, Ausderau KK, Adluru N, Kirk GR, Guerrero-Gonzalez J, Skaletski EC, Kecskemeti SR, Dean DC, Weismer SE, Alexander AL, Travers BG (2022) Brainstem white matter microstructure is associated with hyporesponsiveness and overall sensory features in autistic children. Mol Autism 13:1–18. https://doi.org/10.1186/s13229-022-00524-3
20. Ouyang Z, Zhang N, Li M, Hong T, Taohui Ouyang WM (2023) A meta-analysis of the role of diffusion tensor imaging in cervical spinal cord compression. J Neuroimaging. https://doi.org/10.1111/jon.13093
21. Wang R, Benner T (2007) Diffusion toolkit: a software package for diffusion imaging data processing and tractography. Proc Int Soc Magn Reson Med 15:3720
22. da Silva JA, Almada RC, Falconi-Sobrinho LL, Pigatto GR, Hernandes PM, Coimbra NC (2023) Neostriatum neuronal TRPV1-signalling mediates striatal anandamide at high concentration facilitatory influence on neostriato-nigral dishinhibitory GABAergic connections. Brain Res Bull 192:128–141. https://doi.org/10.1016/j.brainresbull.2022.11.014
23. Nashold BS, Wilson WP, Slaughter DG (1969) Sensations evoked by stimulation in the midbrain of man. J Neurosurg 30:14–24. https://doi.org/10.3171/jns.1969.30.1.0014
24. Tracey I, Ploghaus A, Gati JS, Clare S, Smith S, Menon RS, Matthews PM (2002) Imaging attentional modulation of pain in the periaqueductal gray in humans. J Neurosci 22:2748–2752. https://doi.org/10.1523/jneurosci.22-07-02748.2002
25. Paschoalin-Maurin T, dos Anjos-Garcia T, Falconi-Sobrinho LL, de Freitas RL, Coimbra JPC, Laure CJ, Coimbra NC (2018) The rodent-versus-wild Snake paradigm as a model for studying anxiety- and panic-like behaviors: face, construct and predictive validities. Neuroscience 369:336–349. https://doi.org/10.1016/j.neuroscience.2017.11.031
26. Ullah F, dos Anjos-Garcia T, Mendes-Gomes J, Elias-Filho DH, Falconi-Sobrinho LL, de Freitas RL, Khan AU, de Oliveira R, Coimbra NC (2017) Connexions between the dorsomedial division of the ventromedial hypothalamus and the dorsal periaqueductal grey matter are critical in the elaboration of hypothalamically mediated panic-like behaviour. Behav Brain Res 319:135–147. https://doi.org/10.1016/j.bbr.2016.11.026
27. Ullah F, dos Anjos-Garcia T, dos Santos IR, Biagioni AF, Coimbra NC (2015) Relevance of dorsomedial hypothalamus, dorsomedial division of the ventromedial hypothalamus and the dorsal periaqueductal gray matter in

the organization of freezing or oriented and non-oriented escape emotional behaviors. Behav Brain Res 293:143–152. https://doi.org/10.1016/j.bbr.2015.07.013

28. Falconi-Sobrinho LL, dos Anjos-Garcia T, Hernandes PM, de Rodrigues BMP, Almada RC, Coimbra NC (2023) Unravelling the dorsal periaqueductal grey matter NMDA receptors relevance in the nitric oxide-mediated panic-like behaviour and defensive antinociception organised by the anterior hypothalamus of male mice. Psychopharmacology 240:319–335. https://doi.org/10.1007/s00213-023-06309-7
29. de Freitas RL, Salgado-Rohner CJ, Hallak JEC, de Souza Crippa JA, Coimbra NC (2013) Involvement of prelimbic medial prefrontal cortex in panic-like elaborated defensive behaviour and innate fear-induced antinociception elicited by $GABA_A$ receptor blockade in the dorsomedial and ventromedial hypothalamic nuclei: role of the endocannabinoid. Int J Neuropsychopharmacol 16:1781–1798. https://doi.org/10.1017/S1461145713000163
30. Almada RC, dos Anjos-Garcia T, da Silva JA, Pigatto GR, Wotjak CT, Coimbra NC (2021) The modulation of striatonigral and nigrotectal pathways by CB1 signalling in the substantia nigra pars reticulata regulates panic elicited in mice by urutu-cruzeiro lancehead pit vipers. Behav Brain Res 401:112996. https://doi.org/10.1016/j.bbr.2020.112996
31. Almada RC, Falconi-Sobrinho LL, da Silva JA, Wotjak CT, Coimbra NC (2022) Augmented anandamide signalling in the substantia nigra pars reticulata mediates panicolytic-like effects in mice confronted by *Crotalus durissus terrificus* pit vipers. Psychopharmacology 239:2753–2769. https://doi.org/10.1007/s00213-022-06127-3
32. dos Anjos-Garcia T, Coimbra NC (2020) Anandamide in the anterior hypothalamus diminishes defensive responses elicited in mice threatened by *Epicrates cenchria* constrictor serpents. Acta Neurobiol Exp (Wars) 80:179–191. https://doi.org/10.21307/ane-2020-017
33. dos Anjos-Garcia T, Coimbra NC (2019) Opposing roles of dorsomedial hypothalamic CB1 and TRPV1 receptors in anandamide signaling during the panic-like response elicited in mice by Brazilian rainbow *Boidae* snakes. Psychopharmacology 236:1863–1874. https://doi.org/10.1007/s00213-019-5170-2
34. Almada RC, Coimbra NC, Brandão ML (2015) Medial prefrontal cortex serotonergic and GABAergic mechanisms modulate the expression of contextual fear: intratelencephalic pathways and differential involvement of cortical subregions. Neuroscience 284:988–997. https://doi.org/10.1016/j.neuroscience.2014.11.001
35. Coimbra NC, Brandão ML (1993) GABAergic nigro-collicular pathways modulate the defensive behaviour elicited by midbrain tectum stimulation. Behav Brain Res 59:131–139. https://doi.org/10.1016/0166-4328(93)90159-N
36. Veenman CL, Reiner A, Honig MG (1992) Biotinylated dextran amine as an anterograde tracer for single- and double-labeling studies. J Neurosci Methods 41:239–254. https://doi.org/10.1016/0165-0270(92)90089-V
37. Felix-Ortiz AC, Burgos-Robles A, Bhagat ND, Leppla CA, Tye KM (2016) Bidirectional modulation of anxiety-related and social behaviors by amygdala projections to the medial prefrontal cortex. Neuroscience 321:197–209. https://doi.org/10.1016/j.neuroscience.2015.07.041
38. Smith SM (2002) Fast robust automated brain extraction. Hum Brain Mapp 17:143–155. https://doi.org/10.1002/hbm.10062
39. Jenkinson M, Smith S (2001) A global optimisation method for robust affine registration of brain images. Med Image Anal 5:143–156. https://doi.org/10.1016/S1361-8415(01)00036-6
40. Behrens TEJ, Woolrich MW, Jenkinson M, Johansen-Berg H, Nunes RG, Clare S, Matthews PM, Brady JM, Smith SM (2003) Characterization and propagation of uncertainty in diffusion-weighted MR imaging. Magn Reson Med 50:1077–1088. https://doi.org/10.1002/mrm.10609
41. Behrens TEJ, Berg HJ, Jbabdi S, Rushworth MFS, Woolrich MW (2007) Probabilistic diffusion tractography with multiple fibre orientations: what can we gain? NeuroImage 34:144–155. https://doi.org/10.1016/j.neuroimage.2006.09.018
42. Jenkinson M, Bannister P, Brady M, Smith S (2002) Improved optimization for the robust and accurate linear registration and motion correction of brain images. NeuroImage 17:825–841. https://doi.org/10.1006/nimg.2002.1132
43. Duvernoy HM (1999) The human brain: surface, blood supply, and three-dimensional sectional anatomy. Springer-Verlag, Wien, 491 pp
44. Johansen-Berg H, Behrens TEJ, Robson MD, Drobnjak I, Rushworth MFS, Brady JM, Smith SM, Higham DJ, Matthews PM (2004)

Changes in connectivity profiles define functionally distinct regions in human medial frontal cortex. Proc Natl Acad Sci USA 101: 13335–13340

45. Jenkinson M, Beckmann CF, Behrens TEJ, Woolrich MW, Smith SM (2012) NeuroImage 62:782–790. https://doi.org/10.1016/j.neuroimage.2011.09.015
46. Andersson JLR, Sotiropoulos SN (2016) An integrated approach to correction for off-resonance effects and subject movement in diffusion MR imaging. NeuroImage 125: 1063–1078. https://doi.org/10.1016/j.neuroimage.2015.10.019
47. Mori S, Crain BJ, Chacko VP, van Zijl PC (1999) Three-dimensional tracking of axonal projections in the brain by magnetic resonance imaging. Ann Neurol 45:247–250. https://doi.org/10.1002/1531-8249(199902)45:2<247::AID-ANA16>3.0.CO;2-U
48. Wilent WB, Oh MY, Buetefisch CM, Bailes JE, Cantella D, Angle C, Whiting DM (2010) Induction of panic attack by stimulation of the ventromedial hypothalamus: case report. J Neurosurg 112:1295–1298. https://doi.org/10.3171/2009.9.JNS09577
49. Falconi-Sobrinho LL, dos Anjos-Garcia T, de Oliveira R, Coimbra NC (2017) Decrease in NMDA receptor-signalling activity in the anterior cingulate cortex diminishes defensive behaviour and unconditioned fear-induced antinociception elicited by GABAergic tonic inhibition impairment in the posterior hypothalamus. Eur Neuropsychopharmacol 27: 1120–1131. https://doi.org/10.1016/j.euroneuro.2017.09.002
50. Falconi-Sobrinho LL, dos Anjos-Garcia T, Coimbra NC (2021) Nitric oxide-mediated defensive and antinociceptive responses organised at the anterior hypothalamus of mice are modulated by glutamatergic inputs from area 24b of the cingulate cortex. J Psychopharmacol 35:78–90. https://doi.org/10.1177/0269881120967881

INDEX

Kazuhito Toyooka (ed.), *Neuronal Morphogenesis: Methods and Protocols*, Methods in Molecular Biology, vol. 2831, https://doi.org/10.1007/978-1-0716-3969-6,

D

E

W

X

Y

Z

SPRINGER NATURE

GPSR Compliance

The European Union's (EU) General Product Safety Regulation (GPSR) is a set of rules that requires consumer products to be safe and our obligations to ensure this.

If you have any concerns about our products, you can contact us on ProductSafety@springernature.com

In case Publisher is established outside the EU, the EU authorized representative is:

Springer Nature Customer Service Center GmbH
Europaplatz 3
69115 Heidelberg, Germany

FSC
www.fsc.org
MIX
Papier | Fördert
gute Waldnutzung
FSC® C083411

Kolibri 360 GmbH
Ferdinand-Jühlke-Straße 7
99095 Erfurt, Deutschland
produktsicherheit@kolibri360.de